Integrated Optics

Physics and Applications

NATO Advanced Science Institutes Series

A series of edited volumes comprising multifaceted studies of contemporary scientific issues by some of the best scientific minds in the world, assembled in cooperation with NATO Scientific Affairs Division.

This series is published by an international board of publishers in conjunction with NATO Scientific Affairs Division

A	**Life Sciences**	Plenum Publishing Corporation
B	**Physics**	New York and London
C	**Mathematical and Physical Sciences**	D. Reidel Publishing Company Dordrecht, Boston, and London
D	**Behavioral and Social Sciences**	Martinus Nijhoff Publishers The Hague, Boston, and London
E	**Applied Sciences**	
F	**Computer and Systems Sciences**	Springer Verlag Heidelberg, Berlin, and New York
G	**Ecological Sciences**	

Recent Volumes in Series B: Physics

Integrated Optics

Physics and Applications

Edited by

S. Martellucci

University of Naples
Naples, Italy

and

A. N. Chester

Hughes Aircraft Company
El Segundo, California

Plenum Press
New York and London
Published in cooperation with NATO Scientific Affairs Division

Proceedings of a NATO Advanced Study Institute on
Integrated Optics: Physics and Applications,
held August 17–30, 1981,
in Erice, Italy

Library of Congress Cataloging in Publication Data

NATO Advanced Study Institute on Integrated Optics: Physics and Applications (1981:
Erice, Italy)
  Integrated optics.

  (NATO advanced science institutes series. Series B, Physics; v. 91)
  "Proceedings of a NATO Advanced Study Institute on Integrated Optics: Physics and
Applications, held August 17–30, 1981, in Erice, Italy"—T.p. verso.
  Includes bibliographical references and index.
  1. Integrated optics—Congresses. I. Martellucci, S. II. Chester, A. N. III. Title. IV. Series.
TA1660.N37 1981   621.36′93   82-25074

ISBN-13: 978-1-4613-3663-1   e-ISBN-13: 978-1-4613-3661-7

DOI: 10.1007/978-1-4613-3661-7

© 1983 Plenum Press, New York

Softcover reprint of the hardcover 1st edition 1983

A Division of Plenum Publishing Corporation
233 Spring Street, New York, N.Y. 10013

This volume contains the proceedings of a two-week NATO A.S.I. on Integrated Optics: Physics and Applications, held from August 17 to August 30, 1981 in Erice, Italy. This is the 8th annual Course of the "International School of Quantum Electronics" presented under the auspices of the "E. Majorana" Centre for Scientific Culture.

The subject was chosen in order to satisfy the demand for a course on integrated optics which is relevant to the expanding use of fiber optics for communication and signal processing.

Integrated Optics, encompassing all of the optical waveguide circuits which are the optical analog of integrated circuits, is finding its way into a variety of applications involving communications, high speed signal-processing, and sensors of many kinds. However, because the technology is still changing very rapidly, the development of these exciting applications relies heavily upon the physics of the integrated optical circuits themselves and the processing techniques used to fabricate them. This NATO A.S.I. provided not only a thorough tutorial treatment of the field, but also through panel discussions and additional lectures treated topics at the forefront of present work.

Therefore the character of the Course was a blend of current research and tutorial reviews. "The Physics and Applications of Integrated Optics" could hardly be a more appropriate title to be chosen for this volume. Many of the worlds' acknowledged leaders in the field have been brought together to review and speculate on the accomplishments of integrated optics.

The papers in these proceedings give a fairly complete accounting of the Course lectures with the exception of the informal panel discussions. The lectures were presented in five major sections:

I INTRODUCTION: A tutorial treatment and overview of the field of integrated optics;
II THEORY OF GUIDED-WAVE STRUCTURES: A review of waveguide theory and the physical principles of optical couplers, modulators, and switches;

III PROCESSING TECHNIQUES: Both theoretical and practical aspects
     of the techniques by which integrated optical circuits and
     devices are fabricated:
 IV DEVICE TECHNOLOGY: A review of device principles and a look at
     the state of the art today;
  V SYSTEMS AND APPLICATIONS: The use of integrated optics
     technology in practical applications, including communications,
     signal processing and sensor systems.

In particular, two areas of application were stressed during the
Course: optical communication and signal processing. Accordingly
there is a large selection of papers relevant to both topics. Papers
are presented in addition describing single mode fibers which are
driving the use of integrated optics in telecommunications. Similarly
the topic of "Geodesic Optics" is represented because of its application to the integrated optical R.F. spectrum analyzer.

The NATO A.S.I. took advantage from the very active audience;
most of the students were experts in the field of Integrated Optics
and contributed with panel discussions and seminars. Some student
seminars are also included in the "contributed papers" section of
these proceedings.

This volume would, of course, not be possible without the
considerable effects put forth by the authors represented here.
Among them, R.M. de la Rue, R.V. Schmidt and G.L. Tangonan volun-
teered to edit the manuscripts in Erice and are cordially acknowledged. We also wish to acknowledge with thanks the invaluable help
of Mrs Maria Fiorini for much of the organization connected with the
Course and her timely editorial work on the entire collection of
papers. In addition we must mention the most qualified cooperation
offered by Maria Teresa Casati and Penny van Landingham in Erice,
Italy and in the USA, respectively.

Finally, but not the least of all, we are also indebted to
R.H. Andrews of Plenum Press for his assistance in the preparation of
this volume.

The Directors of the I.S.Q.E.

         A. N. Chester   S. Martellucci
         Hughes A. C.   Engineering Faculty
         El Segundo, Cal. (USA)   Naples (Italy)

CONTENTS

LECTURES AND SEMINARS

CONTRIBUTED PAPERS

INTEGRATED OPTICS DEVICES FOR OPTICAL COMMUNICATIONS

H. Kogelnik

Bell Laboratories

Holmdel, NJ 07733 USA

INTRODUCTION

The field of integrated optics covers the exploration of guided-
wave techniques for the construction of new or improved optical devices.
Dielectric waveguides, usually in the form of a planar film or strip
of higher refractive index than the substrate, are used to confine
the light to very small cross sections over relatively long lengths.
The goal is to accomplish compact and miniaturized devices of improved
efficiency and reliability, better mechanical and thermal stability,
and lower drive voltage and power consumption. In addition, one hopes
that several guided-wave devices can be combined and connected on a
common substrate to form more complicated optical circuits in analogy
with electronic integrated circuits. The branch of integrated optics
which is trying to serve optical fiber telecommunications has been in
the research laboratory for more than a decade. It has witnessed the
acceptance of a first generation of lightwave technology near 0.8
micrometer wavelengths, and the emergence of a second generation
technology near 1.3 micrometers. Both of these technologies use multimode fibers, yet integrated optics is essentially a single-mode technology, even though there have been several efforts to make contributions to multimode systems. Thus, integrated optics had to wait
in the research laboratory. However, now, there is serious interest
in developing a third generation lightwave technology for high-data-
rate long-distance transmission based on single-mode fibers. Now integrated optics has a good chance of contributing devices and circuits
such as switches, spare switches or switching networks, high-speed
modulators, filters and wavelength multiplexing circuits, and many
others.

1

In recent years there has been remarkable progress in research efforts focussed on two materials systems. The first is based on Ti-diffused waveguides in the electro-optic crystal LiNbO$_3$, and the second uses III-V semiconductors such as GaAs, InGaAs, and InGaAsP. The latter two are long-wavelength materials lattice-matched to InP. Integrated optics work in these materials is rather recent, but of considerable interest because of the wave-length compatibility with the new 1.3 and 1.6 micrometer lightwave technologies. The examples used here to illustrate recent research trends are arranged in three groups. The first includes switches, high-speed modulators and filters. The second group deals with the polarization problem posed by single-mode fibers, and the third with integrated optics in semiconductors, particularly with the technique of opto-electronic integration.

SWITCHES AND MODULATORS

A schematic of an optical switching network is shown in Figure 1. The sketch shows eight input ports and eight output ports that could be connected to optical fibers. Optical switching devices are shown as boxes. One of the research goals is to demonstrate devices which allow integration of several switches on one chip. Several types of optical switches are under exploration. Many guided-wave amplitude modulator or switching devices are based on Ti:LiNbO$_3$ and the directional coupler element, and several use the alternating phase mismatch ($\Delta\beta$) configuration to achieve low cross talk in switches. An example[1] is a switch, 10 mm in length, that uses six sections of alternating $\Delta\beta$ and operates with drive voltages as low as 3V. More recently a high-speed directional coupler modulator[2] was reported in the same materials systems, which allows a (10 percent to 90 percent) switching

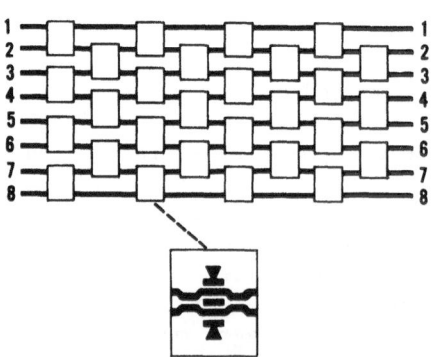

Fig. 1.   Optical switching network.

time of 110 picoseconds. This device has an electrode and interwave-
guide gap of only 1 micrometer and a device length as short as 750
micrometers. The required drive voltage was 6V. A reminder of what
this technology may offer is an earlier experimental integrated optical 4 x 4 switching network[3] where 5 $\Delta\beta$ couplers were integrated on
the same $LiNbO_3$.

FILTERS

A sketch of an optical circuit allowing wavelength division multiplexing is shown in Figure 2. The sketch indicates three input channels operating at different wavelengths and a network consisting of
filter devices which combines the three channels onto one single-mode
fiber, preferably without loss. A variety of suitable filter devices
are under exploration. Two recent research examples of filter devices
are based on $Ti:LiNbO_3$. The first employs, again, a directional coupler configuration, but with two waveguides of different effective refractive index. The dispersion curves of the two guides are designed
to intersect at the wavelength of peak filter response. Efficient
directional coupler filters with a bandwidth of about 200 Angstroms
have been reported[4], and the lowest achievable bandwidths for these
filters have been estimated at about 50 Angstroms. For systems requiring smaller bandwidth, the recent mode-converter filters[5] may be
a good option. Using interdigital electrodes of 7 micrometer period
and single $Ti:LiNbO_3$ guides, filter bandwidths between 5 and 50 Angstroms were demonstrated employing electrode lengths between 6 and
0.5mm. In a further demonstration of the capability of this technique, three of these filter devices were integrated on a single $LiNbO_3$
chip to form an experimental wavelength division multiplexing circuit
with 3 channels of 15 Angstroms bandwidth spaced 80 Angstroms apart.

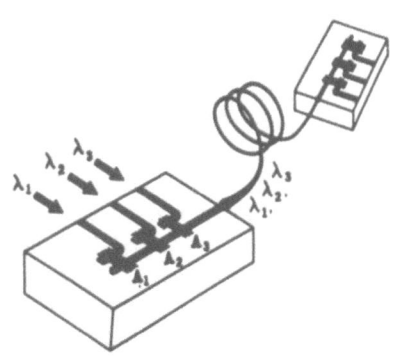

Fig. 2. Optical wavelength division multiplexing circuit.

POLARIZATION CONTROL

The polarization problem posed by the single-mode fiber technology is due to two simple facts: (1) ordinary single-mode fibers
do not preserve the polarization of the lightwave, and (2) ordinary
guided-wave devices of integrated optics behave very differently for
the (orthogonal) TE and TM polarizations. There are at least three
current approaches to solve this problem: (1) fiber makers are trying
to make special single-mode fibers that can maintain polarization,
(2) polarization transformers are used at the fiber output port together with automatic control circuits to compensate for polarization
variations[6-8], and (3) the guided-wave devices are designed to be
polarization independent. Integrated optics research is contributing
to the latter two approaches. Very recently a guided-wave polarization
transformer circuit was demonstrated[9] in Ti:LiNbO$_3$, where two voltage
controllable phase shifters and one controllable TE-TM converter were
integrated on a single chip. The length of the individual devices
was 4mm and 3mm respectively and the required voltages were of the
order of 3V. Several results have been achieved in efforts aiming
at polarization-independent device design. Examples, all in LiNbO$_3$,
are a polarization independent Y-branch interferometer modulator[10],
a polarization independent $\Delta\beta$ switch employing weighed directional
coupling[11], and a polarization independent directional coupler filter
employing the symmetry of the TE-TM interaction[12].

OPTO-ELECTRONIC INTEGRATION

The trend in integrated optics employing semiconductors is characterized by increasing research attention to opto-electronic integration, to high-resolution etching techniques, and to the long-wavelength materials InGaAs and InGaAsP lattice-matched to InP. Both wet
chemical etching as well as reactive ion etching of InGaAsP structures
are under exploration. Recent examples are the use of wet chemical
etches for the preparation of a monolithically integrated laser-detec-
tor structure[13], and the employment of reactive ion etching for the
fabrication of monolithic multi-section laser resonators structures
in InGaAsP[14]. The integration of optical guided-wave devices with
electronic devices, particularly with FET's, is under active investigation both in the GaAs- and the InP-based materials systems. Integration of a semiconductor junction laser with one or two FET's has
been demonstrated[15,16] in GaAs and is considered of promise for high-
speed laser modulation. The integration of a junction laser with a
monitoring detector is being investigated both in the GaAs-[17] and
InP-based[13] systems, and the monolithic integration of preamplifier-
detector structures[16] as well as laser-saturable absorber structures
for Q-switching[18] is under exploration.

INTEGRATED PIN-FETs

The detectors required for lightwave communications are room-
temperature devices of high sensitivity and fast response, that can
detect the attenuated digital optical signal with error rates better
than 10^{-9} (for recent reviews see[19,20]). First-generation systems
use Si p-i-n photodetectors for lower bit rates and Si avalanche photodetectors (APDs) for higher bit rates. This is illustrated in Figure
3. Si p-i-n detectors are followed by low-noise amplifiers such as
Si FETs. The internal gain in APDs is optimized to gain values
ranging from about 10 to 100 in the presence of excess noise due to
the avalanche multiplication process. At bit rates above 10 Mbit/s
the APD receivers offer a sensitivity improvement of about 15dB as
compared to Si p-i-n receivers, for a typical overall sensitivity of
50dBm (see e.g. Reference 21). However, APDs require relatively high
voltages (100-400V) and temperature compensation circuits to stabilize
the gain. There is, therefore, continuing interest in improving p-i-n
FET combinations, and the recent availability of low-noise GaAs FETs
has widened the applicability of p-i-n-FETs to higher speeds (see e.g.
References 22,23).

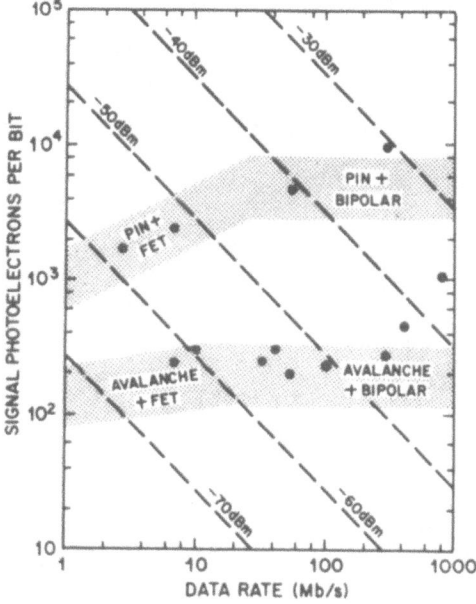

Fig. 3. Calculated sensitivity (in photo-electrons per bit) of
         various optical receivers as a function of bit rate. The
         full circles represent experimental results achieved to
         date. (From Reference 21).

Si detectors are no longer useful at wavelengths longer than
1.1µm, and device research is busy trying to provide detectors for
the long-wavelength range. Ge is a well developed material that is
a good candidate for this purpose. However, Ge is an indirect gap
material with a slow fall-off in response towards longer wavelengths
and with relatively large dark currents. The search for better materials is focusing on InGaAsP and GaAsSb, which are direct gap materials and are sensitive up to about 1.7µm. Figure 4 illustrates
the spectral response of these materials and their sharp cut-off towards longer wavelengths, which can be moved by changing the material's
composition (see e.g. Reference 24). This is a very desirable feature
as leakage currents tend to increase exponentially with the cut-off
wavelength due to the smaller band gaps. For p-i-n FET applications
of these materials the requirements include those for low leakage
current, low capacitance, and high breakdown voltage. Encouraging
results have already been reported where these parameters lie in range
of 10nA, 1pF and 100V (see e.g. References 25, 26). Particularly the
lattice-matched materials system of $In_{.53}Ga_{.47}As/InP$ which has good

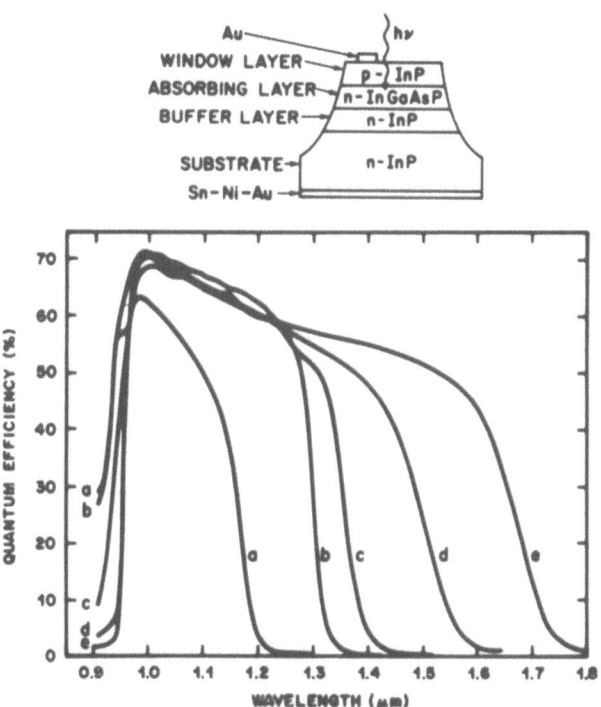

Fig. 4.   (a) Structure and (b) external quantum efficiency of five
          $In_{1-x}Ga_xAs_yP_{1-y}$ photo-detectors with different composition.
          The As content y was: A, 0.47; B, 0.61; C, 0.66; D, 0.88;
          and E, 1.0; respectively. (After Reference 24).

photoresponse up to 1.6 micrometers, appear to be of considerable
practical interest.

Added to these favorable characteristics of low-impurity InGaAs/
InP should be the fact that this material promises mobility values
that should at least be comparable to those of GaAs which is now
favored as an FET material. Theoretical predictions and preliminary
measurements point to potential mobility values as high as 15,000cm^2
V $^{-1}$s^{-1} [27]. This has stimulated efforts directed at integrating PIN
detectors and FET amplifiers on a single chip of InGaAs/InP, and the
first experimental integrated PIN-FET receiver has been recently demonstrated[28]. A sketch of this device is shown in Figure 5. These first
PIN-FETs employ junction field effect transistors with a p-n junction
photodiode as an integral part of the FET gate electrode as shown.
They are fabricated on InGaAs layers prepared by LPE on semi-insulating
InP:Fe substrates. While work is continuing to improve these structures, transconductances of 50 mS/mm have been achieved so far in integrated junction FETs. This last example of opto-electronic integration is somewhat unique in the sense that it is also compatible
with multimode fibers and has, thus, a chance to find earlier application.

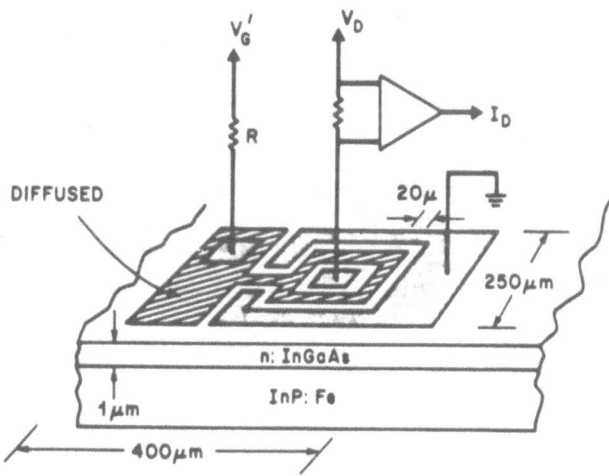

Fig. 5. Diagram of p-i-n FET photoreceiver. Diffused gate electrode
(cross-hatched area) is extended to form p-i-n diode region.
Shaded regions correspond to the metallized contacts. Equivalent electrical circuit is shown inset. (After Reference
28).

CONCLUSIONS

The field of integrated optics is still very young and there are streams of new ideas that are as yet little explored. Among these are the area of optical bistability and the generation and use of ultrashort optical pulses such as in the recently proposed sub-pico-second gates, and many others. In addition, there are many challenging problems awaiting solution in the technology of fabricating proposed devices and structures, and much work is left to be done.

REFERENCES

1. R. V. Schmidt and P. S. Cross, Optics Lett. 2:45, 1978.
2. R. C. Alferness, N. P. Economou and L. L. Buhl, Appl. Phys. Lett. 38:214, 1981.
3. R. V. Schmidt and L. L. Buhl, Electronics Lett. 12:575, 1976.
4. R. C. Alferness and R. V. Schmidt, Appl. Phys. Lett. 33:161, 1978.
5. R. C. Alferness and L. L. Buhl, Opt. Lett. 5:473, 1980.
6. M. Johnson, Appl. Opt. 18:1288, 1979.
7. R. Ulrich and M. Johnson, Appl. Opt. 18:1857, 1979.
8. M. Kubota, T. Oohara, K. Furuya and Y. Suematsu, Electronics Lett. 16:573, 1980.
9. R. C. Alferness, IEEE J. Quant. Electron QE-17:965, 1981; R. C. Alferness and L. L. Buhl, Appl. Phys. Lett. 38:655, 1981.
10. R. A. Steinberg, T. G. Giallorenzi and R. C. Priest, Appl. Optics 16:2166, 1979.
11. R. C. Alferness, Appl. Phys. Lett. 35:748, 1979.
12. R. C. Alferness and L. L. Buhl, Appl. Phys. Lett., to be published.
13. K. Iga and B. I. Miller, Electronics Lett. 16:342, 1980.
14. L. A. Coldren, B. I. Miller, K. Iga and J. A. Rentschler, Appl. Phys. Lett. 38:315, 1981.
15. H. Matsueda, T. Fukuzawa, T. Kuroda and M. M. Nakamura, Proc. 12th Conf. Solid State Dev., Tokyo, 1980.
16. S. Margalit, N. Bar-Chaim, J. Katz, I. Ury, D. P. Wilt, M. Yust and A. Yariv, Laser Focus, p. 76, Sept. 1980.
17. J. L. Mertz, R. A. Logan and A. M. Sergent, IEEE J. Quant. Electron. QE-15:72, 1979.
18. D. Z. Tsang and J. N. Walpole, Digest 100C, San Francisco, p.34, 1981.
19. H. Melchior, Phys. Today. 30:32, 1977.
20. H. Melchior, A. R. Hartman, D. P. Schinke and T. E. Seidel, Bell Syst. Tech. J. 57:1791, 1978.
21. T. Li, IEEE Trans. Commun. 26:946, 1978.
22. S. Hata, K. Kajiyama and Y. Mizushima, Electron. Lett. 13:668, 1977.
23. D. R. Smith, R. C. Hopper and I. Garrett, Opt. Quantum Electron 10:292, 1978.
24. M. A. Washington, R. E. Nahory, M. A. Pollack and E. D. Beebe, Appl. Phys. Lett. 33:854, 1978.

25. C. A. Burrus, A. G. Dentai and T. P. Lee, Electron. Lett. 15:655,
     1979.
26. R. F. Leheny, R. E. Nahory and M. A. Pollack, Electron Lett.
     15:713, 1979.
27. R. F. Leheny, A. A. Ballman, J. C. DeWinter, R. E. Nahory and
     M. A. Pollack, J. Elec. Matls. 9:561, 1980.
28. R. F. Leheny, R. E. Nahory, M. A. Pollack, A. A. Ballman, E. D.
     Beebe, J. C. DeWinter and R. J. Martin, Electron. Lett. 16:353,
     1980.

PLANAR OPTICAL WAVEGUIDE AND COUPLER ANALYSIS

H. G. Unger

Institut für Hochfrequenztechnik
Technische Universität Braunschweig
D-3300 Braunschweig/West-Germany

Planar optical waveguides in form of films on substrates as well as
strips on and in substrates, and various strip derived structures
serve in integrated optics to confine optical waves in components
and connect such components. In many of these applications they must
be small enough to guide light only in one mode of propagation, usually
the fundamental mode.

1. THE TRANSVERSE IMPEDANCE METHOD

To facilitate the design of integrated optics components and
circuits a method of analysis for wave propagation in planar optical
waveguides is to be preferred that is simple, for fast and easy computations, and yet accurate enough for reliable results. The transverse impedance method seems to offer these advantages. It will first
be explained for film waveguides and subsequently be extended to strip
waveguides and strip derived structures.

1.1. Film

Film waveguides such as the thin film of refractive index n_f on
a substrate of refractive index $n_s < n_f$ in Figure 1 may be assumed
to be uniform in the direction z of wave propagation and also in the
transverse y-direction parallel to the film. Waves that the film
guides may then also be assumed to have fields, which are independent
of y, so that $\partial/\partial y = 0$, and which depend on z as $\exp(-j\beta z)$. Under
these conditions Maxwell's equations for the electric field $\vec{E}$ and the
magnetic field $\vec{H}$ have the following Cartesian components where ω is
the circular frequency, μ_o and ε_o are the vacuum permeability and

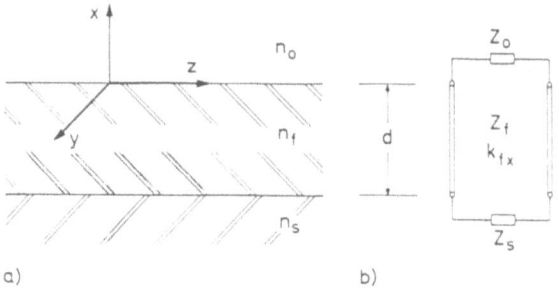

Fig. 1.   Filmwaveguide a) and transverse transmission line model b)

permittivity respectively and n_i is the refractive index of the respective cross-sectional area of the film waveguide.

$$j\beta E_y = -j\omega\mu_o H_x \tag{1.1}$$

$$-j\beta E_x - \frac{\partial E_z}{\partial x} = -j\omega\mu_o H_y \tag{1.2}$$

$$\frac{\partial E_y}{\partial x} = -j\omega\mu_o H_z \tag{1.3}$$

$$j\beta H_y = j\omega n_i^2 \epsilon_o E_x \tag{1.4}$$

$$-j\beta H_x - \frac{\partial H_z}{\partial x} = j\omega n_i^2 \epsilon_o E_y \tag{1.5}$$

$$\frac{\partial H_y}{\partial x} = j\omega n_i^2 \epsilon_o E_z \tag{1.6}$$

By substituting H_x from eq. (1.1) into eq. (1.5) we obtain

$$\frac{\partial H_z}{\partial x} = -j(\omega n_i^2 \epsilon_o - \frac{\beta^2}{\omega\mu_o})E_y \ . \tag{1.7}$$

This equation together with eq. (1.3), is a system of transmission
line equations for the voltage

$$V = E_y \tag{1.8}$$

and the current

$$I = H_z \tag{1.9}$$

on a transmission line with the propagation constant

$$\gamma_{ix} = jk_{ix} = j\sqrt{n_i^2 k^2 - \beta^2}$$

(1.10)

and characteristic impedance

$$Z_i = \omega\mu_o/k_{ix} \cdot$$

(1.11)

The fields for which these transmission line equations describe the x-distribution have no E_z component and will hence lead to TE- or H-waves with respect to the direction of propagation.

In the remaining three of Maxwell's equations we substitute E_x from eq. (1.4) into eq. (1.2)

$$\frac{\partial E_z}{\partial x} = -j(\omega\mu_o - \frac{\beta^2}{\omega n_i^2 \epsilon_o})H_y$$

(1.12)

and together with eq. (1.6) have a system of transmission line equation for the voltage

$$V = E_z$$

(1.13)

and the current

$$I = H_y$$

(1.14)

on a transmission line, whose propagation constant is again γ_{ix} but whose characteristic impedance is given as

$$Z_i = k_{ix}/(\omega n_i^2 \epsilon_o) \cdot$$

(1.15)

The fields for which these lines describe the x-dependence have no H_z-component and will hence lead to TM- or E-waves.

H- as well as E-waves have now the equivalent transmission line model in Figure 1b for their x-dependence. In it Z_f and

$$k_{fx} = \sqrt{n_f^2 k^2 - \beta^2}$$

(1.16)

are the transverse wave impedance and phase constant in the film. In the substrate and in the space above the film the fields for film-modes are transversely evanescent with

$$\gamma_{sx} = jk_{sx} = \alpha_{sx} = \sqrt{\beta^2 - n_s^2 k^2}$$

(1.17)

and

$$\gamma_{ox} = jk_{ox} = \alpha_{ox} = \sqrt{\beta^2 - n_o^2 k^2}$$

(1.18)

as transverse attenuation constants. The transverse wave impedance
Z_s and Z_o of these regions terminate the transverse transmission line.
They follow from eq. (1.11) or eq. (1.15), and because of the evanescent character of the fields in these regions are purely reactive with
$k_{ix} = -j\alpha_{ix}$. For the voltages and currents to be continuous in the
transmission line model, the input imedance

$$Z_{ef} = Z_f \frac{Z_o + jZ_f \tan k_{fx}d}{Z_f + jZ_o \tan k_{fx}d} \tag{1.19}$$

of the film at the substrate interface must match $-Z_s$:

$$Z_{ef} = -Z_s \tag{1.20}$$

This transverse resonance condition when evaluated for H-waves leads
to the following characteristic equation for these waves

$$\tan k_{fx}d = \frac{k_{fx} (\alpha_{sx} + \alpha_{ox})}{k_f^2 - \alpha_{sx} \alpha_{ox}}. \tag{1.21}$$

When evaluated for E-waves it leads to their characteristic equation

$$\tan k_{fx}d = \frac{n_f^2 k_{fx} (n_o^2 \alpha_{sx} + n_s^2 \alpha_{ox})}{n_o^2 n_s^2 k_{fx}^2 - n_f^4 \alpha_{sx} \alpha_{ox}}. \tag{1.22}$$

Together with the transverse phase and attenuation constants according
to eq. (1.16), (1.17) and (1.18) these characteristic equations determine the phase constants of guided modes of the film waveguide. Upon
introducing the following transverse phase and attenuation parameters

$$u = k_{fx}d, \quad v = \alpha_{sx}d, \quad w = \alpha_{ox}d \tag{1.23}$$

the characteristic equation for H-waves appears as

$$\tan u = \frac{u (v + w)}{u^2 - vw} \tag{1.24}$$

and for E-waves it appears as

$$\tan u = \frac{n_f^2 u (n_o^2 v + n_s^2 w)}{n_o^2 n_s^2 u^2 + n_f^4 vw}. \tag{1.25}$$

Another example of a practical film waveguide is the four layer
structure in Figure 2a, which for $n_o \leqslant n_s < n_f < n_p$ represents the

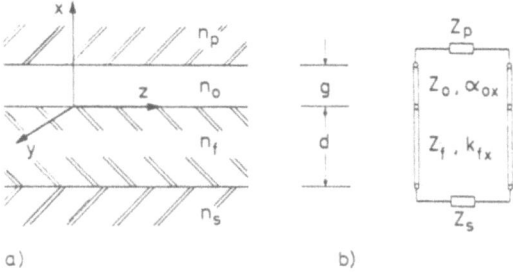

Fig. 2.   Four layer structure of a prismcoupler a) and transverse
          transmission-line model b)

situation in the prism coupler, where the upper most layer with refractive index n_o is the prism-region, separated by the gap of width g from the film. Because of $n_p > n_f$ the structure supports only leaky modes, which serve to couple power in and out of film waves.

Figure 2b shows the transverse transmission line model for this four layer structure. The transverse wave impedance of the prism is

$$Z_p = \omega\mu_o/k_{px}$$

where
$$k_{px} = \sqrt{n_p^2 k^2 - \beta^2} .$$

In the transverse transmission-line model this wave impedance is also the input impedance of the prism at the prism base. The gap transforms this input impedance Z_p into

$$Z_{eo} = Z_o \frac{Z_p + Z_o \tanh\alpha_{ox}g}{Z_o + Z_p \tanh\alpha_{ox}g} .$$

With the reflection coefficient
$$r = (Z_p - Z_o)/(Z_p + Z_o)$$

at the prism base, this input impedance may also be written as
$$Z_{eo} = Z_o \frac{1 + r\ \exp(-2w\delta)}{1 - r\ \exp(-2w\delta)}$$

where $w\delta = wg/d = \alpha_{ox}g$ represents the gap attenuation.

We focus our attention now on the more important case of H-waves and compare Z_{eo} for these waves with the input impedance

$$Z_o = j\omega\mu/\alpha_{ox} = j\omega\mu d/w$$

of the upper space without prism. From this comparison we note that
w must be replaced by

$$w_p = w \frac{1 - r \exp(-2w\delta)}{1 + r \exp(-2w\delta)} \qquad (1.26)$$

in order to obtain relations for the prism coupler from the relations
for H-modes in the film-guide. With w_p instead of w in eq. (1.24)
this equation represents the characteristic equation for leaky H-modes
in the prism coupler.

For a wide gap between film and prism we have $w\delta \gg 1$, and the
exponentials in eq. (1.26) are so small that the prism perturbs the
film modes only slightly. A perturbation analysis for this case can
start from unperturbed values u, v, and w of film modes with small
perturbations added to them in the form

$$u + u_1, \quad v + v_1, \quad w + w_1$$

to account for the effect of the prism. Such perturbations u_1, v_1,
w_1 of u, v, and w cause a perturbation β_1 of the unperturbed phase
constant β in form $\beta + \beta_1$. Equations (1.16), (1.17), (1.18) and
(1.23) relate the perturbation β_1 of β to the perturbations u_1, v_1,
w_1 of u, v, and w:

$$\beta_1 \beta d^2 = -u_1 u = v_1 v = w_1 w \qquad (1.27)$$

The characteristic equation (1.24) now yields the following expression
for β_1 due to its right-hand side perturbation:

$$\beta_1 \beta d^2 = \exp(-2\delta w) \frac{w-jt}{w+jt} \frac{2u^2 vw^2}{(u^2+w^2)(v+w+vw)} \qquad (1.28)$$

where $t = k_{px} d$. The perturbation β_1 of β is complex. Its real part
indicates a slight shift in phase constant of the film mode underneath
the prism. The negative of the imaginary part of equation (1.28) is

$$\alpha_1 \beta d^2 = \exp(-2\delta w) \frac{4t\, u^2 w^2}{(t^2+w^2)(u^2+w^2)(1+\frac{1}{w}+\frac{1}{v})} \qquad (1.29)$$

it gives the attenuation constant α_1 of the film mode underneath the
prism due to its power leaking into the prism beam. If we introduce
the effective width d_e of the film mode according to

$$d_e = d (1 + \frac{1}{v} + \frac{1}{w}) \qquad (1.30)$$

and replace the remaining factors of the right-hand denominator by
$u^2 + w^2 = k^2 d^2(n_f^2 - n_o^2)$ and $t^2 + w^2 = k^2 d^2(n_p^2 - n_o^2)$, the leaky mode
attenuation is given by

$$\alpha_1 = \exp(-2\delta w) \; \frac{4t^2 u^2 w^2}{(n_f^2 - n_o^2)(n_p^2 - n_o^2)k^4 \beta d^5 d_e} \; . \qquad (1.31)$$

Due to the exponential factor it depends most critically on the relative gap width $\delta = g/d$ and the transverse attenuation parameter $w = \alpha_o d$ of the evanescent fields in the gap. For moderate to large values of δw the exponential tails of gap fields are quite weak at the prism base and transfer power of the film mode to the prism beam at a very low rate.

1.2 Strip waveguides

Strip waveguides, in their simplest form, consist of a narrow strip of film deposited on a substrate of somewhat lower refractive index. They may also be embedded into the substrate with their cover boundary flush with the substrate surface. Figure 3a shows the basic strip guide structure, which will be a raised strip if $n_i = n_o$ and an embedded strip if $n_i = n_s$.

Optical waves guided by such strips may be regarded as film modes in the strip region which propagate at an angle θ with respect to the strip axis and experience total reflection at the side walls of the strip. With such total reflection the film mode travels on the zig-zag path of Figure 3b inside the film. It forms a mode of propagation of the strip guide if after two reflections it repeats in phase in order to interfere constructively with itself.

When the film mode has the phase constant β_f in θ-direction it has

$$\beta_y = \beta_f \sin\theta \qquad (1.32)$$

as its phase constant in the transverse y-direction.

A TE- or H-film mode with an electric field E has

$$E_z = -E \sin\theta \qquad (1.33)$$

as its transverse electric field with respect to the y-direction. The transverse magnetic field with respect to this direction consists of

$$H_x = -\beta_f/(\omega\mu_o)E \qquad (1.34)$$

from eq. (1.1) and of

$$H_z = (j/\omega\mu_o)(\partial E/\partial x) \cos\theta \qquad (1.35)$$

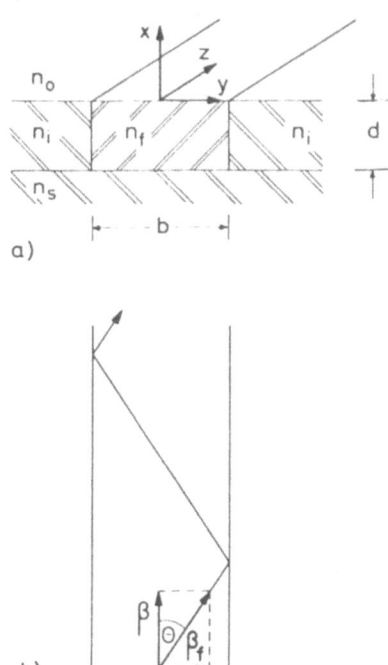

a)

b)

Fig. 3.   Strip guide model for raised and embedded strips (a) and
          zig-zag-path for a film wave in the strip (b)

from eq. (1.3). Usually the strip has a refractive index n_f that is
not much larger than n_s. H-modes in films with $n_f - n_s \ll n_s$ have a
longitudinal magnetic field that is small compared to the transverse
magnetic field. H_z from eq. (1.35) may therefore be neglected compared to H_x from eq. (1.34). For H-film modes in the strip a transverse wave impedance in y-direction may then be calculated from

$$Z_f = \frac{E_z}{H_x} = \frac{\omega\mu_o \sin\theta}{\beta_f} = \frac{\omega\mu_o \beta_y}{\beta_f^2} . \tag{1.36}$$

An equivalent transmission line for this transverse wave propagation has therefore the propagation constant

$$\gamma_y = j\beta_y = j\beta_f \sin\theta = j(\beta_f^2 - \beta^2)^{1/2} , \tag{1.37}$$

where β is the phase constant in z-direction, that is the phase constant of the stripguide mode, and it has the characteristic impedance
Z_f according to eq. (1.36).

A TM- or E-film mode with a magnetic field H has

$$H_z = -H \sin\theta \tag{1.38}$$

as its transverse magnetic field with respect to the y-direction. The transverse electric field with respect to this direction consists of

$$E_x = \left[\beta/(\omega n_f^2 \epsilon_o)\right] H \tag{1.39}$$

from eq. (1.4) and of

$$E_z = -j\left[\cos\theta/(\omega n_f^2 \epsilon_o)\right] \partial H/\partial \mathbf{x} \tag{1.40}$$

from eq. (1.6). For a strip with $(n_f - n_s) \ll n_s$ the E-film mode just as the H-film mode is nearly TEM and as a consequence E_z may be neglected compared to E_x. The transverse wave impedance in y-direction is then given by

$$Z_f = -\frac{E_x}{H_z} = \frac{\beta_f}{\omega n_f^2 \epsilon_o \sin\theta} = \frac{\beta_f^2}{\omega n_f^2 \epsilon_o \beta_y} \tag{1.41}$$

It represents the characteristic impedance of the equivalent transmission line for transverse wave propagation of E-film modes in the strip; with $\gamma_y = j\beta_y$ in eq. (1.37) as the propagation constant.

For total reflection of the film modes at the side walls of the strip their fields beyond are transversely evanescent. They decay exponentially in y-direction with an attenuation constant

$$\alpha_y = (\beta^2 - \beta_f^2)^{1/2} = (\beta^2 + k_{fx}^2 - n_i^2 k^2)^{1/2} \tag{1.42}$$

This y-dependence results from the separation condition

$$k_{ix}^2 + k_{iy}^2 + k_{iz}^2 = n_i^2 k^2$$

in which $k_{iy} = -j\alpha_y$ and $k_{ix} = k_{fx}$ and $k_{iz} = \beta$ from the excitation of the evanescent fields by a particular film mode with its specific k_{fx} and $\beta = \beta_f \cos\theta$. The transverse wave impedance in y-direction for these evanescent fields is

$$Z_i = -j\omega\mu_o \alpha_y/\beta_i^2 \tag{1.43}$$

in case of a H-film mode and

$$Z_i = j\beta_i^2/(\omega n_i^2 \epsilon_o \alpha_y) \tag{1.44}$$

in case of an E-film mode. The phase constant

$$\beta_i = (n_i^2 k^2 - k_{fx}^2)^{1/2} \tag{1.45}$$

is that of a film mode in a film of refractive index n_i but with an
x-dependence corresponding with k_{fx} to the x-dependence of the film
mode in the strip, to which the evanescent fields must be matched.
The equivalent transverse transmission lines for these evanescent film
mode fields have Z_i from eq. (1.43) or eq. (1.44) as their characteristic impedance and $\gamma_y = \alpha_y$ from eq. (1.42) as their propagation constant. Figure 4b shows the transverse transmission line circuit for
the basic strip guide structure in Figure 4a.

The strip guide in Figure 4a and its equivalent transmission
line circuits are symmetric with respect to their center plane at
$y = 0$. Its modes of propagation have therefore fields which are also
either even or odd with respect to $y=0$. A H_m-film mode with m+1 half
waves of its standing wave pattern in x-direction forms an HE_{ml}-hybrid-
mode of the strip with $+1$ half waves of its standing wave pattern
in y-direction. For even l its field component E_z has a node at $y=0$
and its transverse line may be shorted there. For odd l its H_x has
a node at $y=0$ allowing an open circuit there. Figure 4c and 4d show
the simplified line circuit for HE_{ml}-modes of even and odd order l.

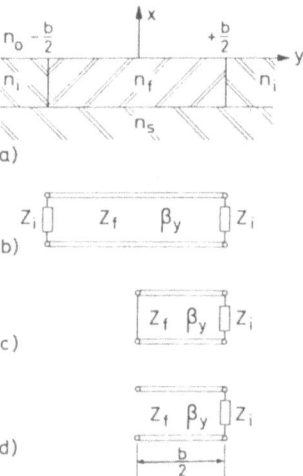

Fig. 4. a) Basic strip guide model
         b) equivalent transverse transmission line circuit
         c) equivalent line circuit for
            HE_{ml}-modes with even l
            EH_{ml}-modes with odd l
         d) equivalent line circuit for
            HE_{ml}-modes with odd l
            EH_{ml}-modes with even l

An E_m-film mode forms an EH_{m1}-hybrid mode of the strip. If it has an even number 1 of half waves in y-direction its H_z is zero at y=0 and the transverse line may be shorted there. If its 1 is odd the component E_x has a node at y=0 allowing an open circuit in the plane of symmetry. These correspondences are indicated in Figure 4c and 4d.

The transverse resonance condition in Figure 4c and 4d leads to the following characteristic equation for HE_{m1}-modes with even 1 and for EH_{m1}-modes with odd 1

$$\tan(\beta_y b/2) = j Z_i/Z_f \tag{1.46}$$

For HE_{m1}-modes with odd 1 and for EH_{m1}-modes with even 1 it leads to

$$\tan(\beta_y b/2) = j Z_f/Z_i \tag{1.47}$$

For a universal formulation we define a transverse phaseparameter in the strip

$$u_s = \beta_y b/2 = (b/2)\left[n_f^2 k^2 - (u/d)^2 - \beta^2\right]^{1/2} \tag{1.48}$$

and a transverse attenuation parameter beyond its sidewalls

$$v_s = \alpha_y b/2 = (b/2)\left[\beta^2 + (u/d)^2 - n_i^2 k^2\right]^{1/2} \tag{1.49}$$

In addition we use $N_f = \beta_f/k$ and $N_i = \beta_i/k$ as the effective indices of refraction for the film modes in the strip and beyond its sidewalls respectively. In terms of these parameters the characteristic equation for HE_{m1}-modes of any order 1 is

$$u_s \tan(u_s - 1\pi/2) = v_s N_f^2/N_i^2 \tag{1.50}$$

and for EH_{m1}-modes of any order 1 it is

$$u_s \tan(u_s - 1\pi/2) = v_s N_f^2 n_i^2/(N_i^2 n_f^2) \tag{1.51}$$

These characteristic equations as we obtained them from the transverse impedance method differ by the factor N_f^2/N^2 from those characteristic equations as they obtain from the effective index method[1].

Figure 5 shows for the HE_{00}-mode of a strip of index n_f surrounded on all sides by a medium of index $n_o = n_s = 1$ the phase parameter

$$B = \frac{(\beta/k)^2 - n_s^2}{n_f^2 - n_s^2}$$

as a function of the film parameter $V = kd\sqrt{n_f^2 - n_s^2}$.

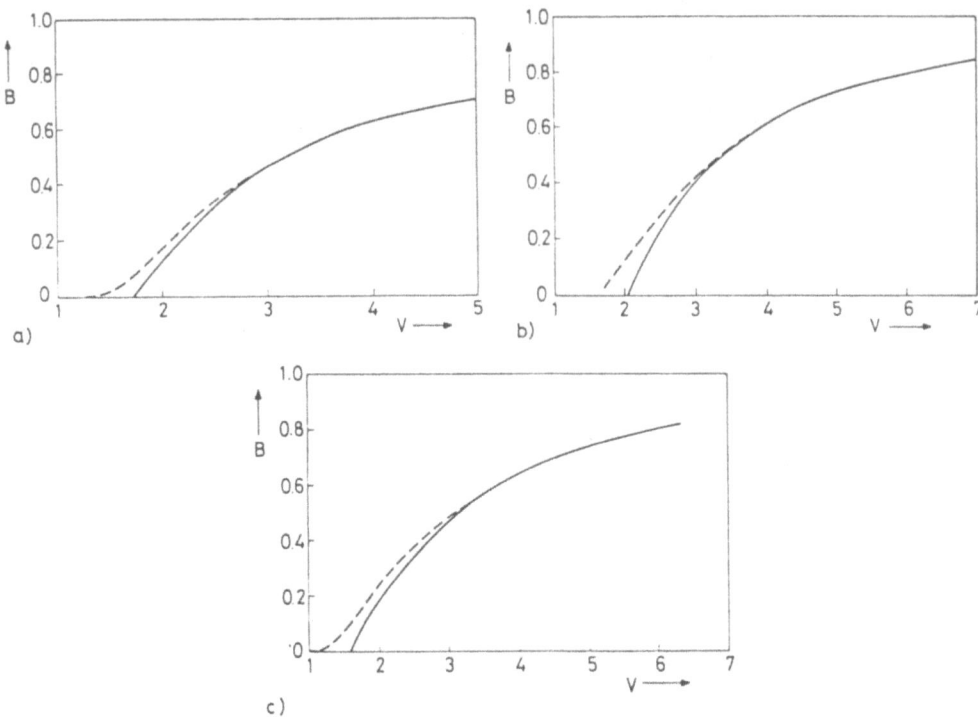

Fig. 5. Phase parameter $B = (\beta^2/k^2 - n_s^2)/(n_f^2 - n_s^2)$ as a function of the frequency parameter $V = kd.\sqrt{n_f^2 - n_s^2}$ of a buried strip-guide

with $n_s = 1$

Solid lines: approximation by transverse impedance method
Broken lines: variational analysis[2]

a) $n_f = 2.5$ $b/d = 0.4$

b) $n_f = 2.25$ $b/d = 0.5$

c) $n_f = 1.1025$ $b/d = 0.5$

The solid lines have been calculated using the transverse impedance method, while the broken lines are from a more accurate variational analysis[2]. Except for low values of the film parameter V the transverse impedance method yields satisfactory results even for numerical apertures as large as $(n_f^2 - n_s^2)^{1/2} = 1.2$.

1.3. Strip guide couplers

The transverse impedance method proves quite helpful for more complicated planar waveguide structures. Figure 6 shows as an

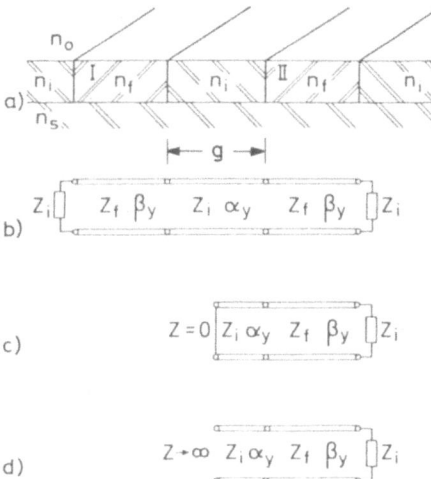

Fig. 6. Strip guide coupler (a) and transverse transmission line
model (b)
c: simplified line model for even HE- and odd EH-modes
d: simplified line model for odd HE- and even EH-modes

example the basic structure of a symmetrical strip guide coupler.
A mode of strip guide I couples to modes of strip guide II by way of
their evanescent fields in the gap between both strips. Coupling
increases with the extent to which evanescent fields of the same kind
and direction overlap. Due to their coupling those modes interact
most strongly that remain in phase synchronism along the coupler. An
HE_{ml}-mode will therefore interact more with the same HE_{ml} of the other
strip then with any other mode.

The interaction is described by the following system of coupled
wave equations for the amplitudes A_I of mode I and A_{II} of mode II

$$dA_1/dz = -j\beta A_I - jcA_{II}$$

$$dA_{II}/dz = -j\beta A_{II} - jcA_I$$
(1.52)

β is the phase constant of each of the degenerate modes and c their
coupling coefficient.

The eigenvector matrix

$$\underline{\underline{V}} = \frac{1}{\sqrt{2}} \cdot \begin{bmatrix} 1 & 1 \\ 1 & -1 \end{bmatrix}$$

$$\underline{V} = \frac{1}{\sqrt{2}} \cdot \begin{bmatrix} 1 & 1 \\ 1 & -1 \end{bmatrix} \tag{1.53}$$

of eqs. (1.52) transforms from modes I and II according to

$$A_1 = \frac{1}{\sqrt{2}} (A_e + A_o')$$

$$A_{11} = \frac{1}{\sqrt{2}} (A_e - A_o) \tag{1.54}$$

to modes e and o with amplitudes A_e and A_o and diagonalizes the coupled wave equations according to

$$dA_e/dz = -j\beta_e A_e \qquad dA_o/dz = -j\beta_o A_o \tag{1.55}$$

the eigenvalues β_e and β_o of the system of equations (1.52) follow from

$$\beta_e - \beta = \beta - \beta_o = c \tag{1.56}$$

and represent the phase constants of mode e and o respectively. Mode e and o are the normal modes of the system of equations (1.52). The inversion of eqs. (1.54) leads to

$$A_e = \frac{1}{\sqrt{2}} (A_I + A_{II})$$

$$A_o = \frac{1}{\sqrt{2}} (A_I - A_{II}) \tag{1.57}$$

It shows that mode e has an even field distribution with respect to the plane of symmetry of the coupler and mode o and odd distribution.

Instead of using eqs. (1.56) to determine the phase constants β_e and β_o of the even and odd normal modes of the coupler we will now regard it as a formula for the coupling coefficient c. To determine c from eqs. (1.56) we need to know β and one of the phase constants β_e or $\beta_o.\beta_e$ and β_o would also allow us to calculate c from eqs. (1.56).

Figure 6b shows the equivalent transverse transmission line circuit for normal modes of the composite strip guide coupler. For the even HE- and odd EH-modes it may be simplified according to Figure 6c. For odd HE- and even EH-modes the simplified line circuit of Figure 6d applies.

In both simplified line models the strip transforms the transverse wave impedance Z_i at the outer side wall to

$$Z_{fi} = Z_f \frac{Z_i + jZ_f \tan 2u_s}{Z_f + jZ_i \tan 2u_s} \qquad (1.58)$$

as the input impedance of the strip as seen by the coupling gap.
Looking in the other direction, half of the shorted gap has the input
impedance

$$Z_{is} = Z_i \tanh(v_s g/b) \qquad (1.59)$$

and half of the open gap has

$$Z_{io} = Z_i \coth(v_s g/b) \qquad (1.60)$$

The transverse resonance conditions

$$Z_{fi} = -Z_{is} \quad \text{and} \quad Z_{fi} = -Z_{io} \qquad (1.61)$$

lead to

$$Z_f(Z_i + jZ_f \tan 2u_s) = -Z_i \tanh(v_s g/b)(Z_f + jZ_i \tan 2u_s) \qquad (1.62)$$

as the characteristic equation for even HE- and odd EH-modes and to

$$Z_f(Z_i + jZ_f \tan 2u_s) = -Z_i \tanh(v_s g/b)(Z_f + jZ_i \tan 2u_s) \qquad (1.63)$$

as the characteristic equation for odd HE- and even EH-modes of the
composite strip-guide coupler.

For a wide gap between the strips, or strongly evanescent fields
in the gap, the strip guide modes couple only weakly through the gap.
For them $v_s g/b = \alpha_y g/2$ is large enough to approximate eqs. (1.59) and
(1.60) by

$$(1.64)$$

and
$$Z_{is} \simeq Z_i \left[1 - 2\exp(-2v_s g/b) \right]$$

$$Z_{io} \simeq Z_i \left[1 + 2\exp(-2v_s g/b) \right] \qquad (1.65)$$

respectively. In these expressions the exponentials represent a small
perturbation of the transverse wave impedance in the gap by the short
or open circuit in its center. This perturbation changes also the
characteristics of the strip guide mode and in particular its phase
constant β. A perturbation analysis based on equations (1.62) and
(1.63) and on the unperturbed strip mode parameter u_s, v_s and β will
yield this change in phase constant $\Delta\beta$ which according to equations
(1.56) represents directly the coupling coefficient for the respective
strip guide mode.

2. A RIGOROUS THEORY FOR COUPLED OPTICAL WAVEGUIDES

To develop an exact coupling theory for optical waveguides we adopt Figure 7 as a suitable waveguide model[3]. It has a cylindrical distribution of linear, isotropic matter with permittivity $\varepsilon_I(x,y)$ which is a function of the transverse co-ordinates x,y only, and which does not change as a function of the axial distance z. This guide is assumed to be confined to a certain region of x and y. Outside this region the permittivity is assumed to be constant and equal to ε. Normally we have $\varepsilon_I(x,y) > \varepsilon$. We enclose the actual optical wave-guide in cylindrical walls of perfect conductivity. These opaque walls are placed at a sufficient transverse distance from the light guiding region of optical waveguide with $\varepsilon_I > \varepsilon$ so that most of the transversely evanescent waves which are guided within the cross-section with $\varepsilon_I > \varepsilon$ have only extremely low fields at the walls. The perme-ability μ is assumed to be constant over the guide cross-section. The guide is assumed to be lossless.

In it, source-free field distributions can be described by normal modes, which propagate independently of one another.

$$E(x,y,z) = \sum_i E_i(x,y)A_i(z_1)\exp\left[-\gamma_i(z-z_1)\right] \qquad (2.1)$$

$$H(x,y,z) = \sum_i H_i(x,y)A_i(z_1)\exp\left[-\gamma_i(z-z_1)\right] \qquad (2.2)$$

For convenience, waves which propagate in the positive z-direction are denoted by positive subscripts i while the respective negative subscripts -i designate waves propagating in the negative z-direction. Thus, if the subscript i represents the forward travelling mode i, the subscript -i represents its replica travelling in opposite direction.

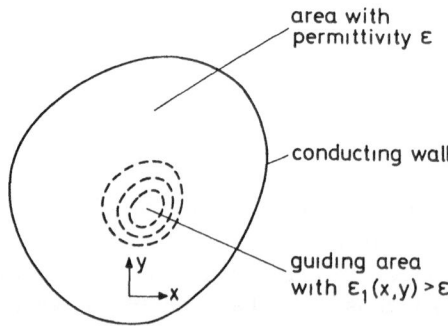

Fig. 7. Cross-section of an optical guide model surrounded by
          conducting walls.

By a suitable normalization of the eigenfields the following orthonormality relation can be satisfied:

$$\frac{1}{2} \iint (\vec{E}_m \times \vec{H}_n + \vec{E}_n \times \vec{H}_m) \cdot \vec{n}_z dS_o = \delta_{mn} \, \text{sign}(m)$$

(2.3)

Here the integration $\iint dS_o$ extends over the total cross-section and δ_{mn} denotes the Kronecker symbol. The value of $\text{sign}(m)$ is given by

$$\text{sign}(m) = 1 \quad \text{for} \quad m > 0$$

$$\text{sign}(m) = -1 \quad \text{for} \quad m < 0$$

(2.4)

The transverse normal mode fields $\vec{E}_{ti}$, $\vec{H}_{ti}$ and the longitudinal field components E_{zi}, H_{zi} of forward and backward waves are related by

$$\vec{E}_{ti} = \vec{E}_{t(-i)} \qquad \vec{H}_{ti} = -\vec{H}_{t(-i)}$$

$$E_{zi} = -E_{z(-i)} \qquad H_{zi} = H_{z(-i)}$$

(2.5)

Referring to the normal mode fields at the conducting walls far away from the guiding area with $\varepsilon_I > \varepsilon$, we can distinguish two types of normal modes of the wave guide model:

(a) There exists a limited number of normal modes which have fields that decrease exponentially as a function of the distance from the guiding area. The field energy of these modes is concentrated in or near the guiding area with $\varepsilon_I > \varepsilon$. The propagation constant γ_i of these modes is purely imaginary for lossless guides.

(b) All the other modes have fields extending over the full cross-section out to the conducting walls. A finite number of them are propagating modes and, in the lossless case, have purely imaginary values of γ_i, while the remaining infinite number is below cutoff and evanescent with purely real γ_i's in the lossless case. If the distance between the guiding area and the conducting wall is extended to infinity, then this set of modes forms the continuous spectrum of space or radiating modes, with which any radiating fields are represented.

Note that, in general, for the rigorous description of an electromagnetic field the normal modes of type (a) as well as the infinite set of modes of type (b) are needed.

We shall denote the complete waveguide model as defined by $\varepsilon_I(x,y)$ and including the conducting wall as waveguide I. In the following we shall assume that the complete set of modes of this guide is known. Let us now define a waveguide II with the same conducting wall in the identical location as for waveguide I, but with a different distribution of matter with permittivity $\varepsilon_{II}(x,y)$. The part of the cross-section of waveguide I with $\varepsilon_I(x,y) \neq \varepsilon$ and that of waveguide II with $\varepsilon_{II}(x,y) \neq \varepsilon$ are assumed to be separated from each other inside the same conducting walls. The complete set of normal modes of waveguide II is supposed to be known as well.

Let any subscript i denote only one mode of either waveguide I or waveguide II. Its field, which according to equations (2.1) and (2.2) is written as

$$\vec{E}_i = \vec{E}_i(x,y) \exp\left[-\gamma_i(z-z_1)\right] A_i(z_1) \qquad (2.6)$$

$$\vec{H}_i = \vec{H}_i(x,y) \exp\left[-\gamma_i(z-z_1)\right] A_i(z_1) \qquad (2.7)$$

will by itself satisfy Maxwell's equation in the corresponding wave-guide I or II

$$\nabla \times \vec{H}_i = j\omega\varepsilon_i(x,y)\vec{E}_i \qquad (2.8)$$

$$-\nabla \times \vec{E}_i = j\omega\mu\vec{H}_i \qquad (2.9)$$

where $\varepsilon_i(x,y)$ is defined by

$$\varepsilon_i(x,y) = \varepsilon_I(x,y) \qquad (2.10)$$

and the ith mode is a mode of waveguide I and by

$$\varepsilon_i(x,y) = \varepsilon_{II}(x,y) \qquad (2.11)$$

if this mode belongs to waveguide II.

These definitions and notation will now be applied to the cylindrical structure of which Figure 8 shows the cross-section. This structure encloses both dielectric waveguides I and II within a common wall of perfect conductivity.

The parallel arrangement of both waveguides within the common wall is described by a cross-sectional distribution $\hat{\varepsilon}(x,y)$ of isotropic permittivity with

$$\hat{\varepsilon}(x,y) = \varepsilon_I(x,y) \qquad (2.12)$$

in the region of waveguide I, and

$$\hat{\varepsilon}(x,y) = \varepsilon_{II}(x,y) \qquad (2.13)$$

in the region of waveguide II, but

$$\hat{\varepsilon}(x,y) = \varepsilon \qquad (2.14)$$

over the remaining part of the cross-section.

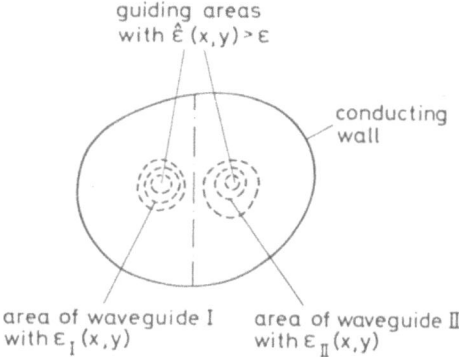

Fig. 8.   Cross-section of waveguide I and waveguide II laid
          in parallel.

The two parallel dielectric waveguides together with their enclosure of perfect conductivity form a uniform cylindrical structure.
A complete set of normal modes exists for this structure which propagate independently of one another.

Usually, however, these modes are not really known to us.   But
we also do not really need them.   In the structure of Figure 8 both
waveguides I and II couple with each other and power converts between
modes of the separate waveguides I and II.   It is therefore, more appropriate to describe the fields in the structure of Figure 8 by the
superposition of the normal modes of waveguides I and II.   In the
following this superposition will be needed for the transverse fields
in any given cross-section with $z = z_1$:

$$\vec{E}_t = \sum_i \vec{E}_{ti}(x,y)A_i(z_1) \tag{2.15}$$

$$\vec{H}_t = \sum_i \vec{H}_{ti}(x,y)A_i(z_1) \tag{2.16}$$

Note that the summations include terms with positive and negative
subscripts i representing modes, which propagate in the positive and
negative z-direction, respectively.

Although this set of normal mode fields is not orthogonal, it
should be recognized that these fields are nevertheless well suited to
describe the field in a given cross-section and that also the series
in equations (2.15) and (2.16) will converge quite rapidly for any
actual field E_t, H_t.

The individual normal mode fields by themselves do not satisfy
Maxwell's equations in the guide of Figure 8. We therefore have
coupling between these modes. To determine this coupling we resort
to the concept of coupled wave equations. We write this coupled wave
equation in the following matrix form

$$\frac{d}{dz} \underline{A} = (-\underline{\gamma} + \underline{\kappa})\underline{A} \qquad (2.17)$$

where the elements of the diagonal matrix $\underline{\gamma}$ represent the undisturbed
propagation constans of the modes in waveguide I or II. The corresponding coupling coefficients κ_{mn} are calculated using the principle
of effective sources.

We can make any mode with subscript n to propagate undisturbed
if we impress electric currents of density $-i_n = -i_n(x,y,z)$ which is
so chosen that Maxwell's equations in the medium with $\hat{\varepsilon}$ are satisfied.
Considering equation (2.8) we obtain

$$\nabla \times \vec{H}_n = j\omega\varepsilon_n(x,y)\vec{E}_n = j\omega\hat{\varepsilon}(x,y)\vec{E}_n - \vec{i}_n$$

and the effective source is given by

$$-\vec{i}_n = j\omega\left[\varepsilon_n(x,y) - \hat{\varepsilon}(x,y)\right]\vec{E}_n \qquad (2.18)$$

This is the only current source which has to be impressed to make
the nth mode to propagate independently. Since μ is unchanged the
second of Maxwell's equations (2.9) is satisfied in any case.

For source-free field distributions the current density $+\vec{i}_n$ can
be regarded as the origin of mode coupling. After introducing equation
(2.6) into equation (2.18) the value of $\vec{i}_n$ at z is obtained as

$$\vec{i}_n(x,y) = -j\omega\Delta\varepsilon_n(x,y)\vec{E}_n(x,y)A_n(z) \qquad (2.19)$$

$$\Delta\varepsilon_n(x,y) = \varepsilon_n(x,y) - \hat{\varepsilon}(x,y) \qquad (2.20)$$

We have

$$\Delta\varepsilon_n(x,y) = \varepsilon - \varepsilon_I(x,y) \text{ if n denotes a mode of waveguide II}$$

$$\Delta\varepsilon_n(x,y) = \varepsilon - \varepsilon_{II}(x,y) \text{ if n denotes a mode of waveguide I}$$

We now consider sources $\vec{i}_n$ according to equation (2.19) in the short
region $z_1 - \Delta z < z < z_1$, and the transverse components $\vec{E}'_{tn}$ and $\vec{H}'_{tn}$ of the
fields which these sources generate in the cross-section at z = z_1.
In our coupled wave representation these transverse fields $\vec{E}'_{tn}$ and
$\vec{H}'_{tn}$ are formed by the superposition of the fields of all normal modes
i which, according to the coupled wave equations, are generated by
their interaction with the nth mode in the region $z_1 - \Delta z < z < z_1$.

In the limit of $\Delta z \to 0$ the coupled wave equations yield for these transverse fields

$$\vec{E}'_{tn} = \sum_i \vec{E}_{ti} \Delta z \kappa_{in} A_n(z_1) \tag{2.21}$$

$$\vec{H}'_{tn} = \sum_i \vec{H}_{ti} \Delta z \kappa_{in} A_n(z_1) \; . \tag{2.22}$$

Note that $\vec{i}_n$ is assumed to radiate only in the positive z-direction. This can in general only be achieved by introducing additional sources S_{na} in the region $z > z_1$ which compensate for the overall field terms propagation into the region $z < z_1 - \Delta z$.

To calculate the coupling from the nth mode to the mth mode we introduce at $z = z_1$ surface currents of magnitude

$$\vec{J}_{(-m)}(x,y) = -\vec{n}_z \times \vec{H}_{t(-m)}(x,y) \tag{2.23}$$

$$\vec{M}_{(-m)}(x,y) = \vec{n}_z \times \vec{E}_{t(-m)}(x,y) \tag{2.24}$$

These sources are assumed to radiate into the negative z-direction only. To prevent them from radiating into the region $z > z_1$ it may be necessary to impress additional sources $S_{(-m)a}$ at $z < z_1 - \Delta z$. The overall arrangement of sources and fields is shown in Figure 9. To determine the field $E'_{(-m)}$, $H'_{(-m)}$ which the sources of equations (2.23) and (2.24) excite in the region $z_1 - \Delta z \leqslant z \leqslant z_1$ in the limit $\Delta z \to 0$ note that the transverse components of this field are equal to the transverse field of the (-m)th mode.

$$\vec{E}'_{(-m)} \times \vec{n}_z = \vec{E}_{t(-m)}(x,y) \times \vec{n}_z \tag{2.25}$$

$$\vec{H}'_{(-m)} \times \vec{n}_z = \vec{H}_{t(-m)}(x,y) \times \vec{n}_z \tag{2.26}$$

The z-components are then obtained from the transverse field components with the help of Maxwell's equations. From them we obtain

$$j \omega \hat{\varepsilon}(x,y) (\vec{E}_{(-m)} \cdot \vec{n}_z) = \vec{n}_z \cdot [\nabla \times \vec{H}_{t(-m)}(x,y)]$$

$$j \omega \mu (\vec{H}_{(-m)} \cdot \vec{n}_z) = -\vec{n}_z \cdot [\nabla \times \vec{E}_{t(-m)}(x,y)] \; .$$

Comparing these equations with equations (2.8) and (2.9) leads to

$$\vec{E}'_{(-m)} \cdot \vec{n}_z = \frac{\varepsilon_m(x,y)}{\hat{\varepsilon}(x,y)} E_{z(-m)}(x,y) \tag{2.27}$$

$$\vec{H}'_{(-m)} \cdot \vec{n}_z = H_{z(-m)}(x,y) \tag{2.28}$$

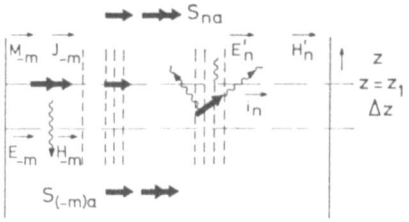

Fig. 9.   Longitudinal section of the coupling structure in Figure
          8 with fields and sources.

where $E_{z(-m)}(x,y)$ and $H_{z(-m)}(x,y)$ denote the z-components of the
field of the (-m)th mode.

   To calculate the coupling coefficients κ_{mn} we can now apply the
reciprocity theorem to the sources of the equations (2.19), (2.21)
and (2.24) and to their fields. The additional sources S_{na} and $S_{(-m)a}$
can be neglected since the fields which have to be multiplied by these
sources are zero in the corresponding regions. We thus obtain for
$\Delta z \to 0$

$$\Delta z \iint \vec{i}_n(x,y) \cdot \vec{E}'_{(-m)} \, dS_o =$$

$$\iint [\vec{J}_{(-m)}(x,y) \cdot \vec{E}'_{tn} - \vec{M}_{(-m)}(x,y) \cdot \vec{H}'_{tn}] \, dS_o \qquad (2.29)$$

Introducing for the fields and sources the results of equations (2.19)
and (2.21) to (2.25) and evaluating the resulting expressions leads
to

$$\kappa'_{mn} = \sum_i W_{mi} \kappa_{in} \qquad (2.30)$$

with

$$\kappa'_{mn} = \frac{j\omega}{2} \iint (\varepsilon_n - \hat{\varepsilon}) \, (\vec{E}_{tn} \vec{E}_{t(-m)} + \frac{\varepsilon_m}{\hat{\varepsilon}} E_{zn} E_{z(-m)}) \, dS_o$$

$$= \frac{j\omega}{2} \iint (\varepsilon_n - \hat{\varepsilon}) \, (\vec{E}_{tn} \vec{E}_{tm} - \frac{\varepsilon_m}{\hat{\varepsilon}} E_{zn} E_{zm}) \, dS_o \qquad (2.31)$$

$$W_{mi} = \frac{1}{2} \iint [\vec{E}_{t(-m)} \times \vec{H}_{ti} - \vec{E}_{ti} \times \vec{H}_{t(-m)}] \cdot \vec{n}_z dS_o$$

$$= \frac{1}{2} \iint [\vec{E}_{tm} \times \vec{H}_{ti} + \vec{E}_{ti} \times \vec{H}_{tm}] \cdot \vec{n}_z dS_o \qquad (2.32)$$

Equation (2.30) when written in matrix form appears as

$$\underline{\kappa}' = \underline{\underline{W}}\, \underline{\kappa} \qquad\qquad (2.33)$$

The elements of $\underline{\kappa}'$ are given by equation (2.31) and those of $\underline{\underline{W}}$ follow from equation (2.32). They can be determined if the normal mode fields are known. The element of $\underline{\kappa}$ can then also be evaluated from

$$\underline{\kappa}' = \underline{\underline{W}}^{-1}\, \underline{\kappa}' \qquad\qquad (2.34)$$

The results of equations (2.33) and (2.34) together with equations (2.31) and (2.32) are rigorous. Their evaluation, however, is very involved. To calculate the coupling between two specific modes it is necessary to consider all modes of type a and b. Thus, usually only an approximate solution can be achieved, which, however suffices for all practical purposes if only the conducting boundaries are far enough removed from the guiding regions.

To find this approximation it is helpful to derive a more suitable relation for the elements of the matrix W. For this purpose we calculate the real part Re(P) of the total power which is carried by a field distribution with an arbitrary set of mode amplitudes A_i. Only the propagating modes contribute to this real part of transmitted power. From

$$
\begin{aligned}
Re(P) &= Re\left\{ \iint (\vec{E} \times \vec{H}^*) \cdot \vec{n}_z \, dS_o \right\} \\
&= Re\left\{ \iint \left[\left(\sum_m A_m \cdot \vec{E}_{tm} \right) \times \left(\sum_n A_n^* \vec{H}_{tn}^* \right) \right] \cdot \vec{n}_z \, dS_o \right\} \\
&= \sum_m \sum_n \frac{1}{2} A_m A_n^* \left[\iint (\vec{E}_{tm} \times \vec{H}_{tn} + \vec{E}_{tn} \times \vec{H}_{tm}) \cdot \vec{n}_z \, dS_o \right]
\end{aligned}
$$

we find

$$Re(P) = \sum_m \sum_n A_m A_n^* W_{mn}$$

which in matrix notation appears as

$$Re(P) = \underline{A}^T \underline{\underline{W}} \underline{A}^* \qquad\qquad (2.35)$$

Note that the components of E_{tm} and H_{tm} are assumed to be real quantities. With the field normalisations of equations (2.3), (2.4) and (2.5) this is possible only for propagation modes. Thus in equation (2.35) we assume that the real power is carried by the fields of propagating modes only. Consequently, in the column vector $\underline{A}$ only the amplitudes of propagating modes are considered. The elements of $\underline{A}^*$ denote the conjugate complex elements of $\underline{A}.\underline{A}^T$ is the transposed (row) vector of $\underline{A}$. Since the guiding structure is lossless, the value of Re(P) must be constant along the coupled waveguides, which requires

$$\frac{d}{dz} \text{Re}(P) = \left(\frac{d}{dz} \underline{A}^T\right)\underline{W}\underline{A}^* + \underline{A}^T\underline{W}\left(\frac{d}{dz} \underline{A}^*\right) = 0 .$$

In this equation we substitute for the rate of change of $\underline{A}$ with z from equation (2.17) and equation (2.34) according to

$$\frac{d}{dz} \underline{A} = \left[-\underline{\gamma} + \underline{W}^{-1}\underline{\kappa}'\right] \underline{A} . \tag{2.36}$$

We note that for all propagating modes in the lossless structure the diagonal matrix has only imaginary elements,

$$\underline{\gamma} = \underline{\gamma}^T = -\underline{\gamma}^*, \tag{2.37}$$

and that for all modes which propagate in the same direction the matrix $\underline{W}$ according to equation (2.32) is symmetrical, and that all its elements are real

$$\underline{W} = \underline{W}^T = \underline{W}^* \tag{2.38}$$

This latter assumption is always justified if a wave travelling in one direction interacts significantly only with waves travelling in the same direction, so that any interaction between waves travelling in opposite directions may be neglected.

    Utilizing all the above relations we obtain the following condition for the conservation of power along the coupled waveguides

$$0 = \underline{A}^T\left[-\underline{\gamma} + \underline{\kappa}'^T\underline{W}^{-1}\right]\underline{W}\underline{A}^* + \underline{A}^T\underline{W}\left[-\underline{\gamma}^* + \underline{W}^{-1}\underline{\kappa}'^*\right]\underline{A}^*$$
$$= \underline{A}^T\left[-\underline{\gamma}\underline{W} + \underline{W}\underline{\gamma} + \underline{\kappa}'^T + \underline{\kappa}'^*\right]\underline{A}^*$$

To satisfy this condition for any combination of wave amplitudes, we must require that the expression within the brackets vanishes, thus

$$-\underline{W}\underline{\gamma} + \underline{\gamma}\underline{W} = \underline{\kappa}'^* + \underline{\kappa}'^T \tag{2.39}$$

holds for all elements m, n with m and n denoting the subscripts of propagating modes, only. In addition, all modes under consideration are assumed to propagate in positive z-direction, so that only positive subscripts m and n are taken.

    For m = n we obtain with

$$W_{nm} = 1 \tag{2.40}$$

the condition

$$R_e(\kappa'_{nn}) = 0$$

This condition may also be inferred from equation (2.31) when the appropriate normalization of fields in equations (2.3) is taken into consideration. For any propagating mode this normalization implies

$$E_{tn} = Re(E_{tn})$$

and

$$E_{zn} = jIm(E_{zn}).$$

Equation (2.39) yields for the off-diagonal terms of W with $m \neq n$:

$$W_{mn} = \frac{\kappa'^*_{mn} + \kappa'_{nm}}{\gamma_m - \gamma_n} \qquad (2.41)$$

This result allows us to very conveniently estimate the interaction terms in the coupled wave equations (2.36). Considering only propagating modes we find that the absolute value of W_{mn} will be the smaller the larger the value of $|\gamma_m - \gamma_n|$ is. Usually one is interested only in those fields which are excited by the modes of type a as described above. The phase constant $\beta_n = \gamma_n/j > k = \omega\sqrt{\mu\epsilon}$ of these modes is the greater the further the particular mode is above its cutoff frequency, while for the propagating modes of type b the phase constants β_n are always smaller than k. Hence, if only the coupling between the modes of type a is to be analyzed then usually only the elements of W which follow from interaction between just these modes need to be considered.

According to equation (2.3)

$$W_{mn} = \delta_{mn}$$

holds in any case in which n and m denote modes of the same dielectric waveguide. If n denotes a mode of one of two waveguides I or II and m a mode of the other waveguide then the value of W_{mn} will in general not vanish. In any case, however, its absolute value is smaller than 1. If the modes m and n are of the previously described type a, then the absolute value of W_{mn} is the smaller the larger the distance between the two guiding areas is and the higher the modes are above cutoff. In this case the fields of modes n and m overlap only little.

Thus, if we assume that the coupling between the modes of type a is small, then we can neglect the off-diagonal elements of W and obtain

$$\kappa_{mn} = \kappa'_{mn} \qquad (2.42)$$

This simple approximation is usually applied when interaction between modes of parallel dielectric waveguides is analyzed.

Results similar to equation (2.42) with equation (2.31) have
been derived previously, for example in[4] and [5] by applying different
methods. These methods do not, however, lead to the rigorous formulation of equation (2.34).

As an example to illustrate the coupled wave theory for parallel
dielectric waveguides we consider the case of two parallel dielectric
slabs in Figure 10. Single dielectric slabs represent the simplest
form of optical waveguides, they also serve as a model for other optical waveguides such as transparent substrates or transparent strips,
which are deposited on or embedded in transparent substrates. By
the same token the two parallel dielectric slabs in Figure 10 represent
a simple form of coupled optical waveguides, and serve well as a model
for other optical waveguide couplers.

The slabs in Figure 10 are assumed to extend to infinity in the
y- and z- direction, and we take the z-co-ordinate for the direction
in which the slab waves propagate. Figure 10 indicates the dimensions
and permittivities of the five-layer structure which forms the coupled
slabs. The conducting enclosure of the guiding structure which our
present coupled wave theory requires is here assumed to be removed
to infinity in the $\pm$x-directions as well as in the $\pm$y-direction.

The guided modes of slab I with permittivity ε_I will be represented in terms of the transverse co-ordinate x_{II}. To unify the representation x_i and ε_i with i = I or II, respectively. All fields are
constant in the y-direction, and hence independent of the y-coordinate.
We will consider here the coupling between TE- or H-waves of the slab
which have only a y-component of their electric field

$$\vec{E}_i = \vec{n}_y E_i \tag{2.43}$$

The index i on fields and on their propagation parameters indicates
that they belong to the mode i in slab i.

To evaluate the coupling coefficients according to equation
(2.30) we need first to determine the field distribution of guided
modes in each of the slabs without the other slab present. Each of
the two slabs by itself represents a symmetrical structure with its
center plane x_i = 0 as the plane of symmetry. Its guided modes have
hence either an even or an odd field distribution about this center
plane and are therefore called even and odd modes, respectively.

The fundamental mode of the symmetrical slab is its even mode.
Actually this mode has zero cutoff. Since this mode is of most practical concern we will consider only coupling between even TE- or H-
modes of the slabs, and list only their field distributions.

From Maxwell's equations follows the wave equation for the only
electric-field component of TE-modes according to

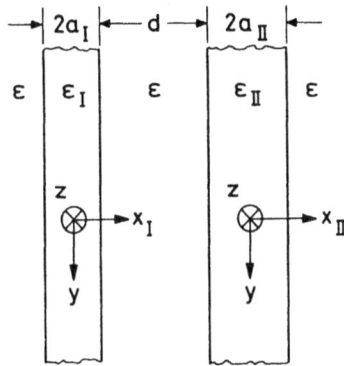

Fig. 10.   Two coupled dielectric slab guides.

$$\Delta E_i + \frac{\varepsilon_i}{\varepsilon} k^2 E_i = 0 \tag{2.44}$$

with $k = \omega\sqrt{\mu\varepsilon}$ as the wave number for the medium outside the slabs.
Its plane wave solution with an even distribution about x_i is for
$|x_i| < a_i$

$$E_i = N_i \cdot \cos(u_i x_i / a_i) \cdot \exp(-j\beta_i z) \tag{2.45}$$

where u_i follows from the separation condition for equation (2.44)
inside the slab

$$(u_i/a_i)^2 = k^2 \cdot \varepsilon_i / \varepsilon - \beta_i^2 \tag{2.46}$$

Plane wave solution for $|x_i| > a_i$, which match the field in equation
(2.45) at $|x_i| = a_i$, are

$$E_i = N_i \cos(u_i) \exp\left[-v_i(|x_i/a_i|-1)\right]\exp(-j\beta_i z) \tag{2.47}$$

where v_i follows from the separation condition for equation (2.44)
outside the slab

$$(v_i/a_i)^2 = \beta_i^2 - k^2 \tag{2.48}$$

One of Maxwell's equations, namely

$$-\nabla \times \vec{E}_i = j\omega\mu\vec{H}_i$$

allows to derive the magnetic field from the above electric field.
Its only components are

$$H_{zi} = \frac{j}{\omega\mu} \frac{\partial}{\partial x_i} E_i \tag{2.49}$$

$$H_{xi} = - \frac{\beta_{|i|}}{\omega\mu} E_i \, \text{sign}(i) \tag{2.50}$$

Matching the tangential components H_z at the slab surface leads to the characteristic equation

$$u_i \tan(u_i) = v_i \tag{2.51}$$

We furthermore obtain from both separation conditions (2.46) and (2.48)

$$u_i^2 + v_i^2 = a_i^2 \, k^2 (\epsilon_i/\epsilon - 1) \equiv V_i \tag{2.52}$$

To normalize the guided mode fields we require it to carry unit power in the z-direction per unit width of the slab in the y-direction

$$\int_{-\infty}^{\infty} (\vec{E}_i \times \vec{H}_i^*) \cdot \vec{n}_z dx_i = - \int_{-\infty}^{\infty} E_i H_{xi} dx_i = \text{sign}(i) \tag{2.53}$$

The normalization factor follows from this condition as

$$N_i = \left[\frac{\beta_{|i|} a_i}{\omega\mu} \left(1 + \frac{1}{v_i} \right) \right]^{-\frac{1}{2}} \tag{2.54}$$

This normalization corresponds to the procedure which equation (2.3) prescribes. Note that the present structure extends to infinity in the y-direction but that the fields are independent of y. When normalizing the fields, and also when evaluating the coupling coefficients we need to integrate only over the transverse x-coordinate.

Equations (2.45) to (2.52) and (2.54) provide us with all the expressions that we require to evaluate the coupling between any of the even TE-modes of two parallel slabs. For a given value of the slab parameter V_i in equation (2.52) the respective slab guides a finite and definite number of such modes. For $V_i \leqslant \pi$ it guides only the fundamental TE-mode.

We designate a TE-mode in one slab with index m and a TE-mode in the other slab with index n. By evaluating equation (2.31) for the fields in these two modes we find for the coupling coefficient between them

$$\kappa_{mn}' = j \sqrt{\frac{\epsilon_m - \epsilon}{\epsilon_n - \epsilon}} \frac{1/2}{(\beta_m a_m \beta_n a_n)^{1/2}} \frac{u_n/a_n}{\{1 + (1/v_n)\}^{1/2}}$$

$$\frac{u_m/a_m}{\{1+(1/v_m)\}^{1/2}} \; \frac{\exp(-v_n d/a_n)}{(v_n/a_n)^2 + (u_m/a_m)^2}$$

$$\left[\frac{v_m}{a_m} - \frac{v_n}{a_m} \; \exp(-2v_n a_m/a_n) + \frac{v_m}{a_m} + \frac{v_n}{a_n}\right]. \qquad (2.55)$$

This formula shows clearly the exponential decay of coupling strength with increasing values of d and v_n/a_n. The two modes couple via their evanescent fields in the gap of width d. Therefore, if d or if the transverse attenuation v_n/a_n of fields increase the coupling strength decreases exponentially.

The coupling coefficients of equation (2.55) do not obey reciprocity. The reason for this discrepancy is that, because of equation (2.34), the formula equation (2.31) only approximates the actual coupling coefficients, and that the guided modes of the single slabs are not orthogonal to each other in the composite structure.

The same result as given by equation (2.42) together with equation (2.55) has also been obtained in[5]. Reference[6] investigates in detail under what conditions these approximate results remain valid. It is found there, that equation (2.42) is rigorous when the two coupled modes m and n are degenerate. The simplification equation (2.42) may therefore be utilized whenever coupled modes do not differ greatly in their propagation constants.

3. WAVELENGTH-SELECTIVE DIRECTIONAL COUPLER

As an example for the application of the theory of coupled optical waveguides a wavelength-selective directional coupler will be analyzed. Directional couplers with a wavelength-selective over-all coupling may be useful in optical communication systems that transmit several optical carriers of different wavelengths over the same transmission medium. A typical case is the bi-directional transmission of optical signals for duplex operation over a single glass fibre in wavelength multiplex. Signals in one direction are transmitted on one wavelength and in the opposite direction on another. Terminals and repeaters for such duplex operation can use wavelength-selective directional couplers as in Figure 11 to transmit on one wavelength λ_1 and receive on another wavelength λ_2 from the opposite direction. Because of the selectivity of the directional coupler nearly all the light power emitted from the transmitter at λ_1 will be launched into the fiber, while nearly all the power arriving through the fiber at λ_2 will be transferred to the receiver. Furthermore, the selectivity

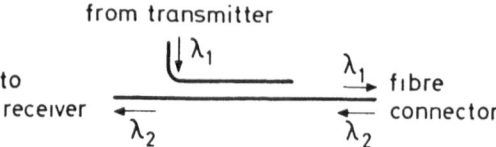

Fig. 11. Wavelength-selective directional coupler as transmit-
          receive duplexer for fiber systems with duplex operation.

of the directional coupler enhanced by its directivity will almost
completely suppress the near-end crosstalk. A very sensitive receiver
will therefore be well protected against high transmitter power. As
an added advantage of this transmit-receive duplexer for monomode
fiber systems the signal received in the lowest order dominant mode
may have any polarization. It will always pass the wavelength-selective directional coupler with little coupling loss.

     To achieve the wavelength-selective power conversion the direc-
tional coupler has the structure shown in Figure 12. The two waveguides 1 and 3 are coupled from z = 0 to z = L to the intermediate
waveguide section 2. The two outer waveguides 1 and 3 may have identical cross-sections, but can also differ one from another. We assume
here only for the analysis that the modes 1 and 3 in waveguides 1 and
3 which are to interact selectively have one and the same phase constant

$$\beta_1 = \beta_3 = \beta \qquad\qquad (3.1)$$

For all practical purposes we need to satisfy this condition for phase
synchronism only for that wavelength λ_1 at which by selective conversion all the power entering in one waveguide is to be transferred
to the other waveguide.

     The mode 1 in waveguide 1 couples to mode 3 in waveguide 3 via
a mode in the intermediate waveguide 2. We assume here for the analysis that mode 1 has the same coupling to this intermediate-wave-
guide mode as mode 1 has it. For all practical purposes also this
condition needs to be satisfied only for the wavelength λ_1.

     Under these conditions, and if heat losses in the coupler may
be neglected, the following system of coupled wave equations holds
for the amplitudes A_1, A_2 and A_3 of the modes in the respective
waveguides

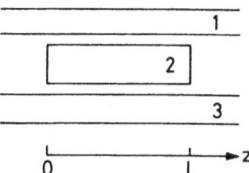

Fig. 12.   Directional coupler with intermediate waveguide for
coupling waves.

$$dA_1/dz = -j\beta A_1 - jcA_2$$

$$dA_2/dz = -jcA_1 - j\beta_2 A_2 - jcA_3$$

$$dA_3/dz = -jcA_2 - j\beta A_3 \qquad (3.2)$$

β_2 is the phase constant of mode 2 in the intermediate waveguide 2
and c the coupling coefficient for coupling between mode 1 and this
intermediate waveguide mode 2 as well as between the latter and mode
3.

The system with 3 coupled waves has three normal modes which
propagate independently from one another in the coupling region.
The phase constants of these normal modes are

$$\beta \quad , \quad \beta - \delta - \sqrt{\delta^2 + 2c^2} \quad , \quad \beta - \delta + \sqrt{\delta^2 + 2c^2} \quad (3.3)$$

where

$$\delta = (\beta - \beta_2)/2 \qquad (3.4)$$

designates half the difference in phase constants between modes 1 and
3 in the outer waveguides and mode 2 in the intermediate waveguide.

The three normal modes with amplitudes w_1, w_2 and w_3 superimpose
to form the following general solution for the amplitudes of the
coupled modes 1, 2 and 3

$$A_1 = -w_1 e^{-j\beta z} + w_2 e^{-j(\beta-\delta-\sqrt{\delta^2+2c^2})z} + w_3 e^{-j(\beta-\delta+\sqrt{\delta^2+2c^2})z}$$

$$A_2 = -\frac{\sqrt{\delta^2 + 2c^2} + \delta}{c} w_2 e^{-j(\beta-\delta-\sqrt{\delta^2+2c^2})z}$$

$$+ \frac{\sqrt{\delta^2 + 2c^2} - \delta}{c} w_3 e^{-j(\beta-\delta+\sqrt{\delta^2+2c^2})z} \qquad (3.5)$$

$$A_3 = w_1 e^{-j\beta z} + w_2 e^{-j(\beta-\delta-\sqrt{\beta^2+2c^2})z} + w_3 e^{-j(\beta-\delta+\sqrt{\delta^2+2c^2})}$$

With respect to the longitudinal plane of symmetry of the coupler in
the center of the intermediate waveguide, the normal modes of the
coupling region are either even or odd. From the general solution
(3.5) we see that w_1 is the amplitude of an odd mode; it has the same
phase constant as the two coupled modes in the outer waveguides, and
does not contribute to the amplitude A_2 of the intermediate waveguide
mode. The normal modes with amplitudes w_2 and w_3 are even modes. They
contribute evenly in a balanced fashion to the amplitudes A_1 and A_3 of
the coupled modes in the outer waveguides.

If unit power is launched into mode 1 of waveguide 1 at $z = 0$
we have as initial conditions

$$A_1 = 1 \qquad\qquad A_2 = A_3 = 0 \qquad\qquad (3.6)$$

Under these initial conditions the amplitudes of the two outer waveguide modes amount to

$$|A_1| = \frac{1}{2}\left|1 + e^{j\delta z}\left(\cos\sqrt{\delta^2+2c^2}\,z - \frac{j\delta}{\sqrt{\delta^2+2c^2}}\sin\sqrt{\delta^2+2c^2}\,z\right)\right|$$

$$ \qquad\qquad (3.7)$$

$$|A_3| = \frac{1}{2}\left|1 - e^{j\delta z}\left(\cos\sqrt{\delta^2+2c^2}\,z - \frac{j\delta}{\sqrt{\delta^2+2c^2}}\sin\sqrt{\delta^2+2c^2}\,z\right)\right|$$

Two limiting cases are of special interest:

1. $\beta_2 = \beta$:

The coupling mode 2 has the same phase constant as the modes 1 and 3.
In this case of $\delta = 0$ the amplitudes of modes 1 and 3 amount to

$$|A_1| = \frac{1}{2}\left|1 + \cos(\sqrt{2}\,cz)\right|$$

$$|A_2| = \frac{1}{2}\left|1 - \cos(\sqrt{2}\,cz)\right| \qquad\qquad (3.8)$$

If the coupling mode 2 is in phase synchronism with modes 1 and 3
the power initially in mode 1 transfers completely to mode 3 at

$$z = (2q + 1)\pi/(\sqrt{2}\,c) \qquad\qquad \text{where } q = 0,1,2\ldots \quad (3.9)$$

while it transfers back to mode 1 at

$$z = 2q\pi/(\sqrt{2}\,c) \qquad\qquad \text{with } q = 0,1,2\ldots \quad (3.10)$$

It thus shuttles back and forth between the outer two waveguides.

For complete power transfer from mode 1 to mode 2 the coupler is best made

$$L = \pi/(\sqrt{2}c) \qquad \text{with } q = 0,1,2... \qquad (3.11)$$

long.

2. $|\delta| \gg c$

Complete power conversion between modes 1 and 3 requires $\delta = 0$, that is phase synchronism for the coupling wave. For $\delta \neq 0$ only part of the input power in mode 1 converts maximally to mode 3. In the limiting case of $|\delta| \gg c$ the converted power fraction remains even rather small. In this case the amplitudes of modes 1 and 3 amount approximately to

$$|A_1| \simeq |1 + j[c^2/(2\delta^2)]\sin\delta z \ e^{j\delta z}|$$

$$(3.12)$$

$$|A_3| \simeq [c^2/(2\delta^2)]|\sin\delta z|$$

According to this approximation only the fraction $c^4/(4\delta^4)$ of the input power in mode 1 converts maximally to mode 3. The rest of the input power remains predominantly in mode 1, but to a small extent also in mode 2. However the fraction which mode 1 will maximally loose to mode 2 under these conditions is only c^2/δ^2 of the input power.

To achieve the desired selectivity, that will lead to complete power conversion at a wavelength λ_1 by a specified difference a suitable intermediate waveguide must be chosen. It s coupling mode 2 must be in phase-synchronism with modes 1 and 3 at λ_1 but at the specified stop wavelength it must differ sufficiently in phase constant from mode 1 and 3 to satisfy the condition $|\delta| \gg c$.

At optical wavelengths these requirements may well be met by planar waveguide couplers with films or strips in the basic form of Figure 13. The outer waveguides 1 and 3 in Figure 13 have the same cross-section and refractive index n_1. The intermediate waveguide 2 has either a wider cross-section than the outer waveguides or a larger refractive index $n_2 > n_1$ or both. The differences in cross-section and in refractive index determine to a large extent the selectivity of the coupler. All three waveguides are deposited on or embedded in a substrate of refractive index $n_0 < n_1, n_2$.

Figure 14 shows the dispersion characteristics of the coupled modes in Figure 13. β is the phase constant of the fundamental mode in the two outer waveguides 1 and 2, while β_2 are phase constants of modes in the intermediate waveguide 2 which can serve as coupling

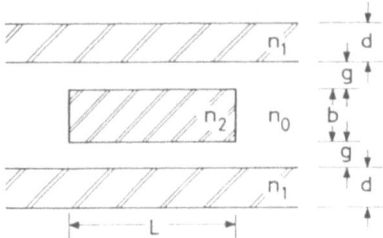

Fig. 13. Planar waveguide coupler with film or strip guides

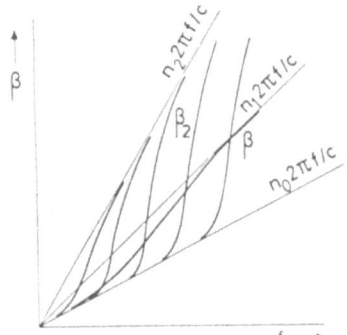

Fig. 14. Phase constant β and β_2 as a function of frequency f in a
          frequency-selective directional coupler.

modes. All phase curves start at the respective cutoff-frequency on
the line $2\pi n_0 f/c$. Far above the cutoff they approach the wavenumber
of the respective waveguide material asymptotically. Except for the
phase curve of the fundamental mode of the intermediate waveguide
the phase curves of all other higher order modes of this waveguide
cross the phase curve of the fundamental mode in the outer waveguides.
Any one of these higher order modes can therefore serve as a coupling
mode between the fundamental modes in waveguides 1 and 3. At the
crossover-points of their phase curves they are able to transfer
power completely between the outer waveguides.

   Which mode to choose as coupling mode and how to design the intermediate as well as the outer waveguides depends on the frequencies
or wavelengths which are to be coupled and those, which should remain
decoupled. For large spacings between the coupled and decoupled wavelengths a low order coupling mode will do, for narrow spacing and high
selectivity it should be a higher order coupling mode. The selectivity

of the coupler may also be enhanced by raising the refractive index
of the intermediate waveguide or by widening its cross-section. These
measures will let the phase curves cross-over at a steeper angle with
respect to each other, and the phase difference δ will increase at a
faster rate with increasing deviation of the wavelength from the crossover point.

In order to investigate the influence of the cross-sectional
dimensions and of the refractive indices on the coupler performance,
coupled film waveguides have been adopted as a model[7]. Interaction
was considered only for H_m-modes with the fundamental H_o-mode in the
two outer films and higher order H_m-modes as coupling modes in the
intermediate film. For the coupling coefficient between these modes
not the approximation (2.55) was used but a more accurate formula
from [6]. Figures 15 and 16 show the power fractions P_1 and P_3 out of
waveguide 1 and 3 respectively for unit power into waveguide 1 plotted
as a function of the wavelength λ relative to the wavelength λ_1 for
maximum power conversion to waveguide 3. In both examples Figure 15
and Figure 16 the H_2-mode of the intermediate film was the coupling
mode. But power conversion due to both modes of adjacent orders in
the intermediate film was also calculated. Power conversion due to
this parasitic coupling remains however so small that it hardly shows
in the diagrams. Only the H_3-mode causes some power conversion near
the short wavelength border of the diagrams as the dash-dotted lines
indicate it.

Figure 15 shows the conversion characteristics for a coupler that
is $L=676\lambda$ long. Its wavelength for P_3 down by 1dB from the maximum
power conversion are spaced by $\Delta\lambda=0.013\lambda_1$. The selectivity when
measured by the wavelength difference from λ_1 where P_1 is 1dB below
the input power amounts to $\Delta\lambda=0.029\lambda_1$ below λ_1. Above λ_1 the output
power P_1 increases only to 71% of the output power. The selectivity
is hence quite poor above λ_1.

Better performance in this respect is shown by the coupler in
Figure 16 with $L=1779\lambda$ it is much longer. But its selectivity for
$P_1 = -1$dB is better than $\Delta\lambda=0.01\lambda_1$ above and below λ_1. As a drawback however, its bandwidth for $P_3 = -1$dB amounts to only $\Delta\lambda=0.003\lambda_1$.

A conversion characteristic with a wider band for nearly complete
power conversion but less selectivity appears in Figure 17. The
coupling mode in this case is the fundamental mode of the intermediate
guide. Its dispersion curve has been tailored to cross the dispersion
curve of the fundamental mode of the outer guide by making the intermediate film quite thin and of relatively high refractive index. The
coupler has a 1dB bandwidth of $\Delta\lambda=0.036\lambda_1$ but its selectivity is so
low that $P_1 = -1$dB only at $\Delta\lambda=0.045\lambda_1$ below λ_1.

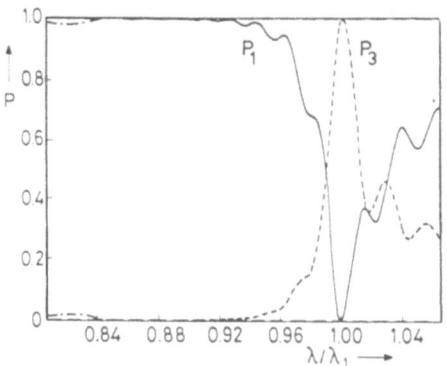

Fig. 15.   Conversion of a wavelength-selective film guide coupler
P_1 output power in film 1
P_3 output power in film 3
$n_1 = n_3 = 1.428$, $n_2 = 1.48$, $n_o = 1.4$
$d = 1.36\lambda_1$, $b = 2.63\lambda_1$, $g = 4.26\lambda_1$, $L = 676\lambda_1$

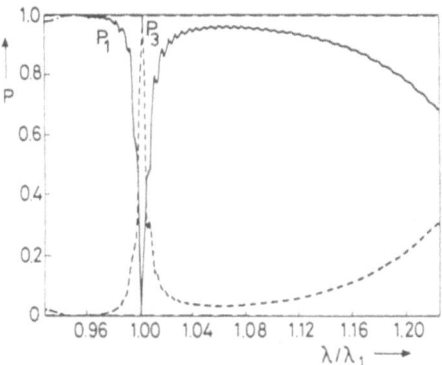

Fig. 16.   Conversion of a wavelength-selective film guide coupler
P_1 output power in film 1
P_3 output power in film 3
$n_1 = n_3 = 1.47$, $n_2 = 1.54$, $n_o = 1.4$,
$d = 1.02\lambda_1$, $b = 2.2\lambda_1$, $g = 5.95\lambda_1$, $L = 1779\lambda_1$

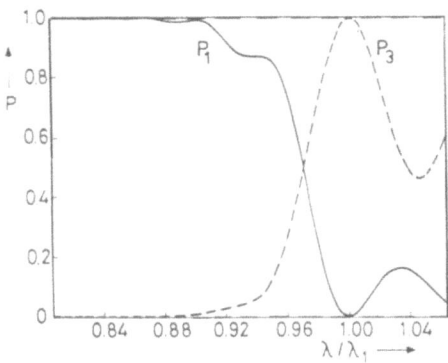

Fig. 17.   Conversion of a wavelength-selective film guide coupler

P_1 output power in film 1

P_3 output power in film 3

$n_1 = n_3 = 1.456$, $n_2 = 1.54$, $n_0 = 1.4$

$d = 0.65\lambda_1$, $b = 0.207\lambda_1$, $g = 5.95\lambda_1$, $L = 1415\lambda_1$

REFERENCES

1. E. A. J. Marcatili, Dielectric rectangular waveguide and directional coupler for integrated optics, Bell Syst. techn. J. 48, pp. 2071-2102, 1969.

2. R. Pregla, A method for the analysis of coupled rectangular dielectric waveguides, Archiv. Elektronik u. Übertragungs-techn. 28, pp. 350-357, 1974.

3. F. Sporleder and H. G. Unger, Waveguide tapers transitions and couplers, P. Peregrinus Ltd., Stevenage, UK, 1979.

4. D. Marcuse, The coupling of degenerate modes in two parallel dielectric waveguides, Bell Syst. Tech. J. 50, pp. 1791-1816, 1971.

5. A. W. Snyder, Coupled mode theory for optical fibers, J. Opt. Soc. Amer. 62, pp. 1267-1277, 1972.

6. E. F. Kuester and D. C. Chang, Nondegenerate surface-wave mode coupling between dielectric waveguides, IEEE Trans, MTT-23, pp. 877-882, 1975.

7. J. Jacob, Studienarbeit, Institut für Hochfrequenztechnik, Technische Universität Braunschweig, 1981.

FABRICATION OF SURFACE OPTICAL WAVEGUIDES

G. Chartier

I. N. P. Grenoble E. N. S. I. E. G.
B. P. 46
38402 Saint-Martin d'Hères, France

1. INTRODUCTION

Review papers[1,2,3] traditionally begin by emphasizing the fact that for centuries, the basic design of optical systems has not been changed and has consisted of bulky and heavy components which require careful alignment and protection.

Guided wave optics is a new development of Optics which avoids the previous disadvantages and will allow the construction of new optical devices for telecommunications, signal processing and instrumentation. This development has been made possible for two reasons: the first is the availability of low loss optical fibers, the second is the possibility of growing thin layers of high optical quality and having a thickness of the order of an optical wavelength.

The idea of repeating, in the optical field, all that had been done in the microwave region appeared soon after the discovery of coherent laser light sources[4]; the problem was then to make "pipes" to guide the electromagnetic energy. The dimensions of those pipes having to be of the same order of magnitude as the guided wavelength (micrometer) it is obvious that it is not possible to make metallic pipes, as in microwaves. It was immediately thought that the solution was in dielectric waveguides, since the design requirements for optical waveguides are within the tolerance limits of existing thin layer technology and of lithographic techniques.

One purpose of my lectures is to review the available techniques that are presently used for fabricating surface optical waveguides.

Ideally one would dream of a single material where the various functions required by optical circuits could be made, this would give totally integrated systems. The only possible materials, at the moment, belong to the III.V semiconductors family, and mainly to the GaAs family; attempts have been made to find new materials to compete with GaAs, especially among II.VI semiconductors[5]; ZnTe seemed very promising but it is almost impossible to make channel waveguide in it[6]. The fabrication of integrated optical devices in semiconductors will be studied elsewhere in this course and I shall limit myself to non-semiconducting materials.

If most functions for optical circuitry can be made in GaAs, this medium is not the best for each function; and it can reasonably be thought that for a long time, if not forever, integrated optics will have to be hybrid.

Many materials have been used for making optical waveguides: most of them had been proposed because they were easy to be elaborated but they will definitely not have industrial importance. Some of them will be described either because of their historical role or because they often allow an easy construction of devices for experiments in the laboratory.

The most important materials for integrated optics are, at the moment, GaAs, $LiNbO_3$, $LiTaO_3$, glass, silicon.

For optical fibres it is well known that there are two kinds of situations: multimode fibres and monomode fibres; in the first one the guiding core is large (50 micrometers or more), whilst the core is very small in the monomode case (5 micrometers). Historically, mainly because of fabrication and connection facilities, multimode fibres were first developed. It is certain that for long range tele-communications monomode fibres will be exclusively used, but still many other multimode devices have an optimistic future. The situation is quite different in the case of integrated optics where the light is guided in a layer; if one is willing to be in a highly multimode situation, the thickness of the layer must be a few tens to a few hundreds of wavelengths (ie. 20 to 200 micrometers). Such thick layers exhibiting a high optical homogeneity and flatness, are not trivial to grow; the only method for making highly multimode planar optical waveguides is ion exchange and diffusion in glass[7].

A thin layer is almost a two-dimensional space vector; in guided optics light energy is trapped in the immediate vicinity of a surface. If one wants a guided light wave to go from one point to another point of the surface, the rule of the game is the "Fermat principle" applied to the surface. Up to now Optics was developed in the R3 space vector, a new kind of optics is to be studied in an R2 space vector[8]. Many good ideas for new experiments can be found when turning over the pages of a text book on classical optics and thinking of transposing experimental situations from R3 to R2; to do so, it is necessary to

be well aware of the technological possibilities for making optical
waveguides.

Most applications for integrated optics are in the signal processing field and do need channel waveguides. When a technology for
making optical waveguides is to be chosen, importance must be given
to its compatibility with masking techniques.

Materials that are used in guided optics can be qualified by
using one, or more, of the following adjectives: linear, non-linear,
active, passive.

Active. A medium is said to be active when its optical properties can be modified from the outside; the characteristics of a wave
propagating in an active material can be modified deliberately. As in
usual electronics the word active has two different possible meanings:

(i)   The shape of the wave remains identical to itself, but its in-
      tensity increases while propagating in the active medium, which
      is also called an amplifying medium.
(ii) The power of the signal remains at best constant, or eventually
      decreases, but the shape of the wave is modified according to an
      externally applied signal; this second sense of the word corres-
      ponds to applications in signal processing.

Most effort has been made on the second kind of active materials.

Non Linear. In non linear media the refractive index and accordingly the dielectric constant, depend on some physical field existing
inside the medium; this physical field can be electric, magnetic or
acoustic; we shall respectively speak of electro-optic, magneto-optic
and acousto-optic activities. The relation between the dielectric constant and the physical field is generally of a tensorial kind.

Integrated optical active devices of the second preceeding kind
mainly use the electro-optic effect, and sometimes the acousto-optic
effect. At the beginning of the 70's an important research effort
was devoted to the elaboration of integrated magneto-optic waveguides
but with rather low success[9].

In so-called "non-linear optics" the interaction takes place
between the fields of optical waves and leads to frequency conversion.
The situation is very favorable in principle for this kind of nonlinear interaction, since we can have a high optical energy density
inside the guides[2]; despite this fact only few results have been obtained on second harmonic generation or optical parametric amplification or oscillation.

2. THE DIFFERENT METHODS FOR FABRICATING OPTICAL WAVEGUIDES

Optical waveguides are of the dielectric kind and are made of a medium having a higher refractive index than the surrounding media. As they must be very thin (on the order of micrometers) one cannot think of making them in the shape of a sheet floating in the air; so optical waveguides necessarily lie on a substrate. Fabrication techniques that are used for making optical waveguides can be roughly classified into two types: in the first one a thin transparent layer is grown on a less refringent substrate; in the second, a transparent substrate is submitted to some physical process which increases its refractive index, starting from the surface.

Besides the fabrication technique, the following criteria are used to characterize and compare different optical waveguides:

-Structure: this can be amorphous (organic or inorganic), poly-crystalline, single crystal.

-Losses: these range from 0.01db/cm to 10db/cm and are usually of the order of 0.5db/cm (at 633nm). The losses are partly due to the bulk absorption of the material which the layer is made of, and mainly due to scattering losses at the interfaces. From this point of view graded index guides are preferable as well as guides buried below the surface of the substrate.

-Value of refractive indices: two refractive indices have to be considered: the index of the substrate and the index of the layer. The more usual substrates are: glass, silicon dioxide, lithium niobate lithium tantalate, semi-conducting materials. For glass and silicon dioxide the index is about 1.5; the index is considerably higher in the case of single crystal materials (2 to 3), the problem of finding a material of higher index is then not an easy one.

-Wavelength region of transparency: for all waveguides described here, the region of transparency goes from the visible to the near infrared; few measurements are available but there is no problem in going as far as 1.5 micrometer, a wavelength which is important for telecommunications. For deposited layers, the transparency is about the same as that of the bulk material; in the case of waveguides induced at the surface of a substrate, the transparency is about that of the substrate.

Value and shape of the gradient index

A very small difference Δn between the refractive indices of he layer and of the substrate is quite enough to observe lightwave guiding. For example in the case of glass Δn can be as low as 0.0001. Of course the smaller Δn is, the thicker must be the layer for having at least one guided mode at a given wavelength.

Table 1. Main Non Semiconducting Optical Waveguides.

Material	Form*	Fabrication technique	Index**	Loss** db/cm	Refer
Soda Lime glass	A	Ag^+ exchange	1.51-1.61	0.5	10-11
		K^+ -	1.51-1.52	0.1	10
		Li^+ -	1.51-1.52	0.1	7
TiF_6 glass (Schott)	A	Ag^+ exchange			12
Aluminosilicate glass	A	K^+ exchange	1.52-1.56	0.1	10
Silica	A	Li implanted	1.46-1.48	0.2	13
7059 glass Corning	A	Sputter	1.43-1.59	2	14
Ta_2O_5	A	Sputter	2.2	1	15
Nb_2O_5	A	Sputter	2.3	1	
Si_3N_4	A	Sputter	2.0	0.5	33
ZnO	PC	Sputter	2	20	16
ZnS	PC	Evapor	2.3	5	16
$LiNbO_3$	MC	Tl exchange	$\Delta n=0.13$		17
		Ion impl.	$\Delta n=-0.1$		18
		Diffusion	$\Delta n=0.1$		19-20
$LiTaO_3$	MC	Ag exchange	$\Delta n=0.05$		21
		Diffusion			18
Polyurethane	A	Organic	1.54-1.58	1	22
Epoxy	A	Films	1.58	0.3	
K.O.R.	A	-	1.78	7	23
Polyphenylsilox	A	-	1.56	2	24
Methylmetacryl	A	-	1.51	0.2	25-2
Vinylmethylsil	A	-	1.54	0.01	26

* A = Amorphous   PC = Polycrystalline   MC = monocrystal

** 633nm

When the guide is induced from outside by some physical process, the index does not vary abruptly with the depth x of penetration inside the substrate. The law of variation of n versus x can be written as: $n(x) = n_s + \Delta n.f(x)$, n_s is the bulk index of the substrate, Δn is the amplitude of the gradient index, $f(x)$ is a normalized dimensionless function which goes to zero when x goes to infinity. The shape of $f(x)$ depends on the physical process used to make the waveguides.

For deposited layers, $f(x)$ is a step-function; in many cases $f(x)$ will be a maximum on the surface, and will have the shape of a half-bell. Gaussian function, complementary error function and Fermi function profiles are usually observed.

3. FABRICATION OF GUIDES BY DEPOSITION OF A LAYER ON A SUBSTRATE

Organic materials

For the purpose of photomasking in integrated electronics a technology had been developed for making thin layers of photopolymerized resins. Such layers appeared to have a thickness quite suitable for waveguiding (0.1 to 2 micrometers): they are geometrically and optically homogenous and have the same flatness as the substrate on which they have been deposited. As they often have a good optical transparency, they have been used since the very beginning of integrated optics. Although they are very convenient, organic layers have limitations which make them obsolete. Their losses can be very low (0.01db/cm[27]) but are usually in the 1-5db/cm range.

Of course plastic films have no electro, magneto or acousto-optic activity. However, they can be doped with the organic molecules which are used in dye lasers (Rhodamine 6G or Rhodamine B for example). When optically pumped by suitable laser light, such a doped layer acts as an optical amplifier[28,29].

The practical use of organic materials may be limited by ageing effects, and this limitation makes them inappropriate for telecommunications applications. Organic layers are still useful for quick construction of a device under study or in prototypes. They can be recommended for practical work for students or for helping non-specialists in integrated optics to become aquainted with optical waveguides. In fact waveguides made by ion exchange in glass are even cheaper and easier to make, they have lower losses and do not suffer from ageing.

A great research effort in this field has been made in the early 70's[22,23,24,25,26,30]. We shall briefly summarize the principle of the method which uses organic compounds, called resins. Resins are made of molecules which remain independent - ie. not bonded - under some physical conditions and which polymerize when some well determined conditions are changed. To obtain polymerization one can use the

evaporation of some solvent, the addition of a catalyst, exposure to
suitable light, bombardment by electrons of well chosen energy; resins
which polymerize according to one of the last two processes are, of
course, well adapted for making laterally bounded waveguides.
Unexposed resins are rather viscous liquids, the viscosity can easily
be controlled using proper dilutant.

One, or a few drops of resin are dropped from a syringe on a one
square inch substrate which is rapidly rotated (spinning speed of
the order of one thousand to ten thousand cycles per minute).

The resultant film thickness depends on the viscosity of the
solution and on the spinning speed, it is usually on the order of 0.1
to 1 micrometer. When thicker films are needed a so-called "controlled
withdrawal technique" should be used: a substrate is immersed in the
resin solution and withdrawn at a controlled speed which is of the
order of a few cm/min. The thickness of the layers decreases when the
withdrawal speed is increased, a typical value is a thickness of about
one micrometer for a speed of 5cm/min, using polyurethane.

Vacuum deposited film

Sputtering: vacuum deposition has been known for a long time to
be one of the best methods for growing thin films and is extensively
used for making optical waveguides. Since the materials to be deposited are dielectrics, simple thermal evaporation is not well adapted; nevertheless it has been used[16] in the case of zinc sulphide,
the heating source being an electron beam. In most other cases sputtering has to be used, sputtered films are very attractive for integrated optical circuits. By employing a radio-frequency sputtering
process films can be produced from substances that cannot be readily
evaporated and repeatable results can be achieved from a source consisting of a compound of materials[14,31,32,33,34].

A disc of the material to be sputtered is mounted on a backing
plate in a vacuum chamber; a small quantity of inert gas (usually
argon) is allowed into the chamber and is ionized by applying an rf
voltage to the backing plate. The difference in mobility of the Ar^+
ions and the electrons in the plasma enables an excess of electrons
to accumulate on the disc during positive half-cycles of the rf
supply voltage. This creates a negative d.c. potential on the disc.
Thus positive Argon ions are accelerated towards the disc which is
struck by one hundred electron-volt argon ions. The "billiard ball"
momentum transfer between the plasma ions and the target disc results
in target atoms, or molecules, being released from the disc; these
neutral particles drift across the plasma and deposit as a thin film
on a suitably placed substrate in the vacuum chamber. For composite
materials, such as metal oxide dielectrics, there may be a loss of
the gaseous components e.g. oxygen, during the sputtering process;
this may be compensated by introducing a few percent of oxygen into
the plasma.

For waveguide fabrication the target is often a sheet of glass (generally a borosilicate or Pyrex type, Corning 7059 glass for example) of slightly higher refractive index than the substrate (a soda lime glass) onto which the film will be deposited. The sputtering process is relatively slow (one micrometer per hour) because the rf power must be limited to prevent thermal cracking of the target; the film thickness is accurately determined by simply timing the process. The refractive index of many glasses may be modified by altering the oxygen content of the constituent. The loss of oxygen in the sputtering process can therefore be used advantageously to control the refractive index of the film. Using an alumino-borosilicate glass film the index can be tuned from 1.53 to 1.59. Typical loss values range from 0.5db/cm to 3db/cm, and most of the loss is due to scattering from isolated defects in the film; the cause of such defects can be attributed to contamination(dust particles, residual cleaning agents ...) on the substrate prior to deposition, flaws in the substrate and also surface defects in the film layer resulting from overheating during the process. If the oxygen deficiency is severe then the films exhibit a pale-yellow-brown appearance and absorption precludes their use as guides, at least in the visible spectrum.

Sputtering is certainly not an easy technology, and if one wants to use it for growing optical waveguides, he must be sure that he is familiar with sputtering difficulties. Nevertheless this technique can be tamed and gives excellent results. The plasma of a sputtering device is a very reactive medium where exotic layers can be grown: such a method is called plasma enhanced Chemical Vapor Deposition; a good example of this technique is given in a work by Tien et al.[27] where polymerization of plastic monomers of vinyltrimethylsilane (or hexamethydisiloxane) are introduced together with Argon in a chamber where the mixture is submitted to an r f discharge, polymer formation is greatly enhanced by the discharge and a very good optical quality film is grown at a rate of 2000 Angstroms/minute; the losses of the films are the lowest known (0.01db/cm).

Among the interesting films that have been grown by sputtering, the following must be cited:

Ta_2O_5[15,33,35,36,38] index of refraction n=2.21.

The best method to make Ta_2O_5 films is reactive sputtering from a Tantalum target in argon-oxygen atmosphere. Deposition rate 5nm/min.

Si_3N_4[36,38] index of refraction n=2.

Silicon nitride films can be deposited from a pressed powder target in nitrogen-argon atmosphere. Deposition rate 5nm/min.

$LiNbO_3$[33] amorphous layers have been made at Naval Research Lab. (Washington).

LiTaO$_3$[37] amorphous layers have been deposited on a glass substrate.

ZnO[16,38] Good quality zinc oxide films have been grown for acoustic surface wave devices, they are well oriented but polycrystalline and not suitable for light propagation.

Chemical Vapor Deposition

Si$_3$N$_4$[39]: according to Stutius the best silicon nitride films are obtained by low pressure chemical vapor deposition (LPCVD). They are grown on a silicon substrate, a buffer layer (SiO$_2$) is at first grown using wet oxygen, then a chemical reaction is made between SiH$_2$Cl$_2$ and NH$_3$ under suitable[40] pressure conditions.

Index of refraction n=2.0 (633nm).
Transparent from 0.4μm to the near-infrared
Losses 0.4db/cm.
Deposition rate 3nm/min. Film thickness less than 400nm.
Silicon nitride films are quite durable and can even be wiped.

As$_2$S$_3$[41]: this compound is very interesting for acoustic-optic applications, Gottlieb and Isaac describe a procedure very close to CVD fabricating.

Index of refraction 2.23 in the visible.
Transparent from 550nm to 7000nm.
Losses of the order of 2db/cm (633nm).

4.   SURFACE MODIFICATION OF A TRANSPARENT SUBSTRATE

The problem is to increase the refractive index of a transparent substrate in the region of its surface. One has to intervene from the outside, through the surface. Two possible methods can be used: chemical diffusion and ion implantation.

General considerations on diffusion

Since a modification of the substrate is needed, starting from its surface, diffusion is a well adapted method. The substrate is put in contact with a chemical phase containing the atoms or ions to be diffused. In the case of ionic diffusion, in order to maintain electrical neutrality, ion exchange must take place between the substrate and the diffusing medium. The physical reason for diffusion is the existence of a gradient of concentration; in the case of ions the diffusion process can be controlled or even enhanced by an applied electric field.

The laws describing diffusion have been formulated a long time ago by Fick - see for example the book of Crank - and relate the chemical concentration C of the diffusion species, the diffusion time t, and the geometrical position (which will be the coordinate x in the simple case of a one dimensional problem). For a semi-infinite problem the solution is of the form:

$$C = C_0 . f(u) \quad \text{with } u = x/ \sqrt{Dt}$$

D is called the diffusion coefficient and follows an Arrhenius type equation.

$$D = D_0 . \exp(-Q/RT)$$

where Q is the activation energy, R is the gas constant, T is the absolute temperature, D_0 is a constant.

The dependence on the square root of the time shows that diffusion processes become slower with time.

The analytic form of the function f(u) is determined by the initial and boundary conditions. There are two extreme situations: either the total amount of diffused material is finite, then f(u) is a gaussian function $\exp(-x^2/DT)$; or diffusion occurs from an infinite reservoir and then:

$$f(u) = \text{erfc} (x/2\sqrt{Dt})$$

The case of ionic diffusion is somewhat more complicated since ion exchange takes place, so that an interdiffusion coefficient must be introduced; the equations of Fick lose linear character[46].

When numerical figures are put into the preceding formulas, it is seen that, for reasonable diffusion times (a few minutes to some hours) the chemical composition of the substrate is disturbed at depths that are measured in micrometers or tens of micrometers.

Ion exchange in glass[7,10,11,12,46]

So far we have discussed only concentration profiles; of more interest, however, for integrated optics is the resulting refractive index modification. Index changes arise for two main reasons, because of exchange between ions with different electronic polarizabilities and also because of internal stresses occuring since the ionic radii of the exchanged ions are different. The first applications of ion-exchange in glass were related to this last property and were aimed at strengthening the glass surface[42,43,48].

It is known that the refractive index of many compound oxide glasses can be predicted by the Gladstone-Dale relation[44] which applies very well for evaluating the refractive index of an exchanged glass.

It is an experimental fact that only glasses having sodium in their composition can exhibit ion exchange. Na ions can easily escape out of the glass and be replaced by other ions; up to 90%, and usually 60%, of the initial Na ions participate in the exchange process.

If a high refractive index change is desired, one must take a glass with a high sodium concentration; the highest index change has been obtained[12] in the case of TiF6 Schott glass where the proportion of Na_2O is 20% by weight and a refractive index change of 0.2 has been measured.

Silica is the main constituent of most glasses, its structure can be described as a random lattice of SiO_4 tetrahedra linked together by their corners (in a crystal of quartz the lattice is perfectly organized). When a metal oxide, such as Na_2O, is added to silica the Na-O bonds are broken, the oxygen atoms join the Si-O lattice which has then a formula SiO_{2+x}, instead of SiO_2. The lattice is still made of SiO_4 tetrahedra where all the Si atoms are tetrahedra coordinated, but where some oxygen atoms are only monobonded. Sodium is present as positive ions, loosely linked to the monovalent oxygen atoms which behave as electrically negative sites. Sodium ions are almost free to move inside the lattice or eventually to escape.

Other metallic oxides (CaO for example) are often introduced in glass and follow about the same mechanism as sodium, but the chemical links with monobonded oxygen atoms are stronger than in the case of sodium.

Ion exchange can be obtained spontaneously or under the control of an electric field. In the first case a glass slide (a microscope slide is quite suitable) is simply immersed in a bath of molten salts; nitrates are often chosen because of their low melting points, but their chemical stability is not very good. In the case of salts having too high a melting point, eutectic mixtures can be used. There are no special requirements for the furnace, the atmosphere is air, the temperature is between $200^\circ C$ and $600^\circ C$ and is not severely controlled ($\pm 1^\circ$).

For electrically controlled ion exchange the ions may come either from a molten salt or from a deposited metallic layer[45,47,49,50]. Most studies have been made for the case of molten salts, and one encounters some difficulties with electrical insulation since the temperature is high and since molten salts are often corrosive media; capillary effects often cause electrical short circuits. The voltages applied across a one millimeter thick glass plate go from 10 to 1000 volts, intensity currents have values from microamps to milliamps.

Ion exchange from metallic layers is a new and promising technique, it will be useful for channel waveguides. But there are still some difficulties in obtaining reproducible layers. The key to the problem is probably in the quality of the evaporated metallic layers.

A great variety of ions can be introduced in glass, either spontaneously or electrically; but only few of them are useful for integrated optics, mainly because the surface is damaged during the ion exchange. The ions used are the following: Ag^+, K^+, Li^+, Tl^+. Thallium can be used but has no significant advantages and is to be avoided because of the toxicity of its salts.

Table 2. Comparison of Different Ions for Ion Exchange.

Ions	Glass	D_o cm^2/s	Q Kcal/mole	Δn	Atom-rad nm
Ag	SLG*	$2.72.10^{-2}$	22.6	0.1	0.27
	Pyrex	$1.01.10^{-3}$	20.5	0.04	
K	SLG*	$3.7.10^{-3}$	26.8	0.01	0.27
	Pyrex	$3.5.10^{-4}$	25.6	0.004	
Li	SLG*	10.2	32.5	0.013	0.12
	Pyrex	11	40.8	0.0011	
Na	-	-	-	-	0.19

*SLG : Soda Lime Glass

Silver ions

Silver ion exchange has been the most extensively studied. Any kind of glass having sodium in its composition can be used. Silver ions usually come from silver nitrate baths (melting point 210°C); the exchange temperature must be lower than about 380°C, otherwise the nitrate begins to decompose and lossy waveguides are obtained. Silver ions do penetrate very easily in glass, although they have a big ionic radius, about the same as potassium; this property is due to the high polarizability of silver ions, which allows a deformation of the electron cloud.

The refractive index change is high (0.1 for soda lime glass; 0.03 for pyrex). If a smaller value is needed, one has to use a mixture of sodium and silver nitrates, the index change is then adjustable from 0 to 0.1. It is worthwhile noting that the silver salt must be very dilute in the sodium salt (10^{-2} mole of Ag/mole of Na), to have a sizable diminution of the refractive index change; this remark may be important for decreasing the cost of experiments, since sodium salts are far cheaper.

The exchange time for making guides having a few guiding modes is about one hour for soda lime glass and five hours for pyrex at a temperature of 300°C. The diffusion constants and the activation energies are given in Table 2.

Silver ion exchange is strongly indicated when high refractive index changes are needed and for making guides with one to a hundred guided modes.

Potassium ions

Potassium ions diffuse remarkably slowly, their diffusion constant is one order of magnitude smaller than for silver. They are of particular interest for single mode guides with low losses. Potassium ions can conveniently come from a potassium nitrate bath. It takes about 24 hours at 360°C to make a monomode guide.

Lithium ions

In contrast to potassium ions, lithium ions diffuse remarkably quickly, with a diffusion constant bigger than in the case of silver by two orders of magnitude. Lithium ion diffusion[7] is probably the only way for making highly multimode planar waveguides. The exchange must be performed at a high temperature, 100° below the softening temperature of the glass. At lower temperatures the glass surface is damaged; in fact the lithium diffusion is so fast and deep that the silica lattice may be broken and that glass devitrification may be readily obtained, giving a nice opalescent appearance which is not very suitable for light propagation. When a sodium ion is replaced by a lithium ion stresses are produced, mainly because of the difference between the ionic radii[48] (0.12nm for Li against 0.19nm for Na). Low loss guides (<1db/cm) are obtained with eutectic mixtures of lithium and potassium salts[11,47], chlorides as well as sulfates have been used, preferably sulfates. Although at a much lower speed than lithium ions, potassium ions penetrate into the glass, and help a rearrangement of the lattice, because of their bigger atomic radius (0.27nm); in the case of an 80 micrometer thick layer, electron microprobe measurements have shown the presence of potassium at a depth of 4μm.

Refractive index profiles

Graded index profiles are always obtained by spontaneous ion exchange in glass. When the exchange temperature is close to the melting point of the molten bath, the refractive index profile is nearly gaussian, as is the case in diffusion problems when the total amount of diffusing material is finite. These results are interpreted by the fact that the thermal agitation is then moderate and once some sodium ions have escaped out of the glass they stay in the vicinity of the surface and act as a screen for further exchanges. At higher temperatures, and/or when the molten bath is stirred, the refractive index profile is nearly a complementary error function, and has there- fore a linear behaviour in the vicinity of the surface.

When an electric field is applied, the refractive index has almost the shape of a step function, its thickness is given by the empirical formula:

$$e = A.j.t^n$$

t is the time in minutes, j is the current density (A/m^2), A and n are characteristic constants of the glass. For soda lime glass we have found A = 0.072, n = 0.89.

Buried profiles and back diffusion[11,47,51,54,55]

In principle buried waveguides can be made by performing two successive ion exchanges; a glass slide is first immersed in a silver nitrate bath and then in a sodium nitrate bath. During the first part of the experiment silver ions penetrate deeply into the slide, when the slide is put in the second bath the silver ions will go from the glass into the melt, producing a concentration profile having a maxi- mum somewhere in the glass. The experiment is not as simple, and two successive field-assisted ion exchanges must be made to obtain a buried waveguide; only multimode thick waveguides can be elaborated by this method[47].

Recently a new method has been proposed by Parriaux[56] for making single mode buried guides: the technique is based on electromigration of ions from a low temperature eutectic of K and Ca nitrates (160°C). At such a low temperature, ion migration merely results from the ap- plied field, thermal diffusion being negligible, the applied field is 800V/mm. Parriaux[57] has shown that the refractive index is de- creased in a thin layer and constitutes a leaky waveguide, he has also observed the existence of a not well explained buried layer of potas- sium ion. This last layer is a buried monomode waveguide.

Ion exchange in lithium niobate and tantalate[58,59,60]

Some of the lithium ions of lithium niobate and tantalate can be ion exchanged, giving optical waveguides. The surface index increase is much more important than in Ti-diffused waveguides ($\Delta n = 0.12$ for thallium-lithium exchange[58], thallium ions coming from thallium nitrate).

Silver-lithium[59,60] exchange is possible in lithium tantalate, only the extraordinary index is increased (0.05) and the losses are small (1db/cm). The exchange is fast and it takes two hours to make an 8 modes guide; from this point of view the ion exchange technique is preferable to the diffusion method that will now be described.

Diffusion in lithium niobate and tantalate

These materials are important since they lead to the most efficient modulators and switches. Most waveguides have been made by in-diffusion of metal or by out-diffusion of lithium.

When heated at about 1100°C lithium niobate and tantalate lose lithium which out-diffuses, the stoechiometry of the crystal is changed, and as a consequence, the index is slightly increased. Although widely investigated this method is now supplanted by metal diffused and ion exchanged guides, since it does not allow the construction of channel waveguides.

Instead of out-diffusing lithium, one can in-diffuse suitable atoms, such as niobium[62], titanium and vanadium[63]. Refractive index profiles usually obey an error function law. This technology is not an easy one.

Ion implantation[64,65]

The principle is rather simple, a collimated ion beam of well defined energy is sent onto the substrate where an optical waveguide is to be made. The energy of the beam is between 50KeV and 2MeV, the doses range from 10^{13} to 10^{15} ions/cm^2.

The ions penetrate into the substrate and interact with the substrate atoms and/or ions, the passage of energetic ions through a solid involves both electronic and atomic interactions. As they propagate the ions have a decreasing energy and will finally stop at a depth which is fixed by the incident energy. Most modifications appear in the region where the ions stop, they are related to the following physical phenomena:

- defects are induced in the lattice, chemical bonds being broken
   while atoms and/or ions are displaced from their original
   position. (Radiation damage).
   In the case of a crystal this may create a local amorphisation.
   In amorphous media one can sometimes observe the appearance
   of a new phase, with different optical properties.
   If the implanted substrate is heated properly, many of these
   defects will be annealed out.

- One definite effect of ion implantation is to introduce new
   energy levels which act as donor or acceptor levels because
   of the damage and because of the introduced impurity ions.
   One frequently chooses an ion of different valence from those
   of the lattice so that the change persists after annealing.

- The stoechiometry of the substrate is, in principle, modified.
   At usual doses this change is not significant; but one could
   think of implanting, at high doses, ions of the lattice and
   thereby change the stoechiometry of materials which are difficult to form. For optical properties this aspect of implantation has not yet been exploited but it could help for elabor-
   ating such as lithium niobate or tantalate which are unstable
   in the melt and cannot be grown with perfect stoechiometry.

Ion implantation is well adapted for making channel waveguides.
Let us examine which parameters control the characteristics of the
optical guide: the kind of implanted ions (in principle any ion can
be used, the most usual are helium, protons and lithium), the energy
of the beam and finally the doses. Thermal annealing is frequently
necessary. The technology is not well understood and no general rules
can be given.

Guides have been made in fused quartz[66], with 300KeV Li ions at
about 10^{15} ions/cm^2. The refractive index change appears to be primarily due to disorder produced by the incident particles rather than
a chemical doping effect. Postbombardment annealing is important,
the losses can be lowered to 0.2db/cm if the substrate is heated
during implantation. The refractive index change can be adjusted by
varying the dose.

Waveguides can be made in lithium niobate[67] and in crystalline
quartz by ion implantation. The following ions have been used: H,
Be,B,N,O,Ne,Ar,Ti and Ag, at energies from 7KeV to 2MeV. Both ordinary and extraordinary indices are decreased. Because the indices
are reduced, ions beams have been used to write low index boundaries
to define waveguiding regions. By using energetic light ions (eg.
2MeV He$^+$) negligible change is produced in the surface layer where
the energy loss is primarily electronic and thus a low index region
is formed deep within the solid. The guides show complex annealing
characteristics but absorption losses are small after annealing at

$200^{\circ}C$; guiding action still exists after heat treatment up to $400^{\circ}C$.
The refractive index change is probably due to some amorphisation,
however the surface layers retain crystallinity and electro-optic
properties.

Of course ion implantation can be used for inducing optical waveguides in semiconductors such as GaAs[68,69] and ZnTe[5].

5. CHARACTERIZATION OF OPTICAL WAVEGUIDES

The characterization of optical waveguides, which is obviously
important, is not easy, because of the extreme thinness of the layers.
It has posed to physicists a set of new interesting problems. Con-
versely, methods for studying optical waveguides are now used for
making measurements on thin film which are not necessarily made for
waveguides purposes[70,71].

The parameters to be measured are the thickness, the refractive
index, the shape of the index gradient and the losses. The refractive
index profile (RIP) has been obtained from reflectivity measurements[72,
73], from interferometric experiments[74], by analyzing the near field
diffraction pattern[75]. Arnaud has proposed a method of determination
of the RIP inside of a multimode optical fiber[76] by simple measurements of the transmitted energy and this method[77] has been used in the
case of planar multimode waveguides.

An electron microprobe can be used to find the ionic or atomic
concentration which induces the gradient index[44,46]. In all cases
there is good agreement between the gradient index and the chemical
concentration.

Refractive index profile measurements

We shall not describe all these methods but focus our attention
on the determination of the RIP from optical measurements of the
propagation constants of the guided modes. The spectrum of the propagation constants can readily be obtained from the so-called m-lines
which appear when a coupling prism is used.

The electric field of a guided mode is obtained from Maxwell's
equations and boundary conditions, it obeys an eigenvalue equation
which is formally identical to the Schroedinger equation for the one
dimensional motion of a particle in a potential well in quantum mechanics. The refractive index profile $n(x)$ plays the role of the potential function, the propagation constants of the guided modes are
the equivalent of the allowed energy levels.

Except for simple laws $n(x)$, the problem cannot be solved rigor-
ously. Starting from the solution obtained for those simple laws,

other results can be obtained for less simple laws, using, as in
quantum mechanics, a perturbation method[78,79,80]. It has been shown
that for certain index profiles the square of the effective indices
of the guided modes has a simple dependence on some function of the
mode number p. A quick method for obtaining an idea of the shape of
RIP consists of plotting n^2_{eff} versus p or versus the p'th root of the
Airy function; it can thus be found whether the RIP is parabolic,
linear or is an error function.

In quantum mechanics the well-known WKB method allows an approximate integration of the Schroedinger equation as long as the variation of n(x) is small over a wavelength.

The WKB method was first used in a direct way to find the mode
spectrum of a given RIP and the computed and measured spectra were
compared[81]. The initial RIP can then be modified to obtain better
agreement between the two spectra.

One can also start from the measured spectrum of the guided modes,
and, using the so-called inverse WKB method (IWKB) determine directly
the refractive index profile[82], provided the index function is mono-
tonically decreasing from the surface. The IWKB method does not work
in the case of buried profiles. Most laboratories now use the IWKB
method in the form of computer programs. The measured spectrum is
fed into the computer and draws the refractive index profile directly.

Loss measurements

The use of low-loss devices in optical telecommunication systems
is an important requirement for success; therefore it is important
to be able to measure accurately the losses of individual modes, as
well as the mode conversion of optical waveguides. The low value of
the absorption coefficient makes it very difficult to measure , the
methods that had been developed by optical physicists for determining
the complex optical constants of thin films are experimentally complex
and do not work for optical waveguides, especially ellipsometry[83] and
Abeles' method[84,85]. On the contrary loss-measurements on thin films
now make use of the methods of guided optics[86].

The most widely used method for measuring the attenuation of a
waveguide is to measure the transmitted power as a function of waveguide length. In this method all the energy flowing through the guide
must be collected in the detector. If one does not want to cut the
guide and polish the end face, one has to use an output decoupling
prism to extract energy from the guide. Various methods have been
used to ensure a 100% extraction efficiency: Weber et al[87] used a
sliding output prism with an index matching liquid between the prism
and the guide. It is difficult to be certain that the extraction
efficiency is 100% or, at least is constant as the prism slides from
one position to another.

The sliding decoupling prism can be avoided by monitoring the light scattered from film imperfections along the light path. It is assumed that the intensity of the scattered light is proportional to the light retained in the film. A photodiode is simply translated above the guide.

Won[88] has modified the sliding prism method using a third prism to give light intensity reference. The method makes use of a coupling prism P_1 and two decoupling prisms P_2 and P_3 disposed one after the other. P_1 and P_3 are fixed, P_2 can slide between them. Let z be a coordinate along the guided beam; light is launched into the guide through P_1 and excites all modes or only desired mode. Let W_2 and W_3 be the output powers of P_2 and P_3. One has:

$$W_2 = k_2 \cdot I(z_2)$$

$$W_3 = k_3 \cdot (I(z_2) - W_2) \cdot \exp(-\alpha(z_3 - z_2))$$

where k_2 and k_3 are output coupling constants for P_2 and P_3, $I(z)$ is the intensity of the light inside the guide, z_i is the position of prism i, and α is the attenuation of the guide which is to be measured.

Let us make $k_2 = 0$ in a first experiment, then the output power takes some value W_{30} which is measured; then the second prism is coupled so that k_2 is no longer equal to zero, the output power of P_3 decreases from W_{30} to $(W_{30} - \Delta W_3)$ and it is easy to show that:

$$I(z_2) (W_{30} - \Delta W_3) = (I(z_2) - W_2) W_{30}$$

$$I(z_2) = W_2 \cdot W_{30} / \Delta W_3$$

The result is independent of the coupling efficiencies k_2 and k_3, which may be small, thus inducing only low mode conversion. Using this method Won[87] has measured the attenuation of K^+ ion exchanged waveguides and has found, at a wavelength of 514.5nm, the value 0.36 ± 0.02 db/cm.

Mode-coupling

In the preceeding method, by choosing a proper incident angle for prism P_1, we can then measure the power carried by the different modes. For the sake of simplicity let us examine the situation of a guide having two modes. Let $I_0(z)$ and $I_1(z)$ be the powers in the two modes at point z; coupling equations can be written as:

$$dI_0/dz = -a_{00}I_0 - a_{01}I_0 + a_{10}I_1$$

$$dI_1/dz = -a_{11}I_1 - a_{10}I_1 + a_{01}I_0$$

If coupling is weak, and if mode 0 only is excited, I_1 remains small and we have:

$$I_0 = A.\exp(-(a_{00} + a_{01})z)$$

$$I_1 = (a_{01}/(a_{00} + a_{01})).A.(1-\exp(-(a_{00} + a_{01})z))$$

where A is the total initial intensity in mode 0.

Using a K^+ ion waveguide, it has been found that:

$$a_{00} = 6.5.10^{-2} cm^{-1} \qquad\qquad a_{11} = 7.8.10^{-2} cm^{-1}$$

$$a_{01} = 2.10^{-4} cm^{-1} \qquad\qquad a_{10} = 6.3.10^{-4} cm^{-1}$$

6.   CHANNEL WAVEGUIDES

From the point of view of components, microguides with a limited width are more interesting than planar guides; those guides are some-times called three dimensional guides, channel waveguides or strip guides. According to Unger's terminology[93], the strip may be depo-sited on a substrate surface (raised strip), it may also be embedded in the substrate or buried more or less deeply inside of the substrate. In fact only a few papers have been devoted to channel optical wave-guide fabrication; in most cases their fabrication uses one of the previously cited methods for planar waveguides together with a masking technique. The width and depth are between 2 and 10 micrometers in the majority of cases.

It is easy to make channel waveguides[89] in plastic films, the exposure to the polymerizing light or to the electron beam being made through a photodeposited mask. Polymerization[90] can also be obtained by focusing the beam in the resin. The guide is then written by moving the film relative to the exposing beam.

Channel waveguides (with 2-8μm width) have been simply made by evaporating the glass from a substrate with the aid of a focused argon laser beam[91].

Diffusion methods for increasing the refractive index can also be used, the diffusion of atoms or ions being also made through a mask[92,93,46,47,11]. Because of lateral diffusion under the mask the optical guide is always wider than the mask. Roughly speaking it can be said that the difference between the guide and the mask widths is about twice the penetration depth.

In the case of ion-exchange in a molten bath, one could imagine that there will be some difficulties in choosing the material for the

mask. In fact, although the medium is very corrosive, aluminium is
well adapted for nitrate salts as well as for sulfates at temperatures
of 600°C and for 15min immersion durations. Even if an external electric field is applied, lateral diffusion occurs. The reasons for
lateral diffusion are not clearly established.

Buried and unburied[47,95] multimode channel waveguides have been
made with very low losses (1db/cm), they are well adapted for mass-
production of multimode devices (beam splitters for example).

Ion implantation is in principle well adapted to the fabrication
of channel waveguides since lateral diffusion is not expected. Ion
implanted channel waveguides have been made in lithium niobate[67],
GaAs and glass. In the case of ZnTe, although ion implantation gives
planar waveguides, it has not been possible to obtain channel waveguide[40].

To make narrow ridge waveguides, Kawabe et al.[94] have used an
ion-milling technique for etching a two micrometers broad raised guide
on a titanium out diffused lithium niobate substrate.

7. CONCLUSIONS

In conclusion it can be emphasized that a large variety of methods
and technologies exist for making optical waveguides, involving many
different fields of physics. All the observed phenomena are not completely explained and an important research effort has still to be
made. Nevertheless integrated technology is now almost adult and
components are now ready for use in the fields of instrumentation,
data processing and telecommunications (both mono and multimode).

REFERENCES

1. P. K. Tien, Rev. Mod. Phys. 49.2, Apr. 1977.
2. P. K. Tien, Appl. Opt. 10.2395, 1971.
3. P. D. Townsend, Journ. Phys. E. Scient. Inst. 10, 1977
4. S. E. Miller, Bell Syst. Tech. Journ. 48.2059, 1969.
5. S. Valette, G. Labrunie, J. C. Deutsch, J. Lizet, Appl. Opt. 16.
      5.1289, May 1977.
6. P. Meyrueiz, Microguides de lumiere par implantation ionique dans
      ZnTe. These Grenoble, Nov. 1979.
7. G. Chartier, P. Jaussaud, A. De Oliveira, O. Parriaux, Electron.
      Lett. 13.25, Dec. 1977.
8. H. Harris, R. Schubert, Journ. Opt. Soc. Am. 62.2.154, Feb. 1972.
9. P. K. Tien, R. Martin, S. Blank, S. Wemple, L. Varnerin, Appl.
      Phys. Lett. 21.207, 1972.
10. T. Giallorenzi, T. West, R. Kirk, R. Gunther, R. Andrews, Appl.
      Opt. 19.7.1092, Apr. 1973.

11. G. Chartier, P. Collier, A. Guez, P. Jaussaud, Y. Won, Appl. Opt. 19.7.1092, Apr. 1980.
12. J. L. Coutaz, P. Jaussaud, G. Chartier, 2nd Top. Meet. on Graded Index Optical Systems. Hawai, May 1981.
13. R. Standley, W. Gibson, J. Rodgers, Appl. Opt. 11.6.1313, Jun. 1972.
14. J. E. Goell, R. Standley, Bell Syst. Tech. Journ. 48.3445, 1969.
15. D. H. Hensler, J. D. Cuthbert, R. J. Martin, P. K. Tien, Appl. Opt. 10.1037, 1971.
16. P. K. Tien, R. Ulrich, R. J. Martin, Appl. Phys. Lett. 14.291, 1969.
17. J. L. Jackel, Appl. Phys. Lett. 37.739, 1980.
18. D. T. Wei, W. W. Lee, L. R. Bloom, Appl. Phys. Lett. 25.329, 1974.
19. R. V. Schmidt, I. P. Kaminov, Appl. Phys. Lett. 25.458, 1974.
20. G. Y. Chin, A. A. Ballman, P. K. Tien, S. San Severino, Appl. Phys. Lett. 26.637, 1975.
21. J. L. Jackel, Appl. Opt. 19.12.1996, June 1980.
22. H. P. Weber, R. Ulrich, Appl. Phys. Lett. 19.38, 1971.
23. G. C. Righini, V. Russo, S. Sottini, Opt. and Quant. Electron. 7.447, 1975.
24. D. Fay, D. Ostrowsky, A. M. Roy, J. Trotel, Opt. Comm. 9.4.424, Dec. 1973.
    J. C. Dubois, M. Gazard, D. Ostrowsky, Opt. Comm. 7.3.237, Mar. 1973.
25. E. Chandross, C. Pryde, W. Tomlinson, H. Weber, Appl. Phys. Lett. 24.2.72, Jan. 1974.
26. J. E. Goell, Appl. Opt. 12.4.729, Apr. 1973.
27. P. K. Tien, G. Smolinsky, R. J. Martin, Appl. Opt. 11.3.637, Mar. 1972.
28. M. S. Chang, P. Burlamacchi, C. Hu, J. R. Whinnery, Appl. Phys. Lett. 20.8.313, Apr. 1972.
29. S. Sriram, H. E. Jackson, J. T. Boyd, Appl. Phys. Lett. 36.9. 721, 1980.
30. J. C. Dubois, M. Gazard, Rev. Tech. C. S. F. Thompson, 6.4.1169, Dec. 1974.
31. J. E. Goell, Appl. Opt. 12.4.737, Apr. 1973.
32. C. Pitt, F. Gfeller, R. Stevens, Thin Solid Films, 26.25, 1975.
33. R. H. Deitch, E. J. West, T. G. Giallorenzi, J. F. Weller, Appl. Opt. 13.4.712, Apr. 1974.
34. W. J. Coleman, Appl. Opt. 13.4.946, Apr. 1974.
35. H. Terui, M. Kobayashi, Appl. Phys. Lett. 32.10.666, May 1978.
36. W. Paulson, F. Hickernell, R. Davis, Journ. Vac. Sci. Techn. 16.2.307, Mar. 1979.
37. D. Russo Gia, C. Kumar, Appl. Phys. Lett. 23.229, 1973.
38. F. Hickernell, R. Davis, F. Richard, 1978 Ultrasonic Symposium Proc. IEEE.
39. W. Stutius, W. Streifer, Appl. Opt. 16.12.3218, Dec. 1977.
40. S. Valette (LETI Grenoble, private comm.)
41. M. Gottlieb, T. J. Isaacs, Appl. Opt. 17.16.2482, Aug. 1978.
42. P. Acloque, C. Guillemet, C. R. Acad. Sc. Paris. 250.4328, 1960.

43. C. Guillemet, J. M. Mery, A. Bonnetin, Symposium sur la surface des verres et ses traitements modernes. Luxembourg USCV, June 1967.
44. D. Gladstone, T. Dale, <u>Phil. Trans. Roy. Soc. London</u> 153, 1893.
45. G. Chartier, P. Jaussaud, A. De Oliveira, O. Parriaux, <u>Electron. Lett</u>. 14.5.134, Mar. 1978.
46. G. Stewart, C. Millard, P. Laybourn, C. Wilkinson, R. Delarue, <u>IEEE Journ. Quant. Electr</u>. QE 13.192, 1977.
47. G. L. Tangonan, Fiber Optics Couplers. <u>Final techn. report</u>, <u>Air Force System Command RADC</u>. TR. 80.358, Jan. 1981.
48. S. S. Kistler, <u>Journ. Amer. Ceram. Soc</u>. 45.2.59, Feb. 1962.
49. T. Izawa, H. Nakagome, T. Kimura, <u>Proc. Quantum Electron. Conf. Montreal 1972</u>, p545.
50. T. Kaneko, H. Yamamoto, <u>Proc. 10th Intern. Cong. on Glass</u>. Kyoto Japan, 1974.
51. G. Stewart, Refractive index modification by ion exchange Ph.D. Thesis, Glasgow, Jan. 1979.
52. P. Collier, Realisation de guides d'ondes lumineuses par diffusion d'ions sous champ electrique. Thèse Grenoble, 1978.
53. Y. Won, Elaboration de guides optiques et mesure des pertes et du couplage entre modes. Thèse Grenoble, 1980.
54. J. L. Coutaz, Gradient d'indice de refraction par échange ionique dans le verre.Thèse Grenoble, 1981.
55. A. Guez, Etude et realisation de dispositifs d'optique intégrée multimode. Thèse Grenoble 1979.
56. O. Parriaux, Buried single mode waveguide in glass. To be pub.
57. G. Chartier, V. Neuman, O. Parriaux, C. Pitt, Low temperature ion substitution in glass. <u>Electron. Lett</u>. To be published.
58. J. L. Jackel, <u>Appl. Phys. Lett</u>. 37.739, Oct. 1980.
59. J. L. Jackel, <u>Appl. Opt</u>. 19.12.1966, Jun. 1980.
60. H. F. Taylor, W. Martin, D. Hall, V. Smiley, <u>Appl. Phys. Lett</u>. 21.95, 1972.
61. I. P. Kaminov, J. Carruther, <u>Appl. Phys. Lett</u>. 22.326, 1973.
62. W. Philips, J. Hammer, <u>Journ. Electron. Mater</u>. 4.549, 1975.
63. R. Schmidt, I. Kaminov, <u>Appl. Phys. Lett</u>. 25.458, 1974.
64. See ref. 99.
65. P. Townsend, <u>Journ. of Physics E</u> 1977. Vol. 10.
66. R. Standley, W. Gibson, J. W. Rodgers, <u>Appl. Opt</u>. 11.6.1313, June 1972.
67. G. De Stefanis, J. Gaillard, E. Ligeon, S. Valette, B. Farmery, P. Townsend, A. Perez, <u>Journ. Appl. Phys</u>. 50.12.7898, Dec.1979.
68. E. Garmire, H. Stoll, A. Yariv, R. G. Hunsperger, <u>Appl. Phys. Lett</u>. 21.3.87, Aug. 1972.
69. T. Nishimura, H. Aritome, K. Masuda, S. Namba, <u>Japan. Journ. Appl. Phys</u>. 15.2883, 1976.
70. W. D. Westwood, J. S. Wei, <u>Canad. Journ. Phys</u>. 57.9.1247, Sept. 1979.
71. M. Olivier, J. C. Peuzin, J. S. Danel, D. Challeton, <u>Appl. Phys. Lett</u>. 38.2, Jan. 1981.

72. J. Heibi, E. Voges, IEEE Journ. Quant. Elec. QE 14.7.501, July 1978.
73. M. Ikeda, M. Takeda, H. Joshikyio, Appl. Opt. 14.814, 1975.
74. W. E. Martin, Appl. Opt. 15.2112, 1976.
75. F. M. Sladen, D. N. Payne, M. J. Adams, Appl. Phys. Lett. 28.5, Mar. 1976.
76. J. A. Arnaud, R. M. DeRosier, Bell Syst. Techn. Journ. 55.10, Dec. 1976.
77. J. A. Arnaud, P. Facq, Private Commun.
78. P. Jaussaud, G. Chartier, Journ. Phys. D 10.645, 1977.
79. Y. Ayant, G. Chartier, P. Jaussaud. Journ. de Physique 38.1089, 1977.
80. J. Lotspeich, Optics Comm. 18.567, 1976.
81. A. Gedeon, Optics Commun. 12.329, 1974.
82. J. White, P. Heidrich, Appl. Opt. 15.151, 1976.
83. R. J. Archer, Journ. Opt. Soc. Amer. 52.970, 1962.
84. F. Abeles, C. R. Acad. Sc. Paris 228.553, 1949.
85. D. Male, Journ. Physique-Radium 11.332, 1950.
86. K. Sasaki, H. Takahashi, Y. Kindo, N. Susuki, Appl. Opt. 19. 17.3018, Sep. 1980.
87. H. Weber, F. A. Dunn, W. Leibolt, Appl. Opt. 12.4.755, Apr. 1973.
88. W. Won, P. Jaussaud, G. Chartier, Appl. Phys. Lett. 37.3.269, 1980.
89. H. F. Taylor, W. Martin, D. Hall, V. Smiley, Appl. Phys. Lett. 21.95, 1972.
90. E. Chandross, C. Pryde, W. Tomlinson, E. Weber, Appl. Phys. Lett. 24.72, 1974.
91. T. Pavlopoulos, K. Crabtree, Journ. Appl. Phys. 45.11.4694, Nov. 1974.
92. P. Simova, I. Savatinova, L. Tsonev, Appl. Phys. 22.237, 1980.
93. H. G. Unger in ref. 96.
94. M. Kawabe, S. Hirata, S. Namba, IEEE Trans. on Circ. and Syst. CAS 26, 12th Dec. 1979.
95. H. Lilienhof, K. Heidman, D. Ritter, E. Voges, Optics Comm. 35. 1.49, Oct. 1980.
96. D. Ostrovski. Fiber and integrated optics. Nato Adv. Study Inst. Series. Plenum Press.
97. K. Barnoski. Introduction to integrated optics. Plenum Press.
98. Physics of thin films. Academic Press.
99. J. Crank. The mathematics of diffusion. Clarendon Press.
100. P. Townsend, S. Valette. Optical effects of Ion Implantation in Treatise on mater. Sc. and Techn. Vol. 18, Edited by J. Hirvonen Acad. Press 1980.

SINGLE-MODE OPTICAL FIBER TECHNOLOGY I.   PROPAGATION PROPERTIES AND

MATERIAL DISPERSION IN SINGLE-MODE FIBERS

P. J. Severin

Philips Research Laboratories
Eindhoven
The Netherlands

INTRODUCTION

In this and the following two lectures the state of the art in
using and manufacturing single mode optical fibers will be discussed
in so far as it is relevant in an integrated optics course. In this
first lecture I shall examine the physical aspects of the propagation
of optical power in fibers stressing single mode (SM) operation and
contrasting it to multimode (MM) operation. This links up automati-
cally with the material aspects of pulse dispersion.

In their pioneering paper on optical communications Kao and
Hockham[1] concentrate on SM fiber propagation, stating that for high
information flow capacity a waveguide is desirable which propagates
an SM with only one polarization. A bandwidth of 1GHz.km can then be
easily attained, once the material loss has been sufficiently reduced.
Low loss was achieved in 1970 in a SM fiber showing 20dB/km at 633nm
wavelength (Kapron et al.[2]). It started a decade of experimental and
theoretical research and development work on MM fibers with a large
bandwidth due to a carefully graded refractive index profile. The
difference between the group velocities of all modes can then be made
small enough to produce, in combination with an appropriately chosen
material dispersion, a fiber with extremely large bandwidth. Though
this holds in the idealized model situation and MM fibers appear
preferable, mainly because the large core leads to simpler manipu-
lation, it is gradually becoming evident that this is experimentally
not the easy path to zero dispersion. Not all modes are launched,
and relaunched at perturbations, with equal efficiency. They are
exposed to different loss coefficients and it is turning out difficult
for the manufacturer to maintain a uniform refractive index profile
with the desired precision over lengths of many kilometers. The

73

disadvantage involved in manipulating SM fibers in the field are gradually being overcome[3].

The main argument put forward for shifting attention to SM fibers is that the large bandwidth desired for telecommunication applications can thus be obtained more readily, both technically and economically. This is proved by the best result[4] reached up to now in fiber optics : a 100km SM fiber with a bandwidth of 100GHz.km and a loss of 0.5dB/km at 1.3μm wavelength.

The next two lectures will be devoted to the specific requirements and the manufacturing technologies by which these may be satisfied. Most papers reporting advances in SM fibers also state the importance of their application in sensors and integrated optics. Optical fiber sensors are generally interferometric arrangements in which one branch is perturbed and this perturbation is measured as a change of phase. Though MM fibers do show the desired effect to some extent with appropriate modulation techniques, SM fibers with unambiguously defined phase are preferable or even necessary, particularly for the three types of sensor studied most intensely: the fiber gyroscope, the fiber hydrophone and the fiber current measuring device. Integrated optics circuits will also generally operate on coherent optical signals of defined phase and polarization. These cannot be delivered by MM fibers. At least one of the dimensions of the integrated optics circuit will demand sizes of the order of the wavelength. Therefore only SM fibers can be matched directly to integrated optics circuits. These are two of the reasons why SM fibers are a topic of this course.

The physical aspects of propagation discussed in this lecture have a direct bearing on technology in two ways. First, it is from this understanding, particularly related to dispersion, that the appropriate wavelength profile, core and cladding materials and V-value can be chosen. Secondly, it is then necessary to test whether the product satisfies the specifications. In the very near future, methods will be required by which the radius, cutoff wavelength, refractive index difference and profile can be measured, all within a few percent precision and on a scale of fractions of a micron. It is not unrealistic to expect that these measurement methods will present greater problems than the SM fiber technology itself.

OPTICAL FIBER PROPAGATION THEORY : MODES

The core and cladding geometry we shall be dealing with, is drawn schematically in Figure 1. In theoretical approaches the cladding radius r_2 is often assumed infinite. The core refractive index n_1 can be of a step or graded nature, $n_1(r)$, as shown in Figure 1 for a "parabolic" and "diffusion profile". Because $n_1 > n_2$, electromagnetic power propagates along the fiber, being trapped in the core by total reflection against the core-cladding interface. As can be seen from

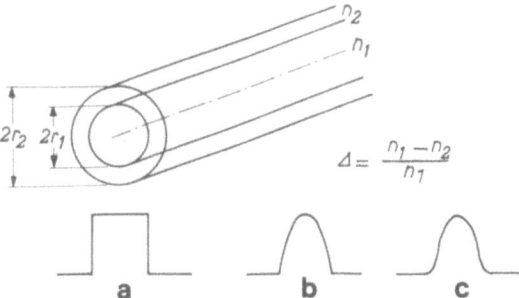

Figure 1.   The core and cladding geometry and three refractive index
            profiles most commonly used: step, parabolic and diffusion-
            controlled.

Figure 2a for a step-index and from Figure 2b for a graded-index
profile, where basically Snell's law is applied to a thin layer approximation of the graded profile, this is clear enough as regards
meridional rays. In the case of skew rays, which do not propagate
in a plane containing the axis, this is less easy to visualize in the
geometrical optics approximation. Through geometrical optics, ray
tracing or the WKB technique are powerful methods, none of these approximate methods is really useful for treating SM fiber propagation
because fundamentally they do not meet the assumptions $\lambda \to 0$ or at least
$\lambda \partial n / \partial r << 1$.

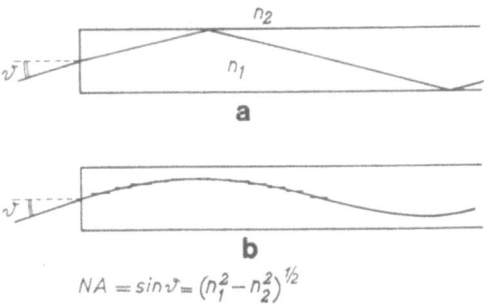

Fig. 2.   The propagation of meridional rays along a step (a) and a
          graded-index (b) optical fiber.

We cannot avoid considering the solution of Maxwell's equations.
These are discussed at length, in Unger's[5] and Marcuse's[6,7] books,
among others. Briefly the solution runs as follows. For a wave of

frequency ω and free space wave number $k = \omega/c$ all field components
depend on time t and axial coordinate z as exp $\{j(\omega t - \beta z)\}$ and
expressions are sought for the dependence of the components of E(r,ϕ)
and H(r,ϕ) on the transverse coordinates. It then turns out that the
components E_r, E_ϕ, H_r and H_ϕ can be expressed in E_z and H_z and that
the latter components satisfy the wave equation. Assuming for E_z a
radial and azimuthal dependence E(r) exp $j\ell\phi$, the wave equation reads

$$\frac{1}{r} \frac{\partial}{\partial r} r \frac{\partial E}{\partial r} + (k^2 n^2(r) - \beta^2 - \frac{\ell^2}{r^2}) E = 0 \qquad (1)$$

For a step-index fiber, n(r) = n_1, there follows a solution for E(r)
in terms of Bessel functions. Hence all components can be found in
core and cladding material in terms of normal modes. A mode is a
field distribution which propagates along the fiber axis without any
change in the cross-sectional intensity pattern. Each mode has a
different propagation constant β which describes the periodicity along
the z-axis. Any actual field is a linear superposition of all modes.
By applying the appropriate boundary conditions at the core-cladding
interface we obtain the characteristic equation which relates the
propagation constant β to the frequency ω for each combination of
azimuthal and radial mode numbers ℓ and p. Gloge[8,9] postulated that
the modes are almost transverse with two perpendicular polarizations.
He derived the longitudinal components from the transverse components
and found them to be smaller of order $\Delta^{\frac{1}{2}}$.

Repeated differentiation of these longitudinal components again
leads to the transverse components which are found to be equal to
the postulated fields apart from a correction term, an order Δ smaller
than the postulated fields. Making use of the fact that $\Delta \ll 1$, this
is called the weakly guiding approximation. The exact mode solutions
HE $_{\ell+1,p}$ and EH $_{\ell-1,p}$ combine to give a linearly polarized mode LP $_{\ell p}$.
The lowest order mode HE$_{11}$ is called the LP$_{01}$ mode, whereas the next
higher order mode, LP$_{11}$, really consists of four modes. The solution
to the characteristic equation for a step-index fiber is plotted
schematically in Figure 3. In order to discuss propagation problems
independently of size, the dimensionless parameters V and b have been
introduced

$$V = kr_1 (n_1^2 - n_2^2)^{\frac{1}{2}} \quad , \quad b = \frac{\beta^2/k^2 - n_2^2}{n_1^2 - n_2^2} \qquad (2)$$

What is of importance to us from these mode theoretical considerations
is that there is a value of V below which only one mode propagates,
designated HE$_{11}$ or LP$_{01}$. This is the next higher order mode cutoff
value V, which equals 2.4 in a stepindex fiber. The LP$_{01}$ mode can
be propagated independently in two linear polarizations and a SM fiber
is effectively a bimode fiber. It is also called the dominant mode,
for at any V-value it is the mode in which β is closest to $n_1^{\frac{k}{1}}$.

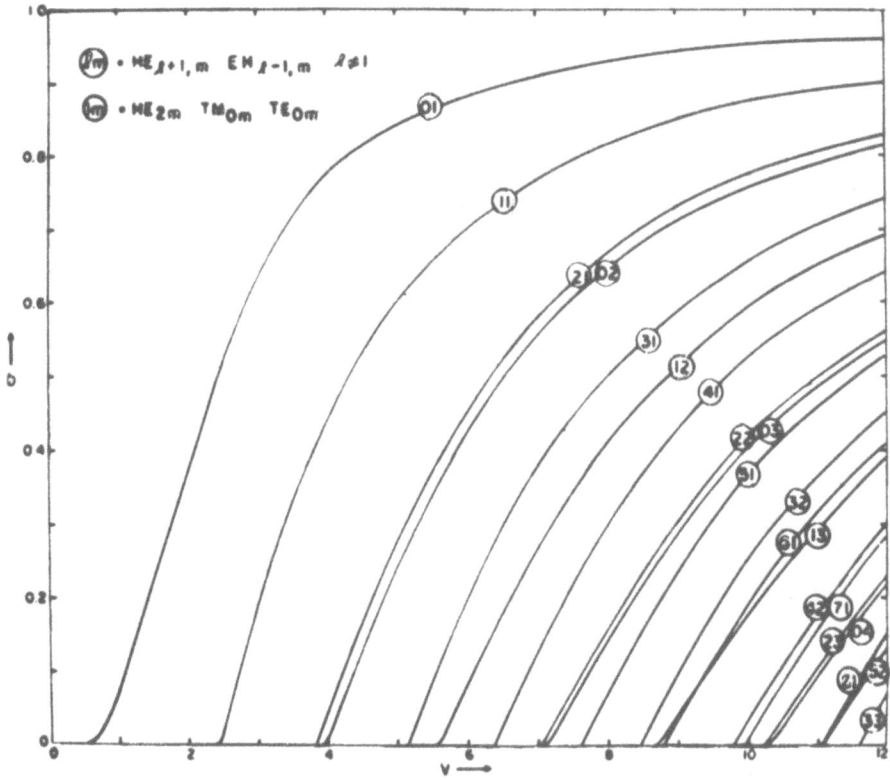

Fig. 3.   Normalized propagation parameter $b = (\beta/k - n_2)/(n_1 - n_2)$ as a function of the normalized frequency V. (from ref. 8 Fig. 3).

In the above derivations it has been assumed that the cladding is of infinite extent. As long as only a minute fraction of the power propagates in the cladding, say 10^{-4} in an MM fiber far from cutoff, a cladding thickness roughly equal to the core radius, $r_1 \approx 25\mu m$, is enough to isolate the core from the outside world. In the V region where only the lowest order HE_{11} mode propagates, the fraction propagating in the cladding is always substantial. At the limit of SM operation, V = 2.4 in a step-index fiber, 80% of the power propagates in the core. However, the propagation constant β/k is closer to the refractive index of the cladding than of the core. In this sense the HE_{11} mode is different from all other modes. It is clear that the large fraction of power propagating in the cladding demands a thick cladding of glass of the same high quality as the core. As far as

material and technology are concerned, this demands special solutions
which will be discussed in the two lectures on technology.

DISPERSION

In optical communications pulses of laser power are transmitted
along the fiber. The envelope of a modulated carrier wave, and hence
also pulses, propagate with the group velocity $v_g = (d\beta/d\omega)^{-1}$. It is
current practice to express this as the group delay time per unit
length

$$\tau = \frac{1}{c} \frac{d\beta}{dk} \tag{3}$$

Using eqs (2) it can be derived[9] that in the weakly guiding approximation

$$\tau = \frac{1}{c}\left[N_2 + (N_1 - N_2) \frac{d b V}{dV}\right] \tag{4}$$

where N_1 and N_2 are the group indices of core and cladding glass which
give rise to material dispersion,

$$N_i = n_i + k \frac{dn_i}{dk} \tag{5}$$

These indices are larger than n_1 and n_2 by about 2% at $\lambda = 0.8\mu m$
for most glasses and the difference decreases with increasing wavelength. The second term determines the pulse broadening in an MM
fiber because power is transmitted at a different velocity for each
mode. The time delay difference between the fastest and the slowest
mode is of the order of Δ/c, that is 50nsec/km. Such large values
allow for a too small bandwidth, about 20MHz. From the very beginning
this deficiency of step-index optical fibers has been recognized and
the obvious solution was to grade the refractive index radially so
that the paths of all rays are almost equal. In a first approximation
an α-profile with $\alpha \approx 2$ satisfies this requirement:

$$n(r) = n_1(1 - \Delta(\frac{r}{r_1})^{\alpha}) \qquad r \leqslant r_1 \quad ,$$
$$= n_2 \qquad\qquad\qquad r \geqslant r_1 \quad . \tag{6}$$

The group velocities of all modes are then almost equal and if all
modes are excited equally the rms pulse broadening can be calculated.
The pulse dispersion is found to be reduced to about $c^{-1} \Delta^2/8$, the
numerical factor depending on the approximation used. In practice,
MM fibers with about 1nsec/km pulse dispersion can be routinely produced[10]. Hence, the actual pulse reduction factor due to graded
profile is an order of magnitude smaller than the theoretically expected value $\Delta/8 \approx 10^{-3}$.

The advantage of using an optical fiber in the SM regime is that there is no intermode dispersion. It is now the intramode dispersion that counts because of the finite bandwidth δk of the carrier, when using a GaAs laser about equal to 1nm. It is called therefore chromatic dispersion. It is found in the weakly guiding approximation by calculating the variation of the group delay over the width δk so that

$$\delta\tau = \frac{\delta k}{k}\frac{k}{c}\frac{\delta^2\beta}{\delta k^2}$$

(7)

$$= \frac{\delta k}{k}\frac{1}{c}\left[k\frac{dN_2}{dk} + \frac{d(N_1-N_2)}{dk}\frac{d\ell V}{dV} + \frac{N_1^2 - N_2^2}{n_1 + n_2}V\frac{d^2 bV}{dV^2}\right]$$

The first term is due to material dispersion, and using equations (4) and (5), equals,

$$\delta\tau_m = \frac{\delta k}{k}\frac{1}{c}k\frac{\delta^2 nk}{\delta k^2} = -\frac{\delta\lambda}{\lambda}\frac{\lambda^2}{c}\frac{\delta^2 n}{\delta\lambda^2}$$

(8)

As shown in Figure 4, for SiO_2 the factor $\lambda^2\delta^2 n/\delta\lambda^2$ decreases almost linearly from a value of 0.04 at 0.7µm to zero at about 1.27µm. For all silicate glasses this factor vanishes at a wavelength somewhere between 1.2 and 1.5µm. The refractive index and the absorption coefficient are coupled by Kramers-Kroning relations so that at any wavelength the refractive index is determined by the absorption coefficient multiplied by a weighting factor integrated over the full wavelength range. In the ultra-pure optical fiber glasses the absorptions that matter are of an intrinsic nature; the electronic absorption in the UV and the vibrational Si-O absorption in the far IR.

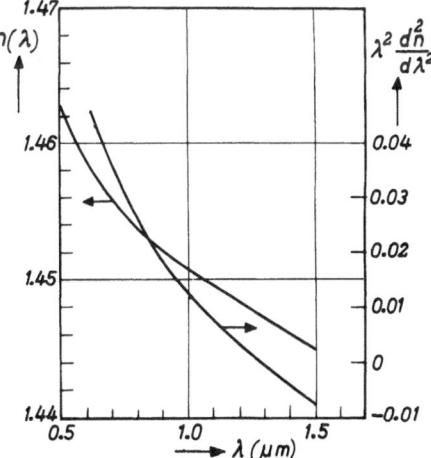

Fig. 4.   The refractive index n and the material dispersion $\lambda^2 d^2 n/d\lambda^2$ as a function of λ for SiO_2.

The inflection point in $n(\lambda)$, occurs where the falling tail of
the former starts being dominated by the rising tail of the latter.
The other components of most silica-based fibers are GeO_2, P_2O_5 and
B_2O_3. With increasing GeO_2 dope the zero material dispersion wavelength moves towards a longer wavelength because both absorption bands
move to higher wavelengths and with increasing B_2O_3 dope it moves to
a shorter wavelength because the vibrational absorption band occurs
at a shorter wavelength. In order to select the wavelength of zero
material dispersion, $n(\lambda)$ should be known very precisely as a function
of composition. The dependence of the refractive index on the wavelength λ or frequency ω is adequately described by a two term Sell-
meier equation based on the model of harmonic oscillators:

$$n^2 - 1 = \frac{E_d E_o}{E_o^2 - \omega^2} + \frac{E_d' E_o'}{E_o'^2 - \omega^2} \qquad (9)$$

where E_d and E_o are due to the UV electronic absorption, E_α' and E_o'
are due to the IR vibrational absorption and the frequency ω is
expressed in eV. Data on various commonly used fiber glasses were
measured by Hammond and Norman[11], Hammond[12], Fleming[13,14], Payne et
al.[15] and Adams et al.[16]. When these data measured on bulk samples
were compared with fiber data it was found that, for instance in
silica-core borosilicate-clad fibers, the refractive index difference
was about twice as large as the bulk samples predicted. These dis-
crepancies are attributed to the highly quenched character of optical
fibers. The nature of the structural difference due to the high degree
of quenching has not yet been fully understood. In an SM fiber, operating at a wavelength close to zero-material dispersion, the second
and third terms are always important. They are plotted in Figure 5
for some α-index profiles. In order to obtain zero chromatic disper-

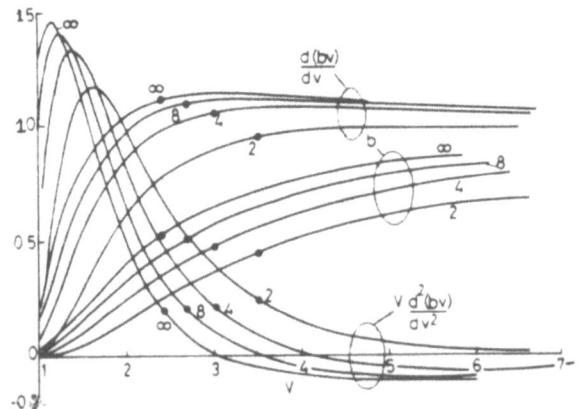

Fig. 5.   The dependence on V of the three basic waveguide parameters,
          used in eqs. (7) and (18), for various values of α in the
          α-profiles defined in eq. (6). The curves with V=∞ refer to
          a step-index profile. The dots mark the values of V_c. (from
          ref. 24 fig. 4).

sion the first terms is chosen negative and of the same magnitude as the positive waveguide dispersion.

The importance of the refractive index distribution in an SM fiber for b(V) will be discussed in the next section.

THE DOMINANT MODE

As explained above, the weakly guiding approximation, $\Delta \ll 1$, allows modes to be described as linearly polarized transverse waves. All components of the HE_{11} or, in weakly guiding terminology, LP_{01} mode can be expressed in the function R(r) which is a solution of the scalar wave equation,

$$\frac{d^2R}{dr^2} + \frac{1}{r}\frac{dR}{dr} + (k^2n^2(r) - \beta^2)R = 0 \tag{10}$$

The longitudinal components E_z and H_z can be neglected and the transverse components are $E_y = R(r)$ and $H_x = -R(r)\beta/\omega\mu_o$.

A refractive index distribution n(r) should be considered, because departures from a step-index profile may occur due to technical deficiencies, or a particular graded-index profile may be chosen deliberately for specific effects. Examples of the first kind are the central dip, which occurs due to evaporation of dopant in the collapsing operation or a diffusion profile arising from inevitable smoothing of an originally abrupt core-cladding interface.

The exact field distributions R(r) for an infinitely large cladding in a step-index profile ($\alpha = \infty$) and in a parabolic-index profile are known

$$R_\infty = \frac{J_o(\mu^r/r_1)}{J_o(\mu)} \qquad , \qquad \mu^2 = r_1^2(n_1^2 k^2 - \beta^2) \tag{11}$$

$$R_2 = e^{-r^2/\omega^2} \qquad , \qquad \omega^2 = 2r_1^2/V$$

Not only the field distribution, but also the cutoff value V_c for single mode operation, the spot radius r_o of the HE_{11} mode, the fraction of power propagating in the core and the characteristic equation from which follows the waveguide dispersion, depend on the shape of n(r). In order to discuss those parameters from one point of view we use the fact, noted by various authors,[17-19] that R_2 in eq. (11) describes the field distribution sufficiently well for practical purposes in any profile n(r). In an attempt to increase physical insight into SM fiber propagation, Snyder[17] derived the relation between the width w of the Gaussian trial function and β. Using a refractive index profile $n^2(r) = n_2^2 + (n_1^2 - n_2^2) f(r/\delta)$ he found that for $f(r/\delta) = \exp(-r^2/\delta^2)$ the spot radius ω is given by

$$\omega^2 = 2\frac{\delta^2}{V-1} \tag{12}$$

which implies that the intensity profile reads exp $(-r^2(V-1)/\delta^2)$.
Hence the light is more concentrated towards the axis for larger V.
With V=2 the light distribution is coincident with the profile which
also turns out to be the situation with maximum concentration of light.
The dispersion equation is found by using this approximation as

$$\delta^2\beta^2 = \delta^2 k^2 n_1^2 - 2V + 1 , \tag{13}$$

from which the SM fiber waveguide dispersion $^2\beta/\omega^2$ can be derived.
He also discusses the step-index profile and an approximation to profiles with smoothed core-cladding interface using the function

$$f\left(\frac{r}{\delta}\right) = \frac{1}{\Gamma(m+1)} \int_{(m+1)r/\delta}^{\infty} t^m e^{-t} dt \tag{14}$$

where the step index corresponds to m = ∞ and the Gaussian index to
m = 0. This approximation could be highly useful for diffusion-
generated profiles in compound-glass multimode fibers, where α-
profiles are inadequate. The waveguide dispersion for these profiles
is shown in Figure 6.

It is clear that the V value at the point where the waveguide
dispersion vanishes, is always beyond the SM regime. The shape of the
profile determines the V value, where waveguide and material dispersion
compensate as well as the precision with which it is predictable. When
material dispersion vanishes at a certain wavelength, given a certain
refractive index difference, the radius can be chosen so that the appropriate V value is reached. Exact compensation is more difficult
with a step than with a Gaussian profile.

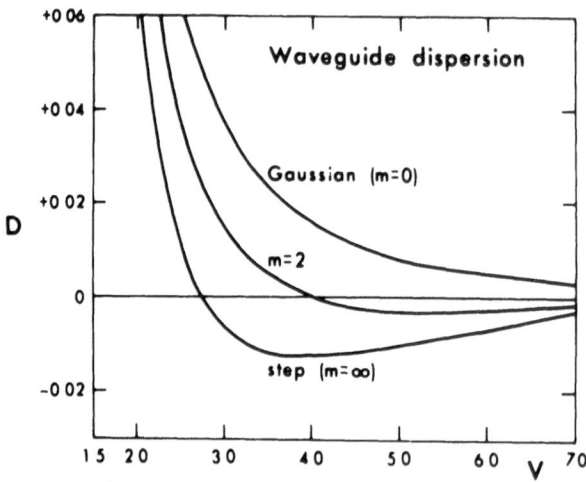

Fig. 6.   The dependence on V of the waveguide dispersion, presented
          in normalized units, for various values of m in the m-profiles
          defined in eq. (14). The shape of the m = ∞ curve is almost
          identical to the α = ∞ curve for Vd^2bV/dV^2 in Fig. 5 because
          waveguide dispersion is mainly determined by this term. (from
          ref. 17. fig. 3).

In order to determine the limit value V_c of SM fiber operation the cutoff frequency of the next higher order mode (TE_{01}) should be determined. For refractive index profiles graded by smoothing of the core-cladding interface Snyder[17] notes that V_c should be between 2.4 and the value for a Gaussian profile 2.59. For α-profiles the situation is more complicated. Gambling et al.[20] found that for an α-profile V_c increases from 2.4 for $\alpha = \infty$ and 3.52 for $\alpha = 2$ to 4.28 for $\alpha = 1$. The central dip is a commonly occurring nonuniformity in silica-based fibers. It is then found that an SM fiber, intended as a step-index fiber, is extremely tolerant even to a dip of full depth and great width. Physically this effect is not unexpected because the TE_{01} mode has a zero intensity in the center.

The parameter V is determined only by the core maximum and cladding values of the refractive index, n_1 and n_2. Gambling et al.[21] introduced a guiding parameter G which averages the number of dope atoms over a nonuniform core,

$$G = \frac{1}{r_1^2} \int_0^{r_1} \frac{n^2(r) - n_2^2}{n_1^2 - n_2^2}\, r\, dr \tag{15}$$

and a new value of V_c valid for any profile

$$V_c = 2.4(2G)^{-\frac{1}{2}} \tag{16}$$

For an α-profile this can be calculated as

$$V_c = 2.4\left(1 + \frac{2}{\alpha}\right)^{\frac{1}{2}} \tag{17}$$

which is the same value as found by Okomoto and Okoshi[22], who used a variation method.

Of course as the aim of the calculation is to find the condition for zero dispersion, any effect which may be neglected with respect to each of the terms can be of the same order as the difference. Sugimura et al.[23] therefore differentiated eq. (2) with respect to λ without any approximation and separated the expression into three terms mainly representing material dispersion, waveguide dispersion and profile dispersion. Each of the terms is a refinement of the three terms presented in eq. (7). The broadening of the pulse of width $\delta\tau$ then equals

$$\delta\tau = \delta\lambda \frac{d\tau}{d\lambda} = \delta\lambda[\delta\tau_m + \delta\tau_g + \delta\tau_p], \tag{18a}$$

where neglecting some terms, it is found that[24]:

$$\delta\tau_m = -\frac{\lambda}{c}\left[\frac{d^2 n_1}{d\lambda^2} H(V) + (1-H(V))\frac{d^2 n_2}{d\lambda^2}\right], \tag{18b}$$

$$\delta\tau_g = \frac{n_2 \, \Delta}{c\lambda} \, V \frac{d^2 Vb}{dV^2} \left[1 - \frac{\lambda}{n_2} \frac{dn_2}{d\lambda}\right]^2 \quad ,$$

$$\delta\tau_p = \frac{n_2}{\lambda c} \frac{d\Delta}{d\lambda} \left[V \frac{d^2 Vb}{dV^2} + \frac{dVb}{dV} - b\right] *$$

$$\left[1 - \frac{\lambda}{n_2 d\lambda} \frac{dn_2}{4\Delta} - \frac{\lambda}{d\lambda} \frac{d\Delta}{d\lambda}\right] \quad ,$$

$$H(V) = \tfrac{1}{2} \left[\frac{dVb}{dV} + b\right]$$

The effect of a profile is fully embodied in the relation b(V); for a step-index fiber the weighting function H(V) is simply the ratio of the power in the core to the total power. This is equal to 0.83 at V=2.4 whereas it is equal to 0.69 for a parabolic-index fiber at cutoff V=3.5. This stresses the importance of the right choice of cladding material.

As soon as it was fully appreciated that the pulse dispersion is a SM fiber can be reduced to the level of a few psec/km for a 1nm bandwidth laser several studies were made preliminary to the actual implementation. South[25], White et al.[26] and Chang[27] present detailed calculations. In graded-index fibers with decreasing in an α-profile the limit V_c increases and the wavelength λ_0 of zero chromatic dispersion can be varied from 1.27 to 1.6μm. Gambling et al.[28] discuss only waveguide dispersion and determine the value of V where zero waveguide dispersion occurs, V_{OD}, in different profiles. They find that V_{OD} is infinite for profiles whose fall-off is parabolic or faster, α≤2, including Gaussian and the well known sech profile. Depressions in the core center affect the profile dospersion. Pal et al.[39] calculate that the usually unavoidable central dip increasingly shifts the value of λ_0 to a shorter wavelength with increasing dip depth.

The zero chromatic dispersion wavelength occurs where negative material dispersion and positive waveguide dispersion, with some minor terms of interest, cancel. The precise experimental profile must be known in order to compute the characteristic equation. Secondly the precise dependence of the refractive index on the wavelength which applies in the fiber, should be known. The first problem can be solved by using the appropriate measurement method on the fiber and mathematical expressions in a close enough approximation. The second problem is more complicated and has to be studied in the near future.

The same problem has also been approached by testing ready-made fibers. Cohen et al.[31] measured pulse transmission over several 1km long fibers in the range 0.6 to 1.6μm. They[32] presented the theory

applied and, using Fleming's Sellmeyer expressions, calculated what
is to be expected. The GeO2 dope profile was measured interferomet-
rically on preform slices and introduced into the wave equation. The
experimental results and the theoretical curves agree very well. The
values of λ corresponding to V_c agree within 25nm around 1.33μm. The
decomposition into independent waveguide and material dispersion clearly shows that assuming core and cladding refractive indices to be independent of wavelength in the waveguide dispersion terms leads to
erroneous results. This is considered theoretically in more detail
by Marcuse[33]. In the above approaches the chromatic dispersion
$^2\beta/k^2$ is made zero. However, in a Taylor expansion the next term
$^3\beta/k^3$ now determines the bandwidth[34]. Furthermore, ellipticity
and other sources of birefringence, uniform along the fiber, give rise
to pulse dispersion if both directions of polarization are excited.
This is included in the calculations by Tsuchiya et al.[35]. Assuming
a spectral width of 5nm at 1.5μm wavelength they determined from the
transfer function a bandwidth as large as 100GHz.km for a 2.5% core
ellipticity. Polarization mixing due to birefringence, nonumiform
along the fiber, and bending will further decrease the bandwidth.

REFERENCES

1. K. C. Kao and G. A. Hockham, Proc. IEE, 1151, 1966.
2. F. P. Kapron, D. B. Keck and R. D. Maurer, Appl. Phys. Lett. 17,
      423, 1970.
3. D. Botez and G. J. Herskowitz, Proc. IEEE, 68, 689, 1980.
4. M. Kawachi, S. Tomaru, M. Yasu, M. Horiguchi, S. Sakaguchi and
      T. Kumura, EL 17, 57, 1981.
5. H. G. Unger, Planar optical waveguide and fibers, Clarendon Press,
      Oxford, 1977.
6. D. Marcuse, Light transmission optics, van Nostrand Reinhold Corp.,
      New York, 1972.
7. D. Marcuse, Theory of dielectric waveguides, Acad. Press, New York,
      1974.
8. D. Gloge, Appl. Opt. 10, 2252, 1971.
9. D. Gloge, Appl. Opt. 10, 2442, 1971.
10. J. G. Peelen and J. W. Versluis, Acta Electr. 22, 255, 1979.
11. C. R. Hammond and S. R. Norman, Opt. Q. Electr. 9, 399, 1977.
12. C. R. Hammond, Opt. Q. Electr. 10, 163, 1978.
13. J. W. Fleming, J. Am. Cer. Soc. 59, 503, 1976.
14. J. W. Fleming, EL 14, 326, 1978.
15. D. N. Payne and W. A. Gambling, EL 11, 176, 1975.
16. M. J. Adams, D. N. Payne, F. M. Sladen, A. H. Hartog, EL 14, 703,
      1978.
17. A. W. Snyder, Proc. IEEE 69, 6, 1981.
18. D. Marcuse, JOSA 68, 103, 1978.
19. A. R. Tynes, R. M. Derosier and W. G. French, JOSA 69, 1587, 1979.
20. W. A. Gambling, D. N. Payne and H. Matsumura, EL 13, 139, 1977.
21. W. A. Gambling, H. Matsumura and C. M. Ragdale, Opt. Q. Electr.
      10, 301, 1978.

22. K. Okamoto and T. Okoshi, _IEEE Trans._ MTT 24, 416, 1979.
23. A. Sugimura, K. Daikohu, N. Imoto and T. Miya, _IEEE J. Q. Electr._ 16, 215, 1980.
24. W. A. Gambling, H. Matsumura and C. M. Ragdale, _IEE J. Micro-waves, Opt. and Acoust._ 3, 239, 1979.
25. C. R. South, _EL_ 15, 394, 1979.
26. K. White and B. P. Nelson, _EL_ 15, 396, 1979.
27. C. T. Chang, _EL_ 15, 765, 1979.
28. W. A. Gambling, H. Matsumura and D. M. Ragdale, _EL_ 15, 474, 1979.
29. A. W. Snyder and R. A. Sammut, _EL_ 15, 269, 1979.
30. B. P. Pal, A. Kumar and A. K. Ghatch, _EL_ 16, 503, 1980.
31. L. G. Cohen, C. Lin and W. G. French, _EL_ 15, 334, 1979.
32. L. G. Cohen, W. L. Mammel and H. M. Presby, _Appl. Opt._ 19, 2007, 1980.
33. D. Marcuse, _Appl. Opt._ 18, 2930, 1979.
34. D. Marcuse, _Appl. Opt._ 19, 1653, 1980.
35. H. Tsuchiya and N. Imoto, _EL_ 15, 476, 1979.

SINGLE-MODE OPTICAL FIBER TECHNOLOGY II. SILICA-BASED SINGLE-MODE

FIBER TECHNOLOGY, LOSS AND POLARIZATION PROPERTIES

P. J. Severin

Philips Research Laboratories
Eindhoven
The Netherlands

INTRODUCTION

In silica-based fiber technology four different processes are
actually used on a routine basis for preform manufacturing: the
outside vapour phase oxidation processes (OVPO), a special version
of which is the vapour phase axial deposition process (VAD), and
the inside vapour phase oxidation process (IVPO), based either on
modified chemical vapour phase deposition (MCVD) or on plasma
activated chemical vapour phase deposition (PCVD). Only the VAD
process yields a massive but porous rod. In the other three methods
a hollow tube is produced, from which the preform is obtained by
collapsing for MCVD and PCVD. With OVPO the hole disappears upon
drawing. Drawing technology will be discussed in the next lecture,
where the double-crucible method will be considered. Each of these
fabrication techniques has its own typical advantages and disadvantages, the relative merits of which can only be assumed with respect
to a specified requirement. In this lecture the yardstick is their
capability to produce single-mode (SM) fibers economically. SM fibers
are attractive as future high-capacity, long-distance transmission
lines with low-loss and hence large repeater distances. The requirement of high capacity implies that the SM fiber operates at longer
wavelengths, $\lambda > 1.27\mu m$, where material dispersion can compensate
waveguide dispersion. This has been discussed at length in the first
lecture. There are two transmission windows, around 1.3 and 1.55μm,
between OH absorption peaks available. The concomitant requirement
of low loss at $\lambda = 1.3\mu m$ implies that the composition of the fiber is
chosen so that any of the fundamental sources of loss due to vibration
tail and OH absorption are avoided. The state of the art is such
that the more trivial transition-metal contamination sources have
been eliminated, while the long wavelength reduces the remaining

term, the Rayleigh ($\sim \lambda^{-4}$) scattering loss, to a very small value.
In this lecture the typical loss properties associated with hydroxyl-
ions and with the different components of the fiber will be examined
in detail when the results of experiments on a series of VAD and MCVD
grown fibers are discussed.

In the next section the OVPO and the closely related VAD methods
will be discussed. No results for SM fibers produced by OVPO are
available to date, but impressive results have been reported with
SM fibers produced by VAD. The MCVD and PCVD methods will be
discussed in the third section, first in general and then applied to
SM fibers. The final section will be devoted to the aims, techniques
and results in the field of polaraization-maintaining SM fibers.

DEPOSITION TECHNIQUES

Chemical Vapour Deposition

The basic process used in preparing silica-based optical fibers
is known in chemical technology as chemical vapour deposition (CVD).
It has been studied intensively because it is used for growing
epitaxial silicon layers, eg. from $SiCl_4$ diluted in H2. This
decomposes on a heated Si surface and the crystal structure of the
slice is continued in the grown layer. The chemical reaction
essentially requires the heated Si surface to behave in this way and
is therefore called a heterogeneous reaction. Due to the catalytic
action of the surface the process runs at a temperature much lower
than that required for fusion of the components concerned. With
increasing $SiCl_4$ pressure it finally decomposes in the gas stream to
form Si clusters in what is therefore called a homogeneous reaction.

This technique has been used in various combinations with known
techniques to grown an optical fiber multimode core layer on the
inside of a glass tube or the outside of a glass, ceramic or carbon
rod. The layer grown can be doped gradually with a variable composition of the various glass former oxides in order to profile the
refractive index: GeO_2, P_2O_5, TiO_2 and several other oxides to
increase the refractive index, and with B_2O_3 and F to decrease it with
respect to SiO_2. The overall reactions are similar to the $SiCl_4$
reaction

$$SiCl_4 + O_2 \rightarrow SiO_2 + 2Cl_2$$

The F atom is embodied in the SiO_2 network in Si-F groups. The effect
of the various dopants X upon the refractive index of SiO_2 is given in
Figure 1, as a function of the molar concentration of the dopant.

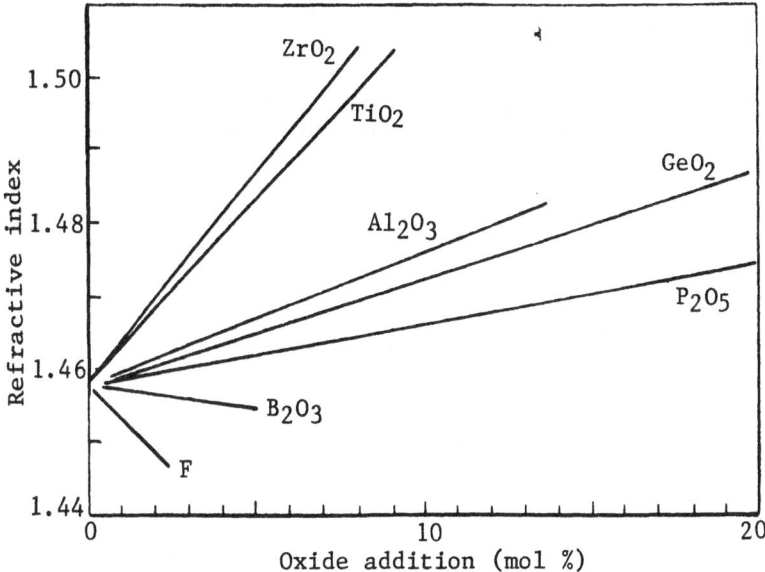

Fig. 1. The refractive index at room temperature and $\lambda = 0.6\mu m$
         as a function of composition.
         (from D. Gloge, Repts. Progr.Phys. 42:1777 (1979) fig.13).

        The most important source of contamination is, in fact, water
which is built into the SiO_2 network as OH. The O-H bond has a
fundamental vibration frequency ν_3 which shows as a strong absorption
peak at 2730nm. Furthermore, the SiO_4 tetrahedron has a fundamental
bending frequency ν_1 corresponding to absorption at $12\mu m$. These two
vibrations and their higher harmonics combine to form a series of
absorption lines listed in Table 1. The fundamental frequency ν_1 is
slightly different in other glass networks; if the OH atom is bonded
to B, P or Ge, a similar series of absorption lines is found. These
fundamental frequencies are also listed in Table 1. The absorption
window at $\lambda = 1.3\mu m$, aimed at for zero-chromatic dispersion, is the
minimum loss between the 1.23 and $1.37\mu m$ absorption peaks. A third
fundamental frequency is due to the Si-O-Si vibration of frequency
ν_0 which gives rise to an absorption peak at $9\mu m$. The corresponding
absorption peaks in B, P, Ge and F are also listed in Table 1. The
second absorption window, at $1.55\mu m$, is the minimum loss between the
$1.37\mu m$ and the $9\mu m$ absorption lines. The first window depends weakly
on the other constituents of the glass, but the second window is
strongly affected by the position of the fundamental absorption
frequency ν_0.

Outside Chemical Vapour Deposition

        Corning produced the first silica-based fibers with a process
called Outside Vapour Phase Oxidation (OVPO)[1,2,3]. The arrangement

Table 1. The positions and intensities of OH-bands in fused SiO_2, combinations of $\nu_3 = 2.8\mu m$, and $\nu_1 = 12\mu m$. The position of the ν_3 O-H vibration depends weakly on the matrix B, P, Ge and ν_1 is an SiO_4 vibration. The fundamental Si-O-Si vibration ν_0 determines infrared absorption loss at $9\mu m$, the equivalent vibrations in B, P, Ge and F are indicated also.

Wavelength (nm)	Intensity (dB/km.ppm)	Identification
2720	8000	
2220	115	$2\nu_1 + \nu_3$
1900	8	$2\nu_3$
1370	50	$\nu_1 + 2\nu_3$
1230	2.4	$3\nu_3$
950	2.2	$3\nu_3 + \nu_1$
880	0.1	
	8	

	Si	Ge	B	P	F
$\nu_3(\mu m)$	2.7	2.8	2.8	3.05	
$\nu_0(\mu m)$	9	11	7.9	8.1	13.6

is shown schematically in Figure 2a. A vapour mixture of suitable starting materials is obtained by bubbling oxygen through or over the surface of liquid $SiCl_4$, $GeCl_4$, BCl_3 and $POCl_3$. It reacts in a flame of CH_4/O_2 or H_2/O_2 mixtures to produce a stream of particles of very high purity and the desired composition.

This jet of glass soot is directed radially at a distance of some 15 cm towards a rotating and traversing ceramic or carbon target rod and a porous glass preform is built up layer by layer. The desired profile is achieved by controlling the $SiCl_4$/dopants flow ratio. The deposition rate increases with the thickness of the preform and equals on the average about 2 g/min at 50% efficiency and a temperature of about 1400-1500°C. The burner traversing speed is about 20cm/min. When the desired thickness is reached the porous preform is slipped off the reusable target and zonesintered to a solid bubble-free glass blank by passing it through a furnace hot zone at 1500°C in a controlled atmosphere, eg. He. The OH impurity introduced by the burner gas is removed during the sintering step by a gaseous chlorine treatment which reduced the hydroxyl level below 0.1ppm. This is facilitated by the large active surface area of the soot particles, 100 to 1000 Å across, and the 75% porosity of the layer. After sintering, the glass blank is heated in another furnace to a temperature of 1800-2200°C, where the viscosity is about $10^5 - 10^6$ poise (= $10^4 - 10^5 Ns/m^2$) and the fiber can be drawn. The hole

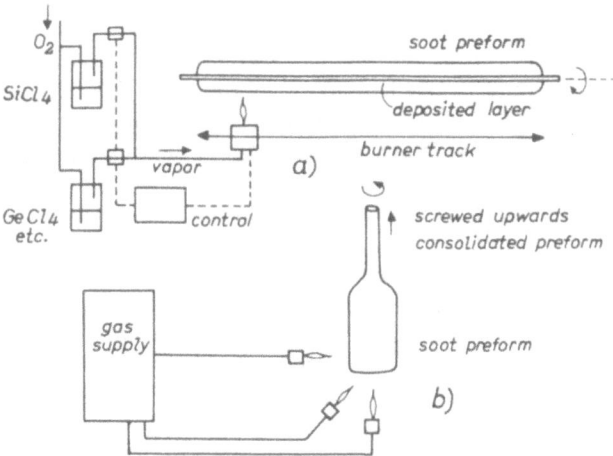

Fig. 2.   The OVPO process (a) and the VAD process (b), shown
schematically.

disappears during the drawing process. A blank yields about 10km of
125μm diameter fiber, the profile being built up in some 200 layers.
This corresponds to about $10^2 cm^3$ or 2.5gr/min = 1cm^3/min, to a
deposition time of 10^2min = 1.5h.

The author is not aware of any published results on SM fibers
made by OVPO. Judging from a distance, however, there appears to be
no reason why OVPO should not be a suitable method for SM fiber
manufacturing.

One of the main drawbacks of the CVD-based processes is that
they are batch processes. The logical conclusion, drawn by the
Japanese[4,5], was that the solution was to modify the OVPO arrangement
slightly, as shown in Figure 2b, and to position an H_2/O_2 burner
directed against the center of the lower end of a vertical silica rod
and pull the rod upwards. It is therefore called the vapour phase
axial deposition process (VAD). In order to obtain a profile a
second burner is positioned close to the first one, directed towards
a slightly excentral position, and the rod is rotated. A thicker
cladding can be formed by using a third burner radially directed. The
porous preform is then dehydrated using O_2 saturated with Cl_2 or
$SOCl_2$, and simultaneously consolidated. The dehydration process
occurs fast at 700°C, the 50-200nm soot particles being small enough
for fast diffusion and offering a large surface area for fast reaction.
The porous preform is consolidated continuously by zone melting in a
carbon ring heater at about 1450°C in He at a rate of 150mm/h. It
has been found that the profile shape depends on the deposition
temperature: GeO_2 deposited at low temperature vaporizes more readily

at the consolidation temperature than GeO_2 deposited at high temperature. The profile also depends critically on raw material flow
rates, exhaust gas flow rates, rotation speed, flame and surface
temperatures and the position of the rod end, which should be
stabilized to within 50μm. Very nearly parabolic profiles have been
realized for multimode fibers[6]. There is a striking similarity
between the VAD method and the OVPO process, both being soot processes.
The growth rates reported are also similar: 2gr/min. The VAD process
has been tested successfully for its capability to produce SM fibers.
Tomaru[7] describes the procedure for preparing the preform, using a
$SiCl_4$, $GeCl_4$ and $POCl_3$-fed core burner, and covering this preform
with a SiO_2 and $POCl3$ barrier-cladding. It is elongated so that the
diameter decreases from 25 to 15mm and pulled into a low OH-content
VAD silica sleeve tube. This composite preform is again elongated
and covered with a commercially available silica rod to a 50cm long
and 26mm diameter preform. This is drawn to a 21km long SM fiber
6.9μm core diameter, $\delta n = 0.0046$ and a deposited core/cladding ratio
of 3.5. The loss is 0.6dB/km at 1.2 and 1.55μm and 0.7dB/km at 1.3μm
wavelength. Because P-OH absorption generates loss at 1.6μm wavelength, in another experiment the core burner is fed with $SiCl_4$ and
$GeCl_4$ and the two barrier cladding burners with $SiCl_4$ only[8]. The
deposited cladding/core ratio is now 7. This preform, 30mm in
diameter and 25cm long, is again elongated and jacketed with a
commercial silica tube. This final preform is 50mm in diameter and
80cm long. This is drawn into a 100km long SM fiber[9]. The loss is
0.3dB/km at 1.55μm and 0.5dB/km at 1.3μm. The core diameter is 8μm,
the refractive index difference 0.0045 and the cut-off wavelength
1.2μm. Because the material dispersion compensates for waveguide
dispersion at 1.3μm, the bandwidth at that point was 100 GHz.km as
against 5 GHz.km at 1.55μm. The conclusion drawn from other experiments was that, for long-wavelength low-loss fibers, B_2O_3 cannot be
used because of the B-O vibrational absorption tail and it now became
clear that if OH is not absolutely absent, P_2O_5 cannot be used because
of P-OH absorption.

Inside Chemical Vapour Deposition

Bell did some pioneering work in applying the original heterogeneous CVD process for building up inside a tube, which was to be
the cladding, a profiled layer which was to be the core. However,
the original CVD process is basically a molecule-by-molecule building
process and therefore intrinsically slow, 24-36h for 1km of fiber, at
$1100^{\circ}C$. This was considerably improved when MacChesney et al.[10,11]
raised the temperature to 1300-1600°C and thus increased the reaction
rate. Both the homogeneous and the heterogeneous processes occur,
but at the higher temperature a homogeneous soot layer is fused on
the wall. This process is called the Modified CVD process and is in
fact a combination of the two processes. The arrangement is shown in
Figure 3a. The layer is sintered to a vitreous glass on the inside

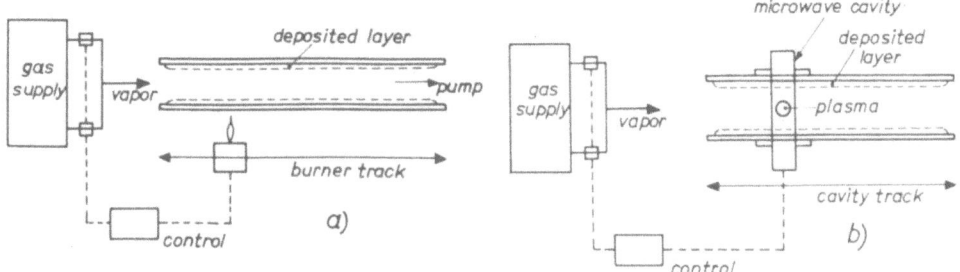

Fig. 3.   The MCVD process (a) and the PCVD process (b), shown
          schematically.

of the tube only where the slowly traversing O_2/H_2 burner (20cm/min)
raises the temperature to about $1500°C$. The burner is rapidly
returned to a low temperature to prevent any deposition during
returning. In this multipass operation typically 50-100 layers are
deposited with thickness typically 5 to 10μm, corresponding to about
0.3g/min.

In order to determine the ultimate low-loss limit, predicted as
being 0.2dB/km at 1.6μm, Miya et al.[12] produced an SM fiber from an
MCVD preform. The most obvious procedure was followed: an SiO_2
cladding and an SiO_2/GeO_2 core layer. The preform rod, 7.8mm across
and 40cm long, was covered with another silica tube and drawn into a
125μm diameter fiber, with a deposited cladding to core ratio of 5.5,
cutoff wavelength 1.1μm, δn = 0.0028, and core radius 4.7μm. The
total loss measured confirmed the expectations. Figure 4 shows the
measured loss broken down into the various contributions. A disadvantage of a pure SiO_2 deposited cladding is that during deposition
and subsequent sintering a very high temperature ($\approx 1700°C$) is
required which endangers the substrate tube. Ainslie et al.[13]
reduced the temperature to $1500°C$ by adding some P_2O_5 ($\approx 1\%$) to
the deposited cladding and some B_2O_3 or F ($\approx 0.5\%$) to compensate
for the increased refractive index. The core consisted of P_2O_5-
GeO_2-SiO_2 deposited in two layers. The preform was sleeved with
another silica tube and 10km of fiber was drawn with 8μm core diameter, 40μm deposited cladding and 100μm outer diameter. With F
instead of B_2O_3 in the cladding the loss at 1.5μm was much less,
which suggests that the tail of the Si-F vibration does not contribute at wavelengths below 1.7μm. Eliminating P_2O_5 in the cladding
as far as the power distribution reaches would reduce the losses
due to P-OH vibrations. Ainslie et al.[15] conclude that with a
SiO_2/GeO_2 core, a layer up to twice the core radius should be pure
SiO_2 and that beyond that radius the cladding could safely contain
P_2O_5 compensation. F. Irven[16] prepared SM fibers using an SiO_2/GeO_2
core and an $SiO_2/GeO/F/P_2O_5$ cladding and losses as low as 0.3dB/km

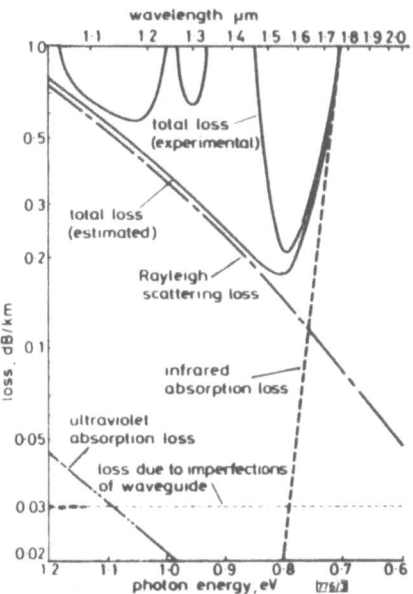

Fig. 4. The total loss measured on a MCVD–SM fiber with a launching
         NA = 0.2 and broken up into the various contributions.
         (from ref.12, fig.3).

at 1.5–1.65μm were reported[15]. The conclusion is that F can be used
as an index reducing dopant, which also lowers the deposition temperature without any adverse effects on loss in the 1–1.8μm region.

Another alternative to the original CVD process was developed
at Philips Research Laboratories in Aachen by Küppers et al.[17–19].
The power needed for the reaction to proceed is not delivered by
heating the reactants to a high temperature but by energizing the
gas to form a plasma. The arrangement is shown in Figure 3b. In
the plasma-activated CVD process a microwave cavity, resonant at
2.4 GHz, moves along the tube through which the reactants flow.
Using about 200 W of microwave power the reactants are converted
100% to the desired deposit. The reaction is completely heterogeneous. The total gas pressure in the tube is between 10 and 25
torr with a three to five fold excess of oxygen. In such a low-
pressure discharge the microwave field only interacts with the
electrons which transfer the energy gained to the molecules of the
gas by collisions. It is therefore called a non-equilibrium plasma.
A plasma containing Cl and O is particularly interesting because
these atoms tend to form negative ions. This reduces the electron
density and may provide a source of long living excited states.

In the PCVD reaction, with gas flows of some hundred cm^3/min
(STP), a substrate temperature of 1100-1200°C, and an inner tube
diameter of 12mm, the gas velocity is 300m/min. For a 1m reaction
tube the residence time of the gas is about 0.1sec. In the absence
of a plasma no thermal reaction products were found even at 1200°C.
With a plasma thermal reactions occur even at 500°C, but the temperature has been increased up to 1200°C to avoid cracking in the
deposit upon cooling. This could be related to the difference in
expansion coefficients and to the chlorine content which was found
to decrease with increasing substrate temperature. It was found
that with a stationary cavity the SiO_2 deposit is not symmetric but
shifts in the direction of the gas flow. It is steeper at the side
of the incoming gas, evidently because there the concentration is
100%, becoming fully depleted after a few cm. This profile is
different for other gases. With a moving cavity, there is therefore
a variation in the SiO_2/GeO_2 ratio over the thickness of a single
layer. Actually the deposition rate is 0.3gr/min but because the
cavity can be moved as fast as 10m/min a large number of layers can
be deposited. Typically a layer of 0.5mm is deposited in a 12mm
diameter tube so that the core can be built up of 500-1000 individual
layers each 0.5-1μm thick. A theoretically required concentration
profile may therefore be approximated experimentally with great
precision. SEM studies have revealed the steps in an MCVD core built
up of 50 layers, whereas in PCVD cores no discrete steps could be
distinguished. This is important for obtaining low pulse dispersion:
fibers with a pulse dispersion of 1nsec/km and a loss of 2.5dB/Km at
0.85μm are being produced routinely[20].

Up to now no results on SM fibers grown with the PCVD method
have been reported. The ability to deposit very thin layers could
be expedient for growing a well defined SM profile to obtain a
defined value of the waveguide dispersion for zero chromatic dispersion. The PCVD method has been found to yield satisfactory results
with any of the dopants commonly used i.e. GeO_2, B_2O_3, P_2O_5 and F.
The second advantage of PCVD, therefore, is that any core/cladding
dopant can be used without any danger of substrate tube softening
which limits MCVD.

BIREFRINGENCE AND CONSERVATION OF POLARIZATION

In a straight SM fiber of perfectly circular symmetry two
polarizations propagate independently with identical dispersion
equations and hence pulse dispersion. In the other extreme situation
the fiber is made strongly birefringent and waves of the two
polarizations along the two main axes x and y propagate independently
but with widely different dispersion equations, $\beta_x(\omega)$ and $\beta_y(\omega)$
and hence pulse dispersion. Aiming at perfectly circular symmetry
is unrealistic because, even if it has been achieved, always to a
limited extent and at great pains, the birefringence will be

determined by the unavoidable curvature. Instead of the birefrin-
gence $\delta\beta = \beta_x - \beta_y$ or $\delta_s = 360° \delta\beta/2\pi$ the more directly accessible
parameter of birefringence length $L = 2\pi /\delta\beta$ is used. In fact this
is the beat length with which waves of propagation constants β_x and
β_y are in phase. Any mechanical perturbation of that period couples
the two polarizations strongly. The mechanical perturbations
corresponding to the birefringence length L are of very small
amplitude in a good fiber in any case and particularly if L < 1mm
or L > 10m. Drawing-induced effects of a periodic nature are very
prone to occur in the range between these two limits. The two
extreme situations to be achieved correspond to these two conditions
for L.

     Aiming at a low-birefringence SM fiber at Siemens, Schneider
et al.[21,22] described how first an SiO_2-B_2O_3 cladding is grown on
the inside wall of a fused silica tube followed by a pure SiO_2 core,
so that the possibility of dip formation is excluded. The low birefringence was evident: $\delta_s = 6°/m$ and the light incident along one
of two principal axes maintained polarization very well also showing
that no higher-order mode conversion occurred. Another approach to
drawing an SM fiber with low birefringence is described by Norman
et al.[23]. In a B_2O_3/SiO_2 cladding and a GeO_2/SiO_2 core fiber with
matched expansion coefficients and $\Delta n = 3.4 \cdot 10^{-3}$, the birefringence
was found to be as low as $2.6°/m$ corresponding to L = 140m.

     Results on SiO_2 core, SiO_2/B_2O_3 cladding SM fibers fabricated
with the MCVD techniques, were reported by Tasker et al.[24]. In a
first type of low B_2O_3 content (2%) the deposition temperature was
too high (1600 - 1700°C) to maintain the substrate tube undamaged.
With increasing B_2O_3 content (10%) the deposition temperature
decreased (1450 - 1600°C) and from such preforms SM fibers could be
drawn with a deposited cladding-to-core thickness ratio of about 8
and $\Delta n = 2.10^{-3}$. In order to produce a birefringent fiber with a
non-circular core, Ramaswany et al.[25,26] collapsed a tube, deposited
with such layers, until the axial core was about 1mm in diameter.
The bore was then evacuated and the collapse was completed. This
produced in a fairly irreproducible way a non-circular core with a
peanut shape. The beat wavelength $L = 2\pi/\delta\beta$ was measured by exciting
the two modes β_x and β_y equally and determining the power ratio
present in both polarizations as a function of length by cutting-
back. It was found to be about 5.5cm whereas in circular fibers
10cm is typical. Then they applied Marcatili's equations[27] for a
rectangular core fiber to the actual dimensions and calculated L.
The beat wavelength should be about 50cm. They explained this
discrepancy by the stress-birefringence δn induced during the manufacturing process. If the anisotropy between the two fundamental
modes is included in the equations, it is found that $L \approx \lambda/\delta n$ so
that $\delta n \approx 10^{-3}$ is required. This term due to anisotropic birefringence, completely dominates the other two terms which are due to
geometry and core-cladding index difference. In order to achieve

the high value of δn required, a large strain has to be built into
the preform and this has to be converted to a large birefringence
by breaking the circular symmetry. An analysis by Kaminow et al.[29]
of a slab geometry with two different thermal expansion coefficients
for the core/cladding combination and for a surrounding jacket shows
that a beat length $L \approx \lambda/\delta n$ can be reached close to 1mm.

   Various methods to approximate the slab geometry experimentally
by removing parts of the SiO_2 jacket to expose the cladding in a
non-circularly symmetric way are reported by Ramaswany et al.[29].
In another method, described by Stolen et al.[30], birefringence is
introduced by grinding flats on the substrate tube before depositing
cladding and core layers in the MCVD process. After collapsing
surface tension tends to restore the smallest circular perimeter of
the substrate tube, and the core or the cladding material deform,
depending on which has the smaller viscosity. With a pure SiO_2
core and a B_2O_3/SiO_2 cladding the lower softening temperature of the
cladding causes the core to remain circular and the cladding to
become elliptical. The birefringence δn was measured as being about
4.10^{-5} with 12% B_2O_3 in the cladding and after grinding-off $\delta D \approx 4\%$
of the fiber diameter on both sides. GeO_2 should yield a higher
value of δn due to the higher expansion coefficient besides yielding
a lower viscosity core which will therefore become elliptical or
slab-like. Using these techniques, Kaminow et al.[31] made a strongly
birefringent SM fiber with GeO_2/SiO_2 cladding and a higher ratio for
the core. The beat length yields a modal birefringence of 3.10^{-4}
and because this is an average over dip, core and cladding the peak
material birefringence δn will be somewhat higher. The elliptical
core and cladding dimensions are $7 \times 1.5\mu m^2$ and $38 \times 11\mu m^2$. Ramaswany
et al.[32] also presented a more detailed analysis in which they compared a number of B_2O_3/SiO_2 cladding, SiO_2 core birefringent SM fibers.
In general, it appears that theory and experiment agree reasonably
well. However, it is not clear what the effect of temperature
variations on the dispersion equation will be under practical conditions. One solution might be to achieve the difference $\delta\beta$ by a
non-circularly symmetric refractive index distribution which is not
basically a temperature effect.[33] Secondly, such highly birefringent
fibers require that orientation is carefully observed at splices.
This is remedied in a proposal by Jeunhomme[34] and Monerie[35] to induce
a strong circular birefringence by twisting the fiber. Barlow
et al.[36] calculated and verified that in order to achieve $L = 1mm$
the fiber has to be twisted to the limit of theoretical glass
strength and that this solution is therefore not practical.

   Summing up, a number of proposals for designing polarization-
maintaining fibers has been considered and tried out experimentally.
Each of them entails different drawbacks and none has been tested
for uniformity and reproducibility in manufacturing. It appears
that the best solution has still to be found.

REFERENCES

1. P.C. Schultz, Proc. IEEE. 68:1187 (1980).
2. D. Charlton and P.C. Schultz, Electr. Opt. Syst.Des. 12:12,23 (1980).
3. B.P. Pal, Fiber and Integr.Opt. 2:195 (1979).
4. T. Izawa, T. Miyashita and F. Hanawa, US. Pat. 4:062,665 (1977).
5. T. Izawa and N. Inagaki, Proc. IEEE. 68:1184 (1980).
6. T. Edahiro, S. Sudo, M. Kawachi, K. Jinguji and N. Inagaki, Electr.Lett. 15:726 (1979).
7. S. Tomaru, M. Kawachi and T. Edahiro, Electr.Lett. 16:511 (1980).
8. S. Tomaru, M. Yasu, M. Kawachi and T. Edahiro, Electr.Lett. 17:93 (1981).
9. M. Kawachi, S. Tomaru, M. Yasu, M. Horiguchi, S. Sakaguchi and T. Kimura, Electr.Lett. 17:57 (1981).
10. J.B. MacChesney, P.B. O'Connor, F.V. Di Marcello, J.R. Simpson and P.D. Lazay, x^{th} Int. Congr. Glass, Tokyo. 6-40 (1974).
11. J.B. MacChesney, Proc. IEEE. 68:1181 (1980).
12. T. Miya, Y. Terunuma, T. Hosaka and T. Miyashita, Electr.Lett. 15:106 (1979).
13. B.J. Ainslie, C.R. Day, P.W. France, K.J. Beales and G.R. Newns, Electr.Lett. 15:411 (1979).
14. J. Irven, A.P. Harrison and C.R. Smith, Electr.Lett. 17:3 (1981).
15. B.J. Ainslie, C.R. Day, J. Rush and K.J. Beales, Electr.Lett. 16:692 (1980).
16. J. Irven, Electr.Lett. 17:2 (1981).
17. D. Küppers, H. Lydtin and L. Rehder, Auslegeschrift, 2444100 Sept./US Pat. 852.068 (1974).
18. D. Küppers, J. Koenings and H. Wilson, J.Electrochem.Soc. 125:1298 (1978).
19. D. Küppers, 7th Int.Conf.CVD, Proc.Electr.Chem.Soc. 159: (1979).
20. J.G. Peelen and J.W. Versluis, Acta Electr. 22:255 (1979).
21. H. Schneider, H. Harms, A. Papp and H. Aulich, App.Opt. 17:3035 (1978).
22. A. Papp and H. Harms, Appl.Opt. 19:3729, 3735, 3741 (1980).
23. S.R. Norman, D.N. Payne, M.J. Adams and A.M. Smith, Electr.Lett. 15:309 (1979).
24. G.W. Tasker, W.G. French, J.R. Simpson, P. Kaiser and H.M. Presby, Appl.Opt. 17:1836 (1978).
25. V. Ramaswany and W.G. French, Electr.Lett. 14:143 (1978).
26. V. Ramaswany, W.G. French and R.D. Stanley, Appl.Opt. 17:3014 (1978).
27. E.A. Marcatili, BSTJ. 48:2071 (1969).
28. I.P. Kaminow and V. Ramaswany, App.Phys.Lett. 34:268 (1979).
29. V. Ramaswany, I.P. Kaminow, P. Kaiser and W.G. French, Appl. Phys.Lett. 33:814 (1978).
30. R.H. Stolen, V. Ramaswany, P. Kaiser and W. Pleibel, Appl.Phys. Lett. 33:699 (1978).

31. I.P. Kaminow, J.R. Simpson, H.M. Presby and J.B. MacChesney, Electr.Lett. 15:677 (1979).
32. V. Ramaswany, R.H. Stolen, M.D. Divino and W. Pleibel, Appl. Opt. 18:4080 (1979).
33. K. Kitayama, S. Seikai, N. Uchida and M. Akiyama, Electr.Lett. 17:420 (1981).
34. L. Jeunhomme and M. Monerie, Electr.Lett. 16:921 (1980).
35. M. Monerie and L. Jeunhomme, Opt.Quant.Electr. 12:(1980).
36. A.J. Barlow and D.N. Payne, Electr.Lett. 17:389 (1981).

SINGLE-MODE OPTICAL FIBER TECHNOLOGY III. DOUBLE-CRUCIBLE

DRAWING TECHNOLOGY AND COMPOUND GLASS FIBERS

P.J. Severin

Philips Research Laboratories
Eindhoven
The Netherlands

INTRODUCTION

From the very beginning of optical fiber studies it has been
evident that the optical fiber will modify the electrical communication network either radically or not at all. The large volume needed
is preferably produced in a continuous production process. A second,
and often more important, advantage of a continuous process is that
the product is more uniform. Within this philosophy in 1974
Koizumi et al.[1] took a considerable step forward when they introduced
the double-crucible system for continuous production of compound
glass fibers. Because of the promising results obtained with the
silica-CVD-based methods at about the same time, the double-crucible
technique has been studied in greater detail in only a few centers:
Nippon Sheet Glass, Philips Research Laboratories and, most intensively, by the British Post Office. Initially each of these centers
concentrated for graded-index fibers mainly on proprietary glass
systems. These were chosen with the particular aim of obtaining a
fast enough diffusion rate during the time when the core and cladding
glasses are in high-temperature contact in the double crucible.

From the electromagnetic viewpoint the refractive index profile
is the weak aspect of double-crucible drawn fibers. The profile
shapes which yield a large bandwidth are generally close to a parabola
and always functions with all derivatives negative. From the
materials viewpoint the absorption due to transition metals and water
is the weak aspect. The solid starting materials are available in
about the same pure state as the gases for CVD grown fibers. Basically the problems associated with the removal of water are identical to those encountered in CVD growth. The difference is that CVD
processes start with a distillation step, which leaves behind

undesired components, whereas with the double-crucible technique the solid materials have to be handled and contained. Each of these process steps has to be scrutinized for its potential contamination effect. A third weak aspect is the lifetime determined by the insufficient corrosion resistance. Though glass compositions exist with higher corrosion resistance than silica, multicomponent glasses in general have lower lifetimes[2].

In this review of single-mode (SM) fiber technology it is not appropriate to discuss the glass system which have been investigated as materials for cheap, moderate bandwidth, graded-index or thick core, large NA, step-index compound flass fibers. On the other hand what we do consider worth discussing is the tool and its mode of operation, because the double-crucible can produce some remarkable fiber core shapes easily and continuously. Due to the wide range of materials available and the variations allowed in the double-crucible geometry, it is a very flexible and controllable system.

COMPOUND GLASS PREPARATION

Compound glasses are amorphous materials based on a small number of possible network formers, e.g. SiO_2, GeO_2, P_2O_5, B_2O_3 and a large number of possible network modifiers. The physical parameters and chemical properties are strongly dependent of the composition and can be chosen within a wide range by varying the appropriate component. For optical fibers the most important effect of the network modifiers is a lower melting temperature. The properties of the glass are determined by the network formers as well as by the network modifiers. The network modifiers are generally available as carbonates or nitrates. Only multicomponent glasses with SiO_2 as the main networkformer have been considered. The level of contamination by transition metals is more likely to increase by handling in these solid materials than in the gaseous materials used for silica-based fibers. The effects of adding particular network modifiers are known in general terms and in more detail if the glass system has been investigated for some particular purpose.

Up to now three glass systems have been investigated as potential candidates for graded-index compound glass fibers. The main point here is whether the system allows the preparation of core and cladding glasses, of about 1% refractive index difference, by adding two components which diffuse fast enough at the drawing temperature. The three systems investigated are presented in Table 1, together with some relevant properties, among which low melting temperature, low Rayleigh scattering coefficient, high corrosion resistance, matched expansion coefficient and viscosity ranges are the most important.

Table 1. The three glass systems investigated to date
with some of the most important properties.

Aspect	Philips R_2O-CaO-GeO $-SiO_2$	Nippon Sheet Glass R_2O-B_2O_3-SiO_2	BPO Na_2O-B_2O_3-SiO_2
exchange	Na $\rightleftarrows$ K	Tl(Cs) $\rightleftarrows$ Na	Na,K,Mg,Ba,Ca
diff.rate	Fast	Fast	Fast
Δn	0.02	0.015	0.015
Rayleigh scattering at 1µm (dB/km)	3	0.75	0.75-1
Infrared transmission	1.6µm	-	-
T_m (^{o}C)	1250	1150	1150
α_{min} (dB/km) at 850nm	9	4.5	3.5

In order to prepare a compound glass the starting materials are
mixed thoroughly under well controlled, dust-free conditions and
melted. The melt is homogenized by various techniques and after
some hours a rod can be drawn from the melt. Rods of core and
cladding material are introduced into the double-crucible system
shown schematically in Figure 1. It goes without saying that scaling
this predrawing stage up to a fully continuous process involves a
considerable amount of high-level technology. Though some experience
with conventional glass melting technology is found helpful, most of
it is being newly developed.

DOUBLE-CRUCIBLE FIBER DRAWING

Of course drawing a thin wire from a material with a wide viscosity range is not a new problem. The problems associated with
polymer fiber drawing have been considered widely, and an up-to-date
critical review has been given in Ziabicki's book[3]. The equations
of conservation of mass were applied to double-crucible optical fiber
drawing by Titchmarch[4]. He derived the conditions for drawing a
fiber with constant core/cladding diameter ratio using an approximation which has not been made here.

The geometry of the double-crucible arrangement is shown schematically in Figure 1. The flows are controlled by the resistance
presented by the crucible nozzles operating in the Hagen-Poiseuille

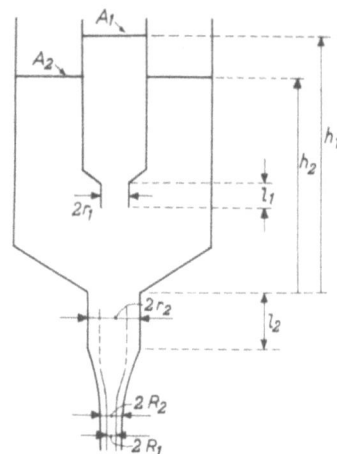

Fig. 1.   The double-crucible geometry.

regime and the pressure difference over the nozzles is determined
by the levels of the molten glasses in the inner and outer crucibles.

In a reasonable approximation the volume flow i_1 through the
inner crucible nozzle responds to the difference between inner and
outer crucible levels as

$$i_1 = \frac{\pi r_1^4 \delta g}{8 l_1} (h_1 - h_2) \tag{1a}$$

Similarly, for the outer crucible nozzle the volume flow i_2 reads

$$i_2 = \frac{\pi r_2^4 \delta g}{8 l_2} h_2 \tag{1b}$$

Drawing a fiber with velocity V, the volume flows can also be
expressed in terms of fiber core and cladding radii R_1 and R_2 and
inner and outer crucible surface areas A_1 and A_2 as

$$i_1 = \pi R_1^2 V = - A_1 \frac{dh_1}{dt} \tag{2a}$$

and

$$i_2 = R_2^2 V = - A_1 \frac{dh_1}{dt} - A_2 \frac{dh_2}{dt} \tag{2b}$$

Combining these eqs. and performing some calculations we obtain

$$h_1 = - k_1 \frac{a}{B_1} e^{B_1 t} - k_1 \frac{1}{B_2} e^{B_2 t} \qquad\qquad (3a)$$

and

$$h_2 = - a(1 + \frac{k_1}{B_1}) e^{B_1 t} - b(1 + \frac{k_1}{B_2}) e^{B_2 t} \qquad\qquad (3b)$$

with

$$2B_{1,2} = - k_1 (1 + \frac{A_1}{A_2}) - k_2 \pm [\{k_1 (1 + \frac{A_1}{A_2}) + k_2\}^2 - 4k_1 k_2]^{\frac{1}{2}},$$
$$\qquad\qquad (4)$$

$$k_1 = \frac{\pi r_1^4 \delta g}{8 \, _1 l_1 A_1} , \qquad k_2 = \frac{\pi r_2^4 \delta g}{8 \, _2 l_2 A_2}$$

and the constants a and b follow from the initial state $h_1 = h_{10}$ and $h_2 = h_{20}$ at t = 0 as follows

$$a = \frac{h_{10}(1 + \frac{k_1}{B_2}) - h_{20} \frac{k_1}{B_2}}{k_1 (\frac{1}{B_2} - \frac{1}{B_1})} \qquad\qquad (4a)$$

and

$$b = \frac{h_{10}(1 + \frac{k_1}{B_1}) - h_{20} \frac{k_1}{B_1}}{k_1 (\frac{1}{B_1} - \frac{1}{B_2})} \qquad\qquad (4b)$$

With $A_1/A_2 << 1$, the solution is identical to the solution obtained by Titchmarch[4].

The ratio of core and cladding radii in the fiber can be expressed using eqs. (1) and (2) as

$$\frac{R_1^2}{R_2^2} = \gamma \, \frac{h_1 - h_2}{h_2} \, , \quad \text{where} \quad \gamma = \frac{k_1 A_1}{k_2 A_2} \tag{5}$$

Eqs. (4a) and (4b) show that this ratio can be made independent of time only if either a = 0 or b = 0. Then the conditions for which the initial heights ratio h_{10}/h_{20} is maintained during the drawing process, are either

$$\left.\frac{h_{10}}{h_{20}}\right)_a = \frac{k_1}{B_2 + k_1} \quad \text{or} \quad \left.\frac{h_{10}}{h_{20}}\right)_b = \frac{k_1}{B_1 + k_1} \tag{6}$$

In practice b = 0 is the condition preferred. In order to keep not only the ratio but also the values of core and cladding diameter themselves constant, the drawing velocity should be reduced according to

$$V = V_o \, e^{B_1 t} \, ,$$

where

$$V_o = k_2 A_2 h_{20} / \pi R_2^2$$

DOUBLE-CRUCIBLE PROFILE TAILORING

In the ideal situation the fiber is drawn at a rate of 1 to 5km/h and the core and cladding glass are long enough in high-temperature contact, in the range with the wide cross-section between inner and outer nozzles, to form a profile. There are two other effects which may modify the profile, the first one decreasing, the second one enhancing the diffusion character of the profile.

The flow through the outer nozzle as formulated in eq.(2b) is really the integral value over the velocity distribution

$$\nu = \nu_o \, (1 - \frac{r^2}{r_2^2}) \tag{7}$$

The velocity ν_o follows from the continuity of flow through the outer crucible nozzle and the fiber mass flow as $\nu_o = V \, 2R_2^2/r_2^2$. This also implies that an arbitrary volume element at radius r in the nozzle moves along a flow line to a corresponding position at radius R in the fiber so that the coordinate transformation experienced by the glass volume in moving from a tubular transport characterized by a parabolic velocity distribution to a transport with a uniform velocity distribution, is

$$\frac{R^2}{R_2^2} = 2 \frac{r^2}{r_2^2} (1 - \frac{r^2}{2r_2^2})$$ (7a)

For an SM fiber with a very small value of r/r_2 there is only a small effect. This effect does not generate a profile but if there is a profile, the coordinate transformation compresses it uniformly close to the axis.

The various glass systems tabulated in Table 1 do not all satisfy the requirement of fast core/cladding dopant diffusion. The outer crucible nozzle has therefore been shaped as a long tube through which the glass flows slowly at high temperature. In this long nozzle the velocity distribution is again parabolic. The velocity $\vec{v}$ does not have any component v_r, but only v_z. Accordingly there cannot be any coordinate transformation of the nature described above, though this again occurs at the output end. Nevertheless, there can be an effect on the profile because the glass that flows down the tube of length l_2 close to the axis is exposed to diffusion for a shorter time than glass flowing close to the outer radius. In a first approximation the effective diffusion parameter ϕ which indicates the degree of grading, after a time at a position r, reads

$$\phi = \frac{Dt}{r^2} = \frac{D}{r^2 v_o} \frac{l_2}{r_2^2 - r^2}$$ (8)

so that for a multimode fiber the diffusion increases towards larger radius, the opposite of what is desired for a profile. In the case of an SM fiber there is again hardly any effect because the profile is in the core where $r \ll r_2$.

Though there is no difficulty at all in drawing a step, or graded-index SM compound glass fiber with several GHz.km bandwidth, there appears to be no aspect which favors the method for SM fiber manufacturing from the technology viewpoint. From the materials viewpoint the fact that the cladding should be of high quality material anyway, is better exploited. From a stability analysis follows which of the parameters of the double-crucible can be chosen from a wide and which from a narrow range of values.

NON-CIRCULARLY SYMMETRIC PROFILES

In this section we shall be concerned with perturbing the core/cladding interface so that it assumes a non-circularly symmetric shape. As discussed with reference to silica-based fibers, such core shapes are desired for single-mode, single polarization fibers.

At the outer crucible nozzle the temperature generally drops rapidly from a high level where η is between 100 and 500Ns/m^2 to a low temperature where the glass can be considered a solid, the glass temperature where $\eta = 10^{12.4}$Ns/m^2. At outer nozzle exit, the high temperature low viscosity and the surface tension make the outer perimeter of the fiber circular. In general, therefore, the shape as offered for the coordinate transformation through drawing, is maintained. Similarly any asymmetry within the double-crucible is manifested in the drawn fiber. Because the core and cladding glass are rheologically almost identical, the lines of equal velocity pass through core and cladding glass without any distinction and the fiber reflects the asymmetry directly. When the inner crucible is positioned slightly out-of-line with the outer crucible, the core is eccentrically positioned in the fiber and remains so.

Positioning the inner crucible slightly eccentrically does not change the total cladding flow. If the gap between the inner crucible nozzle and the outer crucible is reduced further, the velocity distribution, which is zero at the walls, strongly reduces the cladding flow locally through the gap. This effect was found experimentally when the end of the inner crucible nozzle was deformed to an elliptical shape. The core glass assumed an elliptical cross-section but in the remaining gap the cladding glass-flow was reduced to such an extent that core glass moved into replace the missing cladding glass. The result was a fiber with a rectangular core, as shown in Figure 2a. The inner crucible nozzle was then given the shape shown in Figure 2b, a circular inner cross-section for core flow and an elliptical outer cross-section for blocking cladding flow. The result was an elliptical core fiber. The blocking of the cladding glass flow, of course, could also be achieved by making the annular space between the inner crucible and outer crucible elliptical, as shown in Figure 2c. This again yielded a fiber an elliptical core.

Each of these two basic mechanisms, positioning and blocking, naturally produces an even more pronounced electromagnetic asymmetry, than that due to geometry alone, when the core and cladding are not in mechanical equilibrium. Breaking the symmetry of the strain in one of these two ways, the fiber becomes highly birefringent. An SM fiber then maintains polarization.

In an experiment described by Brongersma et al.[5], a fiber was drawn under the conditions described above to yield a rectangular core, of $1.5 \times 30\mu$m^2. The core and cladding glasses differed by only 0.5mol % Nb_2O_5 core dope, which yields about 1% refractive index difference without any strain. The modes could be made clearly visible by varying the angle of the incident beam. The minimum number N of spots decreased with the wavelength λ and it was found, as it should be, that $N \sim \lambda^{-2}$. It was also verified that the polarization could be maintained very well in straight sections. The strain broken in this fiber was not large enough to obtain the

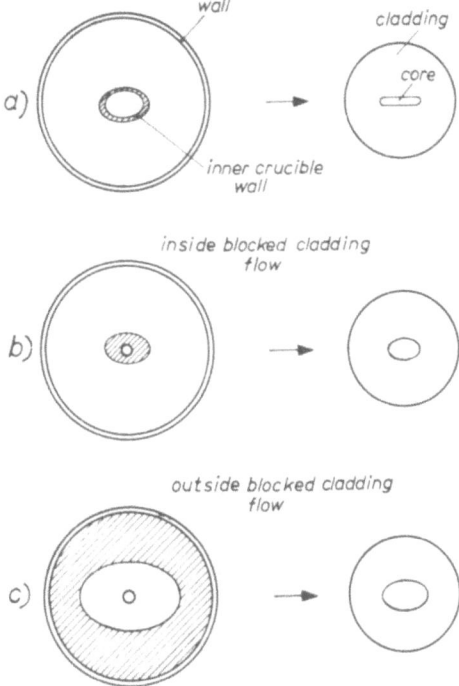

Fig. 2. The double-crucible non-circular geometry and the
         corresponding cross-section of the drawn fiber in
         situation (a), (b) and (c).

desired insensitivity to bending. It is entirely feasible to produce
SM, polarization-maintaining fibers with any arrangement and geometry
of the cores in a controlled process. It would be possible, for
instance, to have four SM fibers in one core. If these were circular,
it would be difficult to avoid cross-talk in the usual dimensions,
but if four slits are made, tangent to a circle, more isolation can
probably be obtained.

REFERENCES

1. K. Koizumi, Y Ikeda, I. Kitano, M. Furukawa and T. Sumitomo,
    Appl.Opt. 13:72 (1974).
2. M. Yoshiyagawa, Y. Kaite, T. Ikuma and T. Kishimoto, J.Non-
    crystall.Sol. 40:489 (1980).
3. A. Ziabicki,"Fundamentals of fiber formation", Wiley, London
    (1976).
4. J.G. Titchmarch, 2nd Europ.Conf.Opt.Fibr.Comm., Paris. 41 (1976).
5. H.H. Brongersma, C.M. Jochem, Th.P. Meeuwsen, P.J. Severin and
    G.A. Spierings, 6th Europ.Conf.Opt.Comm., York. Post dead-line
    papers. 2-5 (1980).

INTEGRATED OPTICAL RECEIVERS FOR TELECOMMUNICATIONS :

A CRITICAL DISCUSSION

B. Daino

Fondazione Ugo Eordoni
Istituto Superiore Poste e Telecomunicazioni
Roma, Italy

INTRODUCTION

The impressive development of first generation optical fiber communication systems has been mainly based on the progress in the technology of fibers and sources. Even though some amount of work has been devoted to the receiver optimization, it mainly concerned the electronic circuitry, that is the optimization of the processing of the signal after the photoelectric conversion. Without exception, in all the systems in use today, the carrier is intensity modulated and direct detection is accomplished at the receiver. The widespread use of direct detection in fiber systems is due to different reasons: the spatial multimode structure of the field to be detected, the lack of stable and spectrally pure sources, the simplicity and in many cases satisfactory properties of the direct detectors. In recent years, however, due to the general development of the field pushing toward the exploitation of better performances of the systems in terms of bandwidths and distances, and given the availability of single-mode fibers and stable single-frequency lasers, new attention has been paid to the possible alternatives for detection, namely quantum amplifiers and heterodyne receivers.

In this paper, a comparison of the different approaches to the low-level optical receiver is presented. Digital angle modulation of the optical carrier, like frequency or phase-shift keying, may offer substantial advantages over the simple intensity modulation. Considering the state of the art of the main components, they seem amenable with the practical realities of the near future, opening new and interesting possibilities for a new generation of optical tranmission systems.

111

THE DIRECT DETECTOR AND THE QUANTUM LIMIT

To begin our analysis, let us refer to a binary digital system, as schematically shown in Figure 1. A single-mode laser emits pulses of light into time slots of duration T. According to a given code, a sequence of "1" and "0" is formed; let E_T be the average energy in each transmitted "1", while no energy is transmitted during the "0"'s. The receiver must recognize, in each time slot, if the light pulse is present or not. In the ideal conditions we assume no background radiation or dark current are present, so that, if E_R is the average received energy in each "1" ($E_R = \alpha E_T$, where $1-\alpha$ is the fractional loss of the link), the detector will on the average emit $\eta E_R/h\nu$ photoelectrons (η is the quantum efficiency and $h\nu$ the photon energy) when a "1" is received and zero photoelectrons in the other case. Since however, as well known, photoelectrons are emitted according to the Poisson's statistics, a nonvanishing probability of detecting an "0" when a "1" is present exists given by

$$P_E = \exp(-\eta E_R/h\nu). \tag{1}$$

The function P_E is the error probability of our ideal system and it is a fundamental limit inherent to the quantum nature of the photodetection process. In this sense, it is often referred to as quantum limit and it is useful as a reference level to which to compare the performance of real systems.

In an analog transmission system, with unitary intensity modulation, a similarly fundamental limit exists for the signal-to-noise ratio SNR, given by[1].

$$SNR = \eta P_R/2h\nu B \tag{2}$$

where P_R is the power at the receiver and B its bandwidth.

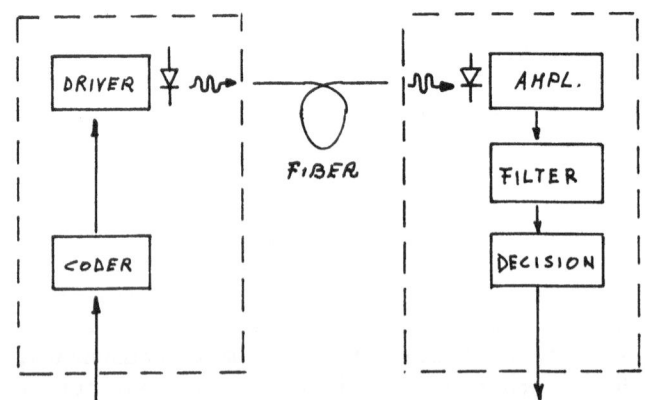

Fig. 1.   Typical optical fiber communication system.

It is common practice in telecommunications to refer to a standard error probability $P_E = 10^{-9}$. According to equation 1, this corresponds to an average number of twentyone photoelectrons per pulse.

In the above considerations, the ideal receiver is supposed to be able to count exactly the photoelectrons (quantum counter). One of the reasons why usual receivers fail to achieve the quantum limit is the way in which photoelectrons are processed. In Figure 2, a very general scheme of the receiver is shown. Electrons emitted in the detector give rise, at the input of the amplifier, to a voltage pulse which is proportional to the total resistance R_T seen by the detector. The pulse is amplified, filtered and sent to a decision circuit. Let v_o be the peak voltage at the output of the filter, due to a single photoelectron; this pulse will be added to the amplified and filtered thermal noise in R_T and excess noise in the input stage of the amplifier. The r.m.s. noise output n_o will depend on the exact details of the amplifier and of the filter; the ratio

$$Z = n_o/v_o \qquad (3)$$

is often used as a parameter to indicate the performance of an amplifier[2,3]. It gives the number of photoelectrons necessary to give a response equal to the r.m.s. value of the noise. For simple (without gain) detectors and nonoptimized wideband amplifiers, usually Z is in the range $10^3 \div 10^4$. Considering that to get an error probability 10^{-9} in detecting a deterministic pulse in a Gaussian additive noise (as it is the output of the filter) of r.m.s. n_o, a peak amplitude of $12n_o$ is required, we see that about 10^4-10^5 photoelectrons, instead of 21, are required in a practical situation. Careful design and optimization of the amplifier and filter may keep these values in the lower range, but still much higher than the quantum limit.

To overcome this situation, photodetectors with internal gain may be used, so as to decrease the relative importance of the thermal and amplifier noise. The drawback with these devices, is that they introduce a new kind of noise, the multiplication noise, which increases with gain. Taking into account this noise too, the SNR may be expressed as[2,3]

$$\text{SNR} = \frac{(\eta P_R/2h\nu B)^2}{(\eta P_R/2h\nu B)<g^2>/<g>^2 + (Z/<g>)^2} \qquad (4)$$

where $<g>$ is the average value of the multiplication factor, $<g^2>$ is its mean square value, and Z, according to equation (3), defines the noise properties of the amplifier. In the absence of multiplication noise ($<g^2> = <g>^2$) and with negligible noise in the amplifier ($<g> >> Z$) equation (4) reduces to equation (2) and the receiver attains the quantum limit. Actually, in semiconductor avalanche devices

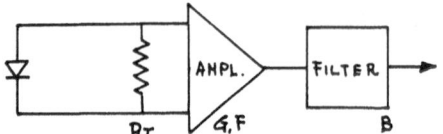

Figure 2.   Optical receiver general schematic.

$$\langle g^2 \rangle = \langle g \rangle^{2+x} \tag{5}$$

where x depends on the material and the device structure. Accordingly
there is an optimum value of <g> giving the maximum SNR, which however
remains lower than the quantum limited value of equation (2). How
much lower, it depends on x and Z, or, more at large, on the devices,
the wavelength range, the bandwidth, etc. Typical calculated performances are shown in Figure 3 (from reference 2), which represents the
minimum power P_{min} necessary to achieve a given SNR. The calculations
refer to a silicon detector and an amplifier having a video bandwidth
B_v = 10 MHz and Z=500; for the other details the reader is referred
to the original paper. The numbers in parentheses indicate the optimum values for the gain of the avalanche photodetector. As is apparent
the quantum limit is approached only at high SNR's; the difference
between the expected characteristics of this example and the ideal
ones become worse at low signal levels and, generally, at higher frequencies. Actually, an optical power 15 ÷ 20 dB higher than the
quantum limit is typical for an optimized receiver.

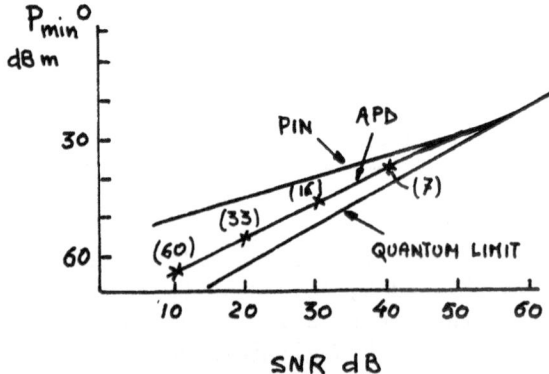

Fig. 3. P_{min} necessary to achieve a given SNR (after ref. 2).

THE HETERODYNE RECEIVER

As we have seen in the previous section, the main problem in the direct detection is to raise the signal to a level comparable or greater than the thermal noise in the associated equipment. Let us now see how heterodyne detection overcomes the problem, or at least alleviates it.

As it is well known, when two waves, having the same wavefront and polarization direction, impinge on a photodetector, they give rise to a current

$$I(t) = (\eta q/h\nu)(P_L + P_R) + (\eta q/h\nu)A_L A_R \cos\{(\omega_L - \omega_R)t + \psi_L - \psi_R\} , \quad (6)$$

where q is the electronic charge, P is the power, A the amplitude, ω the pulsation and ψ the phase of the local oscillator (L) or of the received signal (R) waves respectively. For practical reasons, $P_L \gg P_R$. It may be recognized that the oscillating term in equation (6), containing the amplitude and phase information of the signal, may be increased just by increasing the local oscillator wave amplitude. Taking into account the shot noise associated with the d.c. term of equation (6), the multiplication noise and the thermal noise in the load resistor R_T and following amplifier, the resulting SNR may be written in a way similar to equation (4)

$$SNR = \frac{(\eta/2h\nu B_v)^2 P_R P_L}{(\eta P_L/2h\nu B_v)\langle g^2\rangle/\langle g\rangle^2 + (Z/\langle g\rangle)^2} , \quad (7)$$

where a photodetector with gain $\langle g \rangle$ has been considered and the bandwidth of the intermediate frequency amplifier has been taken to be twice the signal (video) bandwidth.

By comparing equations (4) and (7), it is clear that, due to the presence of P_L instead of P_R in the first term of the denominator, the amplifier noise may be easily overcome also at low detector gains, thus reducing the multiplication noise too. If this is the case, equation (7) reduces to

$$\{SNR_{IF}\}_{ideal} = (\eta/2h\nu B_v)P_R , \quad (8)$$

attaining the quantum limit.

The other important characteristic of the heterodyne detection is that the advantages of angle modulation schemes can be exploited. For example, using an analog frequency modulation, it may be shown[4] that after IF detection, at signal levels above the FM threshold, the signal-to-noise ratio SNR_{PD} is

$$SNR_{PD} = 3\beta^2(B_o/B_v) SNR_{IF} , \quad (9)$$

where β is the modulation index and B_0 the bandwidth of the modulated optical carrier. If the same conditions of equation (8) apply, equation (9) becomes

$$SNR_{PD} = 6\beta^2 (\eta P_R / 2h\nu B_v) \qquad\qquad (10)$$

A detailed comparison of the different modulation scheme sensitivities has been developed in reference 5. Some of these results are shown in Figure 4, which gives the minimum power necessary to achieve a bit error rate BER = 10^{-9}, for values of the parameters typical at 1.5 µm. It may be seen that the best system is the phase-shift-keying (PSK) homodyne, being about 3dB better than the PSK heterodyne (due to the twice larger IF bandwidth of the heterodyne) and more than 20dB better than the amplitude modulated direct baseband system. The gain of the photodetector is supposed optimized in the latter, and unitary in the homodyne and heterodyne systems. The importance of optimizing the gain may be seen in Figure 5, where the minimum power for a BER=10^{-9} is reported versus the local oscillator power, for three values of R_T. According to equation (7), optimization of <g> plays a determinant role when the local oscillator power is not high enough to get the shot-noise limited operation.

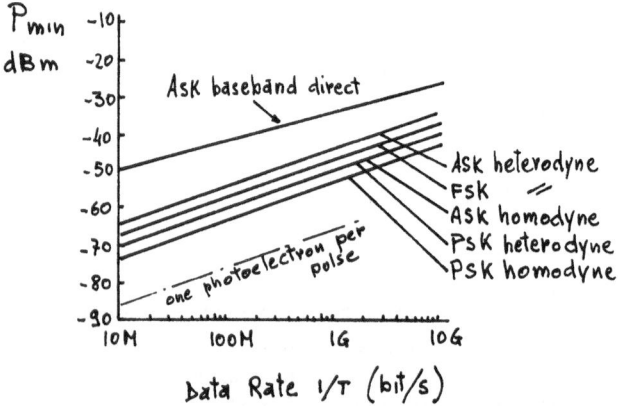

Fig. 4. P_{min} versus Data rate (after ref. 5).

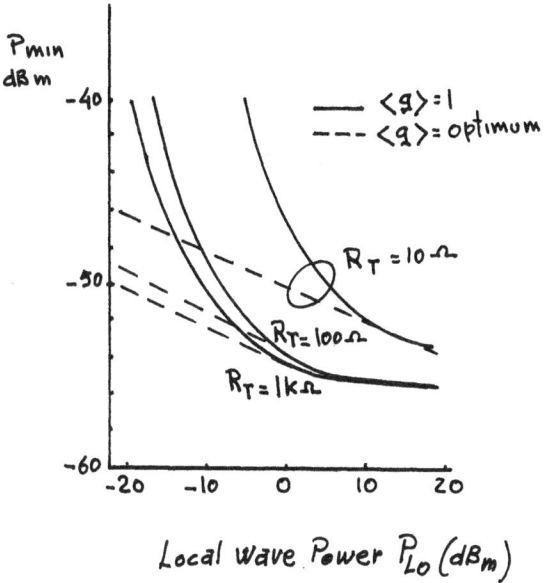

Fig. 5. P_{min} versus local oscillator power (after ref. 5).

THE OPTICAL AMPLIFIER

The use of laser amplifiers in fiber systems has been investigated in the past years[6-9]. Recently, also experimental results on semiconductors diode amplifiers have also been reported[10-11], showing signal power gains greater than 20dB. The limits to the use of such amplifiers come from the spontaneous emission noise they introduce. Since this noise is dependent on the number of modes collected by the detector, an efficient filtering of the optical bandwidth (spatial and temporal) is required. Spatial filtering is easily accomplished in single-mode fibers. Strong frequency filtering demands for coherent detection (homodyne or heterodyne). To be more quantitative, let us say that semiconductor diode laser amplifiers can be divided in two classes, resonant (FP) or travelling wave (TW), according to whether the active medium is enclosed or not in a cavity. Performances of both structures as amplifiers directly coupled to the detector have been calculated in reference 6. As an illustration, some of the results are reported in Figures 6 to 9. Calculations have been performed for typical values of the parameters and the reader is referred to the original paper for the details.

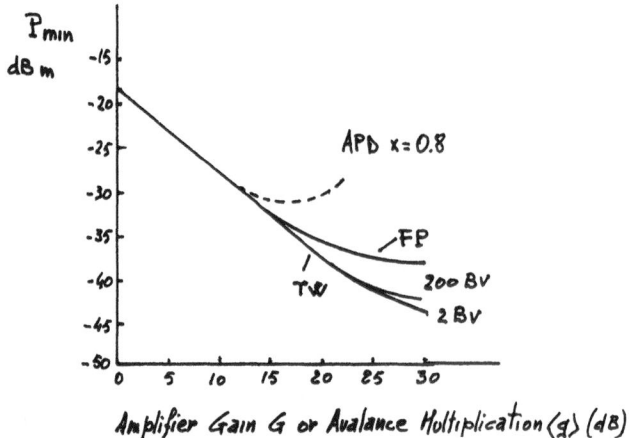

Fig. 6. P_{min} at the optical amplifier input versus gain G or <g>.

In Figure 6, the minimum power P_{min} necessary to achieve a BER= 10^{-9} is plotted versus the optical gain G and the avalanche gain <g> in the absence of optical amplification. Data rate is assumed to be 1 Gbit/s and load resistance of the photodiode $R_T = 50\Omega$.

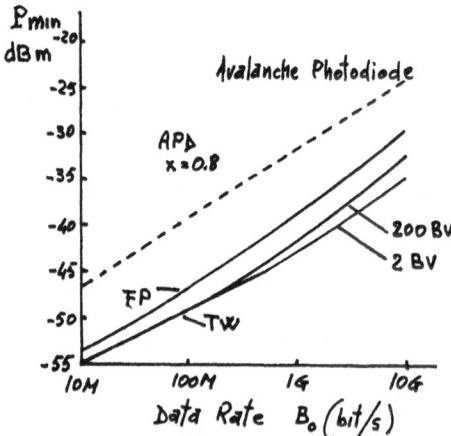

Fig. 7. P_{min} versus Data rate. Optical gain 25dB, APD gain optimum.

In Figure 7, the same minimum power is shown versus the data
bit rate, assuming an optical preamplifier gain of 25dB or, for the
avalanche photodiode, the optimum gain. As easily seen, a marked
improvement is achievable with respect to the simple direct detection.
At high gains and narrow optical bandwidth (the bandwidth in the il-
lustrations is the product of the number of spatial modes times the
optical frequency bandwidth) the improvement is comparable with the
one obtained by using amplitude modulation and heterodyne detection.
The minimum detectable power is however still 6dB higher than in the
PSK heterodyne and 9dB higher than in the PSK homodyne system.

In the case in which many optical amplifiers are cascaded to
compensate the losses of the line (nonregenerative linear repeaters),
the SNR will be degraded along the line due to the noise contribution
of each amplifier. In Figure 8, the dependence of SNR on the number
of repeaters for different values of the optical bandwidth B is shown.
The gain of each amplifier is assumed 25dB with an input power of
−35dBm, at a wavelength of 1.5μm and data rate of 100Mbit/s.

In Figure 9 is reported the maximum number of repeaters R which
can be cascaded preserving a BER=10^{-9}, for different values of the
amplifier input power P_i and optical bandwidth B. Each amplifier is
assumed to have a gain of 25dB.

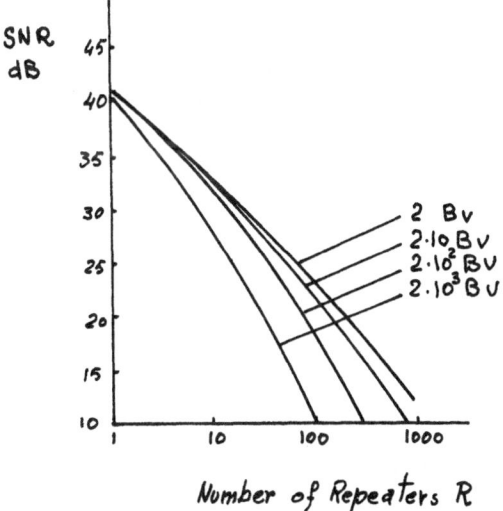

Fig. 8.   SNR degradation after R repeaters.

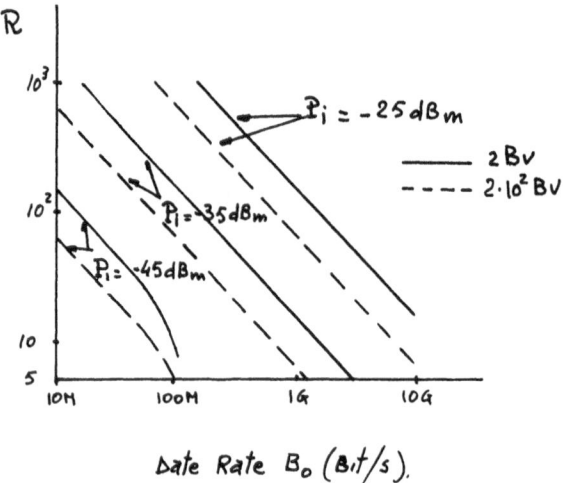

Fig. 9.   Maximum number R of repeaters preserving BER = 10^{-9}.

DISCUSSIONS AND CONCLUSIONS

From the considerations exposed in the previous sections, it is clear that, from the point of view of the optimum use of the received optical power, the PSK homodyne or, with a minor loss, heterodyne modulation formats should be preferred. Simple optical amplifiers used as intermediate linear repeaters could very considerably increase the span between the regenerative repeaters.

For the practical applicability of heterodyne systems, two major achievements must be accomplished : polarization maintaining fibers and high spectral purity lasers. Both fields are steadily progressing[12]. Single polarization fibers of various structures have been reported. For the frequency spectrum of free-running lasers, the best reported results show a linewidth of about 100MHz; locking the laser to an external Fabry-Perot resonator reduces the frequency wandering within a few MHz. This seems to be also the minimum linewidth due to the quantum phase noise[13]. For a binary phase modulation, the requirement is that the phase of the carrier should be stable well within π in the time of a digit; this brings to a maximum linewidth[9]

$$\Delta\omega << \pi/2T \tag{11}$$

where T is the clock period, which shows that for most of the high speed systems the obtained performances could be adequate.

Integration on a single chip of a heterodyne receiver should not present substantially new technological problems: GaAs process compatability of the local oscillator laser, the directional coupler, the detector and, eventually, the IF amplifier has been proven. Feedback control procedures for tuning and power and gain stabilization should be developed. Frequency locking to an external resonator seems inadequate and a stable wavelength reference amenable to planar technology should be found; distributed feedback structures or ring resonators are promising ways. Also compact integrable isolators could be required for cascading a long chain of optical amplifiers, whose stability behaviour is not yet known. These problems do not seem however to be unsolvable; the efforts necessary to solve them will be justified once the advantages that more sophisticated transmission systems can offer will be clearly recognized.

REFERENCES

1. J.P. Gordon, Quantum effects in communication systems, Proc.I.R.E. 50 : 1898 (1962).
2. S.D. Personick, Receiver design for optical fiber systems, Proc. I.E.E.E., 65 : 1670 (1977).
3. W.M. Hubbard, Utilization of optical-frequency carriers for low- and moderate-bandwidth channels, Bell Syst.Techn.J., 52 : 731 (1973).
4. R.M. Gagliardi and S. Karp, "Optical Communications", J. Wiley & Sons, New York (1976).
5. Y. Yamamamoto, Receiver performance evaluation of various digital optical modulation-demodulation systems in the 0.5-10 μm wavelength region, IEEE J.Quantum Electron., QE-16 : 1251 (1980).
6. Y. Yamamoto, Noise and error rate performance of semiconductor laser amplifiers in PCM-IM optical transmission systems, IEEE J. Quantum Electron. , QE-16 : 1073 (1980).
7. S.D. Personick, Applications for quantum amplifiers in simple digital optical communication systems, Bell Syst.Techn.J.,52 : 117 (1973).
8. G. Zeidler and D. Schicketantz, Use of laser amplifiers in glass-fiber communication systems, Siemens Forsch. , 2 : 227 (1973).
9. F. Favre, L. Jeunhomme, I. Joindot, M. Monerie, and J.C. Simon, Progress towards heterodyne-type single-mode fiber communication systems , to appear on IEEE J. Quantum Electron.
10. Y. Yamamoto, Characteristics of AlGaAs Fabry-Perot cavity type laser amplifiers, IEEE J. Quantum Electron. , QE-16 : 1047 (1980).
11. S. Kobayashi and T. Kimura, Gain and saturation power of resonant AlGaAs laser amplifier, Electron.Lett. , 16 : 230 (1980).
12. Proceedings of the third International Conference on Integrated Optics and Optical Fiber Communication (Optical Society of America, Washington , D.C., 1981).
13. Y. Yamamoto, T. Mukai, and S. Saito, Quantum phase noise and linewidth of a semiconductor laser, Electron.Lett. , 17 : 327 (1981).

PRINCIPLES AND PERFORMANCE OF ACCESS COUPLERS FOR MULTIMODE FIBERS

AND DEVICES FOR WAVELENGTH DIVISION MULTIPLEXING

F. Auracher

Research Laboratories of Siemens AG
D-8000 Muenchen 83

INTRODUCTION

Access couplers -- both wavelength selective and non-selective --
play an important role in fiber optic communication. Several examples
are the monitoring the signal level in a transmission system, spatially
distributing optical signals eg. TV-signals in large buildings, and
data bus systems.

Reviews of non-selective access couplers have been published[1-3].
Because of the small dimensions of typical glass fibers (50μm to 200μm
core diameter) it is not easy to fabricate and align these devices
reproducibly. Nevertheless, several promising prototypes have been
built and some are already commercially available. Wavelength selective access couplers are used for wavelength division multiplexing
to increase the transmission capacity of fiber optic transmission
lines for long distance or bidirectional communication. In Part I of
this paper we review the basic principles for non-selective access
couplers and give typical performance data. Part II deals in a similar way with access couplers for wavelength division multiplexing/de-
multiplexing. Review papers have been published previously[4-7].

PART I: NON-SELECTIVE ACCESS COUPLERS

There are two possibilities to tap an optical fiber, namely by
coupling light out via the end face or the cylindrical surface of the
fiber. Figure 1 defines an ideal access coupler with four ports. Input and output fibers are assumed to be multimode fibers of the same
type. Usual specifications of the coupler are its tapping ratio

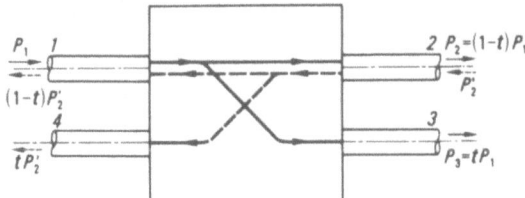

Fig. 1. Ideal access coupler with tapping ratio t.

$$t = P_3/P_2 \tag{1}$$

and its insertion loss

$$L_I = -10 \log (P_2 + P_3)/P_1 \tag{2}$$

The performance of access couplers depends, in general, on the coupler
principle, on the type of fiber (step index or graded index), its acceptance angle, the ratio of the core radius to the cladding thickness
and on the excited modal distribution in the input fiber.

Before the various principles are discussed we list some of the
desirable characteristics of access couplers. These are: low insertion loss, low modal dependence of performance, reproducible and easy
fabrication, small size and light weight.

1. Coupling via fiber end face

Access coupler with branching structure. Perhaps the most in-
tuitive and straightforward approach for access couplers is to use a
branching structure. Figure 2 shows an early version of a planar
branching structure. The tapping ratio is determined by the ratio
a/b of the relative widths of the output waveguides. This early version was fabricated photolithographically in a light sensitive plastic
sheet laminated onto a fused quartz substrate[8]. Simultaneously with
the waveguide structure alignment grooves for the butt coupled glass
fibers were fabricated. While this fabrication process would be ideally suited for mass production of even complicated light distribution
networks, the devices suffered from high insertion loss and poor ageing
characteristics due to the properties of the available materials.
Especially heavy losses occurred in the encircled area of Figure 2.
More recent research activities concentrate on using an ion exchange
process in glass substrates through a photolithographically defined
mask to obtain devices with good aging properties and lower insertion
loss[9]. An all-fiber version of a branching structure is sketched in
Figure 3. The output fibers are ground to a wedge-like shape and

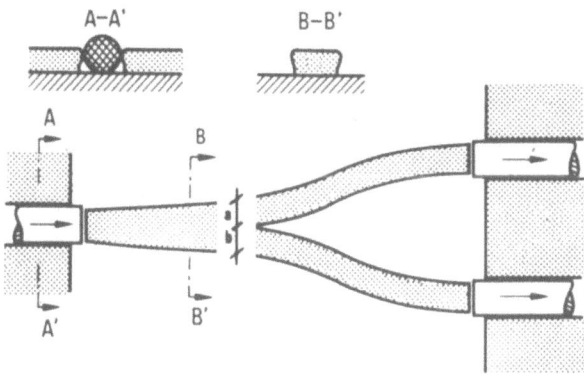

Fig. 2. Branching type access coupler in planar technology after[2].

butted against the end face of the input fiber. Although the principle is simple the fabrication is rather cumbersome. This design is
well suited for both graded index and step index fibers; the mode
selectivity and losses increase with increasing angle ε of the fiber
junction. The inclination angle of rays with respect to the respec-
tive fiber axis will be changed from γ to $\gamma \pm \varepsilon$. Therefore some of
the guided light is lost because some of the rays with inclination
$\gamma + \varepsilon$ exceed the acceptance angle γ_c of the output fibers. Consequently the separation angle ε should be much smaller than the acceptance angle γ_c of the fibers. The theoretical losses can be found
from the literature on splicing losses for an angular misalignment ε
of the spliced fibers[11-13]. Losses of only 0.6dB are reported[14] for
this type of coupler with a tapping ratio t = 0.1.

Access couplers with laterally off-set fibers. Another very
simple and easily fabricated access coupler uses laterally offset
fibers. The principle is shown in Figure 4 together with a picture
of an early version. Obviously the overlap of the fiber cores of the

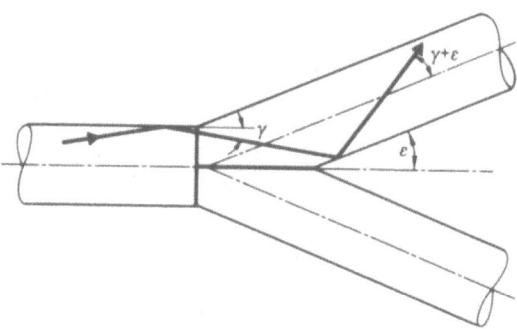

Fig. 3. Branching type access coupler in planar technology after[2].

(a)

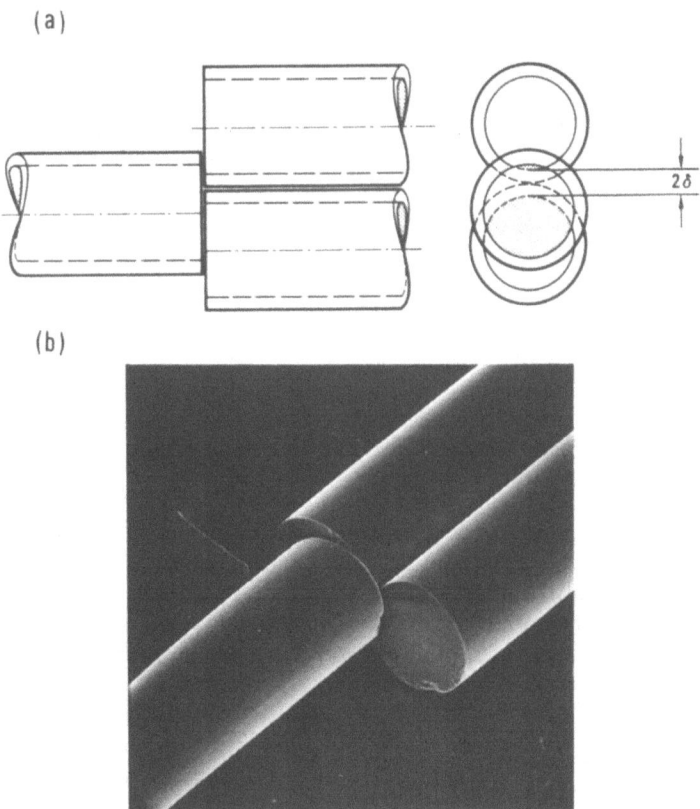

(b)

Fig. 4. Lateral offset fiber coupler: (a) principle, (b) early
         prototype after reference 2.

output fibers 2 and 3 decreases rapidly with increasing ratio of cladding thickness δ to core radius r_c. More recent versions include arrangements, where the input fibers have smaller diameters than the
throughput fibers to improve the overlap at the fiber junction[15].
Theoretically calculated losses for step index fibers as a function
of this ratio and the tapping ratio t are given in[16]. The losses stay
below 1dB for all tapping ratios in the case of step index fibers as
long as $\delta/r_c<0.1$. For graded index fibers additional losses occur
for increasing lateral offset due to the mismatch of the local acceptance angles at the butt joint. The losses can again be found from
the literature on splicing losses for laterally offset fibers [10-13,17].

     Access couplers with beam splitter. Miniature beam splitter
cubes of only 180µm edge length with the fibers directly butted
against the cube were used in an early prototype[18]. A better and
simpler design uses fibers cut under 45° with one side coated with a
partially reflecting metallic or dielectric mirror. Figure 5a shows
the schematic arrangement of this coupler. This coupler achieves

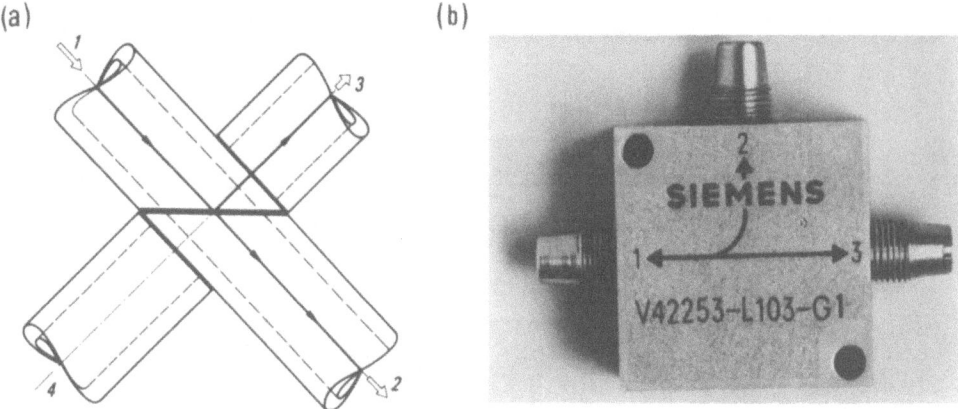

Fig. 5.   Beam splitter in fiber version: (a) principle, (b) finished device.

very low loss, no mode selectivity in the straight through direction (1 to 2) and only small losses and little mode selectivity in the direction 1 to 3 due to the divergence of the unguided reflected beam. An estimate of the theoretical losses is given[19] as a function of the acceptance angle of the fibers and the cladding thickness/core radius ratio δ/r_c. The theoretical losses are very low - below 0.5dB even for a tapping ratio of 1 for all typical fibers. High performance access couplers of this type are being fabricated now using preferentially etched silicon grooves for precision alignment of the coupled fibers[27]. Packaged components with connectors (Figure 5b) have specified insertion losses of 1dB to 1.5dB depending on the tapping ratio.

Access couplers with lenses. A class of access couplers uses graded index rod lenses. The imaging properties of this lens type are sketched in Figure 6a for on-axis and Figure 6b for off-axis input coupling. Several types of access couplers using these imaging properties are shown in Figure 7. The thick lines in the drawings mean partially (Figure 7b to c) or totally reflecting coatings (Figure 7d). The most interesting design is that of Figure 7c because of its simplicity and the nearly perpendicular incidence of the rays of the mirror thus minimizing the polarization effects of the mirror. A measured insertion loss of 0.9dB for a small tapping ratio of 0.03 is cited for a coupler of this kind[20,21].

2. Coupling via fiber surface

Light can also be extracted from the fiber surface by converting guided modes into radiation modes or by coupling to a second fiber via

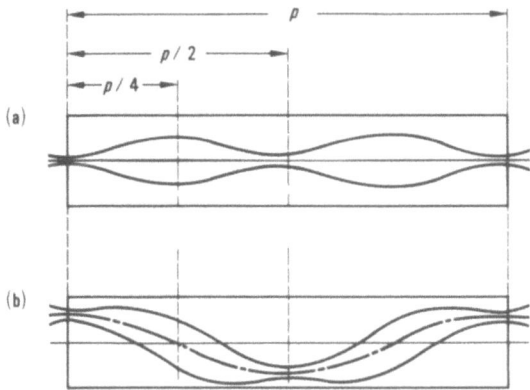

Fig. 6.   Imaging properties of graded index rod lenses (shown are the
          envelopes of guided rays).

the evanescent wave. Conversion into radiation modes can be achieved
by bending the fiber, by stripping the cladding or by tapering down
the core cross-section. The only principle of practical value so far
is the latter using a twisted and fused pair of fibers with biconical
tapers in the fusion region. Figure 8 shows the schematic arrangement
of the biconical taper coupler. In the converging taper the higher
order modes are converted into radiation modes and fill up the total
cross-section (cores plus cladding) of the two fused fibers. In the
expanding taper these radiation modes are mostly converted back to
guided modes, ie. trapped in the cores of the output fibers. Figure 8

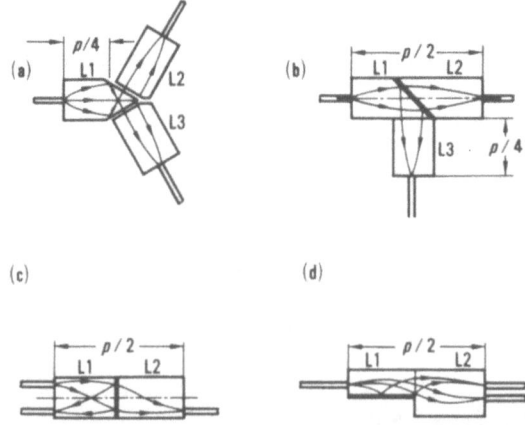

Figure 7.   Access couplers using graded index rod lenses. Heavy lines
            symbolize a partially or totally reflecting mirror.

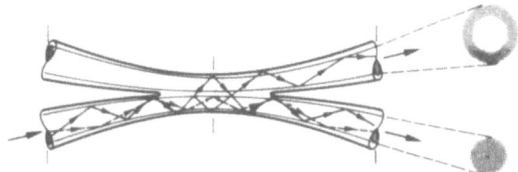

Fig. 8.   Biconical-taper coupler.   Two meridional rays corresponding
          to a low order and a high order mode in a step index fiber
          are shown.

shows this conversion process in a ray optics picture. One would
expect a strong mode selectivity of this coupler type and additional
excess losses in the case of graded index fibers. However, due to
the twisting of the fibers considerable mode scrambling occurs reducing the mode selectivity considerably. In principle this coupler is
easy to fabricate, however reproducibility might be a problem. In
the literature low insertion losses (typically 1dB) are often quoted[22],
but measurement conditions may not be fully adequate for graded-index
fiber couplers.

3.   Conclusion

     The most important types of wavelength unselective couplers have
been reviewed and typical performance data cited. For step index
fibers and small tapping ratio all mentioned principles yield good
performance. For increasing tapping ratio especially in the case of
graded index fibers the performance of the lateral offset fiber coupler and the tapered fiber coupler become somewhat inferior to the
other coupler principles.

PART II: WAVELENGTH SELECTIVE ACCESS COUPLERS

     Wavelength selective access couplers are used for wavelength
division multiplexing (WDM) systems. WDM increases the transmission
capacity of optical fibers and has already undergone field trials in
long distance one-way and shorter distance two-way communication links.
Besides its advantage, WDM leads, however, to higher insertion losses
and requires more transmitters and receivers. The multiplexers
needed for WDM should have low insertion loss. The demultiplexer
should, in addition, also show a high isolation between channels.
The channel separation should be reasonably matched to the spectral
widths of the sources (LED or laser diodes). Until now four different
principles have been used to obtain wavelength selective access couplers. Figure 9a to d is a schematic representation of the four demultiplexer principles used hitherto in experimental setups. These

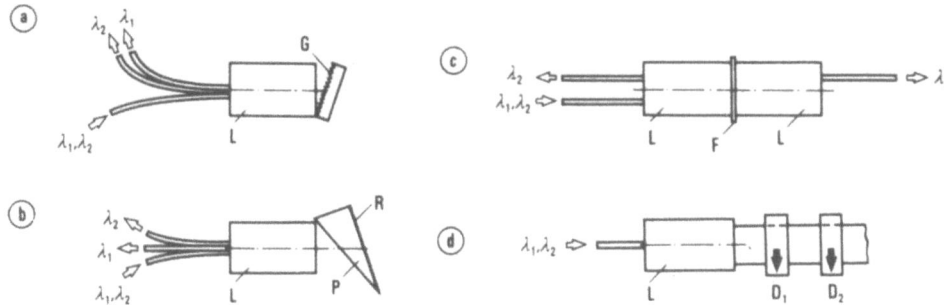

Fig. 9.   Demultiplexing principles[7]: (a) grating, (b) prism, (c)
filter, (d) absorber type. G blazed grating, L graded-
index rod lens, P prism, R reflector, F dichroic filter,
D selective photodetector.

principles rely on the utilization of the three wavelength-selective
effects, viz. angular dispersion, interference and absorptance. All
are shown for free optical wave propagation with beam-shaping by means
of a graded-index lens system. The demultiplexers shown in Figure
9a, b utilize the angular dispersion by means of a grating[23-26] or a
prism. Figure 9c shows a configuration with an interference filter
for channel separation[21,27]. The absorber configuration shown in
Figure 9d so far gives good results only for demultiplexers. The
individual absorbers D are wavelength-selective photodiodes cascaded
according to their band-edge shift[28,29]. The version with a prism
(Figure 9b) is seldom investigated any more because a sufficiently
large angular dispersion is only realizable with materials that are
also lossy whereas low dispersion disallows miniaturization. Configurations with a blazed diffraction grating (Figure 9a), whose 1st-
order angular dispersion

$$\frac{d\phi}{d\lambda} = \frac{2 \tan \phi}{\lambda} = \frac{1}{\sqrt{\Lambda^2 - (\lambda/2)^2}} \tag{3}$$

can be controlled via the grating constant Λ represent on the other
hand an attractive alternative to the configurations with interference
filters (Figure 9c). The collinear absorber cascade (Figure 9d) is
likely to prove very interesting for small numbers of channels, say
two to three, as demultiplexer.

1.   Grating type access couplers

Fig. 10a shows a demultiplexer with a blazed grating and a graded
index rod lens. If the optical power in each channel is purely mono-

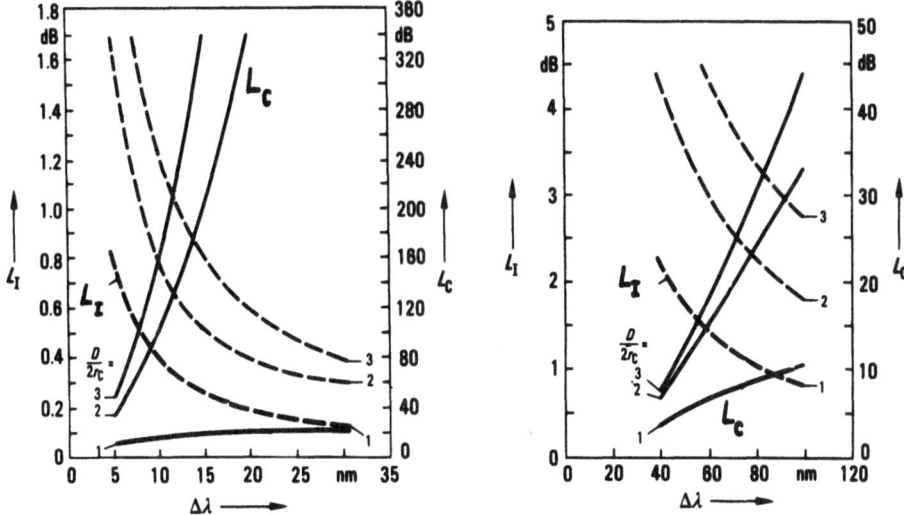

Fig. 10.   Insertion loss L_I and crosstalk attenuation L_C as a function of channel separation $\Delta\lambda$ for a grating demultiplexer according to[7]. (a) for laser diodes with $\delta\lambda = 2$nm, (b) for LED with $\delta\lambda = 40$nm. The parameter is the fiber spacing D over the core diameter $2r_c$.

chromatic, the channel separation is given by

$$\Delta\lambda = (D/f)\sqrt{\Lambda^2 - (\lambda/2)^2} \tag{4}$$

where f is the focal length of the Lens, Λ the grating period and D the spacing of the output fibers. For LED-sources the above assumption does not apply and the spectral widths $\delta\lambda \approx 40$nm of the sources have to be taken into account. Due to the finite $\delta\lambda$ the optical power is diffracted over an angular range $\delta\phi$ causing a spread of the input fiber image. Thus for the same fiber spacing D the cross talk between neighbouring channels is increased as compared to monochromatic sources. As a consequence grating multiplexers are poorly suited for LED sources[7].

This effect has been calculated[7]. The results are shown in Figure 10a for a laser diode with $\delta\lambda = 2$nm and for a LED with $\delta\lambda = 40$nm in Figure 10b. Shown are the insertion loss L_I and the crosstalk attenuation L_C under the detailed assumptions given. With increasing ratio $D/2r_c$, where r_c is the core radius of the fibers, ie. with decreasing fiber packing density, the crosstalk decreases at the cost of a higher insertion loss. Typical achieved data for grating multiplexers and demultiplexers are given in Table 1 at the end of part II.

2.   Interference filter type access couplers

The interference filter demultiplexer shown in Figure 9c uses
graded index rod lenses featuring nearly axially parallel incidence
of the optical rays on the filter. The filter response is corres-
pondingly almost independent of polarisation. The second type shown
in more detail in Figure 11a has an interference filter applied directly to a fiber endface of a 45° cut fiber. This device is identical to the beam splitter in fiber version except for the wavelength
selective dielectric filter. Due to the butt joint of the three
fibers no collimating or focusing lenses are required. The oblique
beam incidence angle, however, results in a slightly different filter
response for the two polarisations. The minimum channel separation
is therefore limited by both the emission spectral bandwidth $\delta\lambda$ and
the splitting of the filter response for the s and p components. An
improved version (Figure 11b) uses a beam incidence angle of 20° instead of 45° and a special filter design[30] to obtain smaller polarizing effects at the cost of slightly higher insertion loss[27]. For
the demultiplexer, narrow band filters are usually added to improve
the cross talk attenuation. Figure 11c shows the WDM coupler used
for a single-mode duplex communication link. In these couplers, 1,2
are single-mode fibers, while 3 is a step-index fiber with a relatively large-diameter core. This type of all-fiber coupler is at
present confined to the more important two-way communication because
the reflex branch 1 → 3 can only be used for coupling out a portion
of the light guided in fiber 1 with low loss; it cannot be used for
coupling in (3 → 1) because serious losses would result. To permit
industrial production of the all-fiber coupler a block design is used
making possible the production of some 30 devices simultaneously.
Figure 12 shows a complete block of 30 WDM couplers, two unfinished
and one finished coupler with attached fiber connectors.

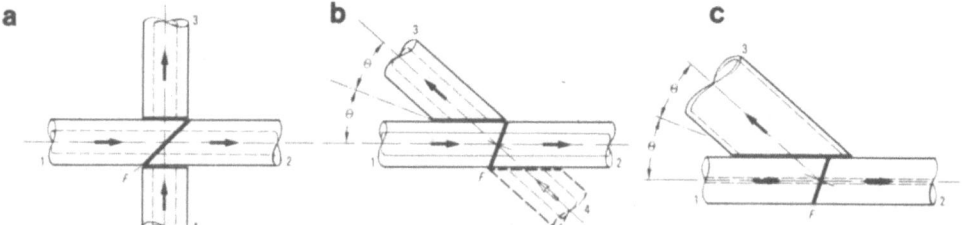

Fig. 11.   All fiber access couplers with interference filter F
           according to[27]. (a) 45° version, (b) improved 20° version,
           (c) single mode fiber version.

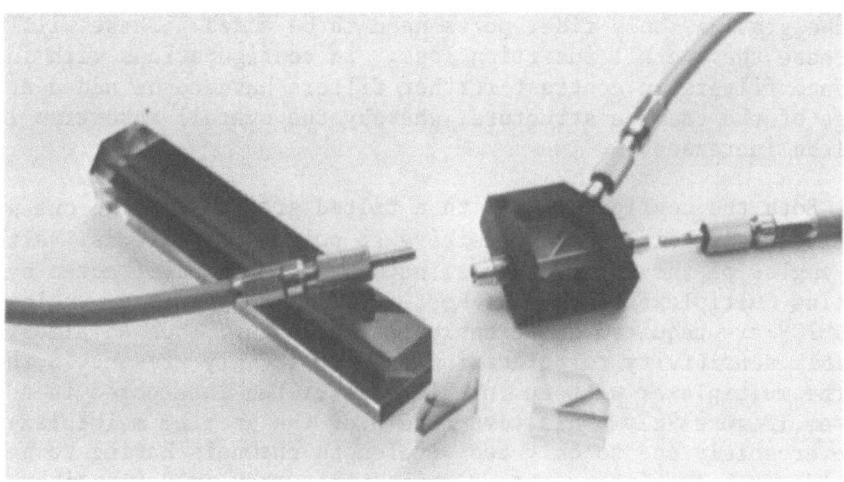

Fig. 12.   Block of 30 access couplers and finished product.

3.   Comparison of multiplexing principles

Only the two most advanced principles - the grating type and
interference filter type access couplers will be compared.   Such an
attempt was made[7] and some of the results are given here.   Figure 13
shows the cross talk attenuation L_c and the insertion loss L_I for
the reflectance (R) and transmission channel (T) of an interference
filter demultiplexer and of a grating demultiplexer (G).   A grating
loss of 1dB and maximum transmission T = 0.6 for the interference
filter were assumed, respectively.   The left diagram is for an LED
source with a spectral width $\delta\lambda$ = 40nm and two channel separations
$\Delta\lambda$ = 50nm and 70nm, respectively.   The right diagram shows the characteristic performance for a laser diode with spectral width $\delta\lambda$ = 2nm
and a channel spacing of $\Delta\lambda$ = 20nm.   With regard to the insertion loss
and crosstalk attenuation ranges under consideration the performance
of the two-channel interference filter demultiplexer operating with
light-emitting diodes is superior to that of its grating counterpart.
The opposite applies to the performance of the two types when operated
with laser diodes.   Very low insertion losses accompanied by very
high crosstalk attenuation can then be realized for the assumed channel
separation with the grating-type version.

With decreasing channel separation $\Delta\lambda$ this situation changes
further in favor of the grating-type version.   With sharply increasing
insertion loss the crosstalk attenuation of the interference filter
demultiplexer reaches saturation upwards of 70dB.   Also the multiplexing principle using a grating will definitely be more favorable in
cases where numerous parallel channels have to be provided because,

optical carrier wavelength separation having already been provided
by the grating, only fiber ports need to be added. These will not
increase the overall insertion loss. In configurations with interference filters in contrast further filters have to be added at each
stage of the cascade structure, whereby the overall insertion loss
will be increased.

Both the configuration with a tilted grating and the one with
a tilted beamsplitter are sensitive to polarizing effects. Although
the angles of the optical beams incident upon and diffracted by the
grating multiplexer should be kept as small as possible, angles of
some 20° are required on technological grounds. Thus an almost negligible sensitivity to polarizing effects is only present in the case
of the multiplexer with an interference filter interposed in a lens
system (Figure 9c). This advantage over the grating multiplexer is
however solely due to only two wavelength channels having to be separated per beamsplitter stage, which means that only two fibers have
to be connected per lens face. Table I gives a comparison of the
reported performance of access couplers for WDM.

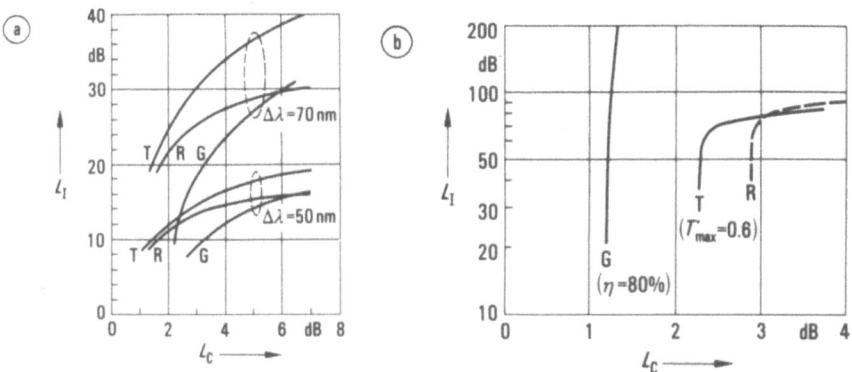

Fig. 13.   Comparison of crosstalk attenuation L_c as a function of the
            insertion loss L_I of 45° demultiplexers with a narrowband
            filter and others with a grating G according to[27]. For the
            crosstalk the influence of one adjacent channel was taken
            into account.   (a) Light-emitting diodes as transmitters,
            $\delta\lambda$ = 40nm, $\Delta\lambda$ = 50nm and 70nm, respectively; (b) Laser diodes
            as transmitters, $\delta\lambda$ = 2nm, $\Delta\lambda$ = 20nm. T transmittance chan-
            nel, R reflectance channel.

Table 1. Performance of Reported Access Couplers for Wavelength Division Multiplexing (MUX) and Demultiplexing (DEMUX).

Device	Type	Source used	Wavelength	Number of channels	Insertion loss	Crosstalk Attenuation		Ref.
						far end	near end	
DEMUX	Concave Grating + Slab Waveguide	Mono-chromator	1.0-1.4 μm	10	2,8 dB*)	25 dB		23
DEMUX	Concave Grating	Mono-chromator	0.7-0.9 μm	10	2,5 dB*)	>30 dB		25
MUX	Planar Grating + Graded Index Rod Lenses	Mono-chromator	1.1-1.5 μm	10	2,2 dB*)			24
MUX	Interference Type	LED	755/825 nm	2	1 dB**) / 2 dB	11 dB / 14 dB	>40 dB / >40 dB	27
DEMUX	Interference Type with add. narrow band filter	LED	755/825 nm	2	2 dB**) / 3,5 dB	30 dB / 30 dB	>60 dB / >60 dB	27
DEMUX	Interference Type, Single Mode Input Fiber	LED	780/858 nm	2	1,5 dB***) / 0,7 dB		>40 dB	27

*)   higher diameter, higher numerical aperture for output fibers

**)   same fiber for input and output

***) output fiber is thick core fiber

ACKNOWLEDGEMENTS

The author gratefully acknowledges the help of his colleagues Dr H. F. Mahlein, A. Reichelt and Dr. G. Winzer in supplying their original work on WDM which is reported in more detail in[7,27,30].

REFERENCES

1. W. Meyer, Verzweigungseinrichtungen in mehrwelligen optischen Datennetzen, Mikrowellenmagazin 2, 153-158, 1978.

2. F. Auracher, Prinzipien und Eigenschaften von Abzweigen fur Multi-modefasern, Frequenz 34, 52-57, 1980.

3. D. Rosenberger, Microoptic Passive Devices for Multimode Optical Fiber Communication Systems, Siemens Forsch. - u. Entw. -Ber. 8, 125-129, 1979.

4. W. J. Tomlinson, Wavelength multiplexing in multimode optical fibers, Appl. Opt. 16, 2180-2194, 1977.

5. T. Miki and H. Ishio, Viabilities of the Wavelength-Division-Multiplexing Transmission System over an Optical Fiber Cable, IEEE Trans. COM-26, 1082-1087, 1978.

6. H. Ishio, Wavelength-Division-Multiplexing Transmission Technology, Opt. Commun. Conf., Amsterdam, Proc. III/1-4, Sept. 1979.

7. G. Winzer and A. Reichelt, Wavelength-Division-Multiplex Trans-mission over Multimode Optical Fibers: Comparison of Multi-plexing Principles, Siemens Forsch.- u. Entw. -Ber. 9, 217-225, 1980.

8. F. Auracher, Planar branching network for multimode glass fibers, Opt. Commun. 17, 129-132, 1976.

9. G. L. Tangonan et al., Planar Multimode Devices for Fiber Optics, Proc. of the Opt. Commun. Conf., Amsterdam, paper 21.5, Sept. 1979.

10. T. C. Chu and A. R. McCormick, Measurements of Loss due to Offset, End Separation, and Angular Misalignment in Graded Index Fibers Excited by an Incoherent Source, Bell Syst. Techn. J. 57, 595-602, 1978.

11. J. Sakai and T. Kimura, Splice loss evaluation for optical fibers with arbitrary index profile, Appl. Opt. 17, 2848-2853, 1978.

12. Y. Daido, T. Iwama, and E. Miyauchi, Estimation of Transmission Losses in Graded Index Fiber Connectors and Splices, Trans-actions of the IECE of Japan, E61, 816-817, 1978.

13. C. M. Miller and S. C. Mettler, Loss Model for Parabolic-Profile Fiber Splices, Bell Syst. Techn. J. 57, 3167-3180, 1978.

14. U. Kalmbach and D. Rittich, Abzweig fur Lichtleitfasern, Nachr. techn. Z. 31, 423-425, 1978.

15. H. -H. Witte and V. Kulich, Branching elements for optical data buses, Appl. Opt. 20, 715-718, 1981.

16. H. -H. Witte and V. Kulich, Planar Input Couplers in Thick-Film Technology for Multimode Optical Fibers, Siemens Forsch.-u. Entw.-Ber. 8, 141-143, 1979.

17. D. J. Bond and P. Hensel, The effects on joint losses of toler-
    ances in some geometrical parameters of fibers, Optical and
    Quantum Electronics 13, 11-18, 1981.
18. Y. Suzuki et al., Concentrated-type directional coupler for op-
    tical fibers, Appl. Opt. 15, 2032-2033, 1976.
19. A. Reichelt et al., Improved Optical Tapping Elements for Graded-
    Index Optical Fibers, Siemens Forsch.-u. Entw.-Ber. 8, 130-135,
    1979.
20. K. Kobayashi et al., Micro-optics devices for branching, coupling
    multiplexing and demultiplexing, Proc. of the Int. Conf. Integr.
    Optics and Opt. Fiber Commun., Tokyo, 367-370, July 1977.
21. K. Kobayashi et al., Micro-Optic Devices for Fiber-Optic Communi-
    cations, Fiber and Integrated Optics 2, 1-17, 1979.
22. F. Szarka, A. Lightstone, J. Lit, and R. Hughes, A Review of Bi-
    conical Taper Couplers, Fiber and Integrated Optics 3, 285-
    297, 1980.
23. R. Watanabe and K. Nosu, Slab waveguide demultiplexer for multi-
    mode optical transmission in the 1.0-1.4µm wavelength region,
    Appl. Opt. 19, 3588-3590, 1980.
24. R. Watanabe, K. Nosu, and Y. Fujii, Optical Grating Multiplexer
    in the 1.1-1.5µm wavelength region, Electron. Lett. 16, 108-
    109, 1980.
25. R. Watanabe and K. Nosu, Optical Demultiplexer using concave
    grating in 0.7-0.9µm Wavelength Region, Electron. Lett. 16,
    106-108, 1980.
26. J. P. Laude and J. Flamand, Un Multiplexeur-demultiplexeur de
    Longeur D'Onde A Reseau, OPTO 3, 33-34, 1981.
27. G. Winzer, H. F. Mahlein and A. Reichelt, Single Mode and Multi-
    mode All-Fiber Directional Couplers for WDM, Appl. Opt. 20,
    in press, 1981.
28. M. J. Sun, W. S. C. Chang and C. M. Wolfe, Frequency demultiplex-
    ing in GaAsInP waveguide detectors, Appl. Opt. 17, 3533-3534,
    1978.
29. W. Eickhoff, P. Marschall and E. Schlosser, Transparent, highly
    sensitive GaAs/(GaAl)As photodiode, Electron. Lett. 13, 493-
    494, 1977.
30. H. F. Mahlein, A high performance edge filter for wavelength-
    division multi-demultiplexer units, Optics and Laser Technol-
    ogy, Febr. 13-19, 1981.

PLANAR TEES AND STAR COUPLERS

E. Voges

FernUniversität, Nachrichtentechnik
P.O. Box 940
D - 5800 Hagen, Germany

1. INTRODUCTION

In multimode fiber-optic systems optical branching networks are of interest for optical data distribution. One principal approach to multimode optical data distribution utilizes microoptic elements or fiber elements (see for example [1]). On the other hand, the fabrication of planar tapping or distributing components by photolithographic techniques has the advantages of reproducibility and batch fabrication. Multimode planar components have been formed preferably by ion exchange processes in glasses (see for example [2-6]).

Here, this planar approach is considered in more detail. We first investigate a multiterminal optical bus system for an assessment of the required components and of important design considerations. The chapter on ion exchanged strip waveguides in glasses is concentrated on the calculation and evaluation of the index profiles. The index profiles of the strip waveguides should be matched to connecting step-index or graded-index fibers to achieve a high coupling efficiency. The design of multimode planar tapping elements or star couplers is based on a statistical ray tracing technique[7]. Some relevant design rules are investigated by this method. Finally, the fabrication and properties of transmission star couplers are considered.

2. SYSTEM EXAMPLE: FIBER-OPTIC DATA BUS

The fiber-optic approach to data communication systems has in particular the advantages of potentially high bit rates, and the lack of electromagnetic interference and ground loop problems. Recently, the concept of an optical multiterminal system or data bus has emerged

as an important application of optical data distribution[8-10]. In such
a system a number of spatially distributed terminals simultaneously
interact with time multiplexed digital signals which are carried by a
common fiber link. The two principal bus configurations are the serial
distribution system (T-system) which employs tee couplers, and the
parallel system (Star-system) which utilizes a Star coupler.

The serial data bus configuration employs a series of N tee couplers (taps) along a common fiber as shown for a ring structure in
Figure 1a. This T-system is convenient when expanding the number of
terminals. However, the series losses through many couplers limit
the number of terminals to at most 10-12. In addition, a high dynamic
range of the receivers is required in order to handle the extremely
different signal levels.

In the alternative parallel Star-bus all connecting fibers are
brought together at a central mixer which distributes an incoming
signal to all other fibers.

The central mixer may operate in a transmission mode as shown in
Figure 1b or in a reflexion mode. In the latter case only one fiber
and an additional 3dB beam splitter are required for each terminal.
In a Star-bus the path loss increases only weakly with the number of
terminals N, and the dynamic range of the receivers can be low.

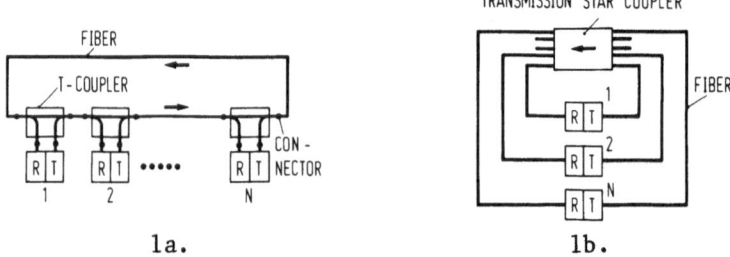

1a.   1b.

Fig. 1a) N-terminal ring-shaped T-bus with tee-couplers (R:receiver,
          T:transmitter)

    1b) N-terminal Star-bus with a transmission Star coupler.

The key devices of these basic configurations - we only note
that the advantages of both systems can be combined in hybrid T/Star-
system - are the coupling component and the mixer, respectively. In
order to assess the relevant requirements for these components we
calculate the path attenuation for both configurations. The maximum
allowable signal loss may then be determined for a system with speci-

fied transmitters and receivers and a specified bit-error-rate (signal-to-noise ratio at the receiver)[9]. In a simplified approach the fiber loss and the internal loss of the tee couplers are neglected. More detailed calculations are given in[9,11].

For the T-system we introduce a connector loss factor C and a tapping factor x of the couplers. The maximum path attenuation occurs between transmitter 1 and receiver N

$$L_{max}^{(T)} /dB = -10 \log \{C^{2N} (1-x)^{N-2} x^2\} \tag{1}$$

and is minimized for the optimum tapping factor

$$x_{opt} = 1 - \frac{N-2}{N} \tag{2}$$

The lower bound of L_{max} then becomes

$$L_{max}^{(T)} /dB = -2N \cdot 10 \log C - (N-2) \cdot 10 \log \frac{N-2}{N} - 20 \log 2/N \tag{3}$$

We note that $L_{max}^{(T)}$ roughly increases in proportion to the number of terminals at a rate given by the connector loss $-10 \log C$. The minimum loss occurs between two adjacent couplers, and is given by

$$L_{min}^{(T)} /dB = -4 \cdot 10 \log C - 20 \log 2/N \tag{4}$$

The required dynamic range of the receivers, the 'optical signal range' OSR is then obtained from

$$OSR = L_{max}^{(T)} - L_{min}^{(T)} \tag{5}$$

In a symmetrical Star configuration with an equally distributing mixer the optical signal range is zero, and the path attenuation is

$$L^{(S)} /dB = -10 \log 1/N - 10 \log C' \tag{6}$$

for a mixer loss including connector losses of $-10 \log C'$. We note the logarithmic dependence of $L^{(S)}$ on the number of terminals.

The results of fig. 2 demonstrate the significant difference in path attenuation of both configurations. We conclude that the connector loss must be very low in a T-system, and that couplers with an adjustable tapping factor are advantageous. In the Star-system the main requirement is a uniform signal distribution of the mixer.

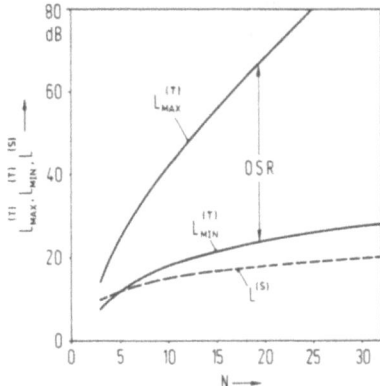

Fig. 2.   Maximum and minimum signal losses $L_{max}^{(T)}$, $L_{min}^{(T)}$ of a T-bus with
          optimized tapping factor and signal loss $L^{(S)}$ of a Star-bus
          in dependence on the number of terminals N. Fiber attenu-
          ation is neglected. A connector loss of 1dB is assumed for
          the T-bus, and 5dB loss of the Star coupler is assumed. OSR:
          optical signal range.

3.   STRIP WAVEGUIDES FABRICATED BY A FIELD-ASSISTED ION EXCHANGE

     The fabrication of planar waveguide structures by a field-assisted
ion exchange in glasses has the advantages of a simple technology, low-
loss wave-guides, and allows the optimization of the index distribution. The Ag-Na ion exchange is most popular for device fabrication[2-6]
and yields an index change of several percent mainly depending on the
type of the glass being used. The geometry of a masked silver ion
exchange for optical strip waveguides is depicted in Figure 3.   One
of the unsolved problems is the calculation of the resulting index
distributions from the technological parameters. We first investigate appropriate theoretical solutions and compare them with experimental results.

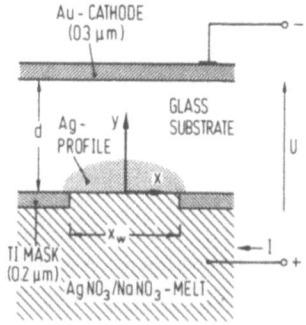

Fig. 3.   Configuration of a field-assisted ion exchange in glasses.

A) Theory

The index profiles of silver ion exchanged waveguides obtain from the distribution of the Ag-ions. This distribution is calculated assuming sodium ions (index n) of equilibrium concentration C_n as the only mobile ion species within the glass, and using diluted silver (index a) salt melts. Because of the lower mobility of the Ag-ions the field-assisted diffusion process depends on the concentration of the Ag-ions due to a space charge formation which couples the migration of the Na- and Ag-ions. The Ag-distribution then is governed by the nonlinear equation

$$\frac{\partial c}{\partial t} + \frac{1}{1-(1-M)c} \, \mu_a \, \vec{E} \, \text{grad} \, c = \frac{D_a}{1-(1-M)c} \, \Delta c \qquad (7)$$

for the normalized concentration $c=C_a/C_n^o$ of the Ag-ions. Here, μ_a and D_a are the mobility and diffusion constant of the silver ions,

$$M = \mu_a/\mu_n < 1 \qquad (8)$$

is the mobility ratio, and the electric field $\vec{E}$, determined via the Poisson equation

$$\text{div} \, \vec{E} = \frac{q}{\varepsilon}(C_a + C_n - C_n^o) \qquad (9)$$

includes the space charge field. Since equation (7), (9) cannot be solved in the two-dimensional case we first investigate approximate one-dimensional solutions where c is a function of depth y and exchange time t, only. For the special boundary conditions at y=0: $c=c_o$ and $\partial c/\partial y=0$, a solution of equation (7) is[12,13].

$$c/c_o = \{1+\exp[vc_o \frac{1-M}{D_a} (y-vt)]\}^{-1} \qquad (10)$$

with a concentration dependent front velocity

$$v = v_o/\{1-(1-M) \, c_o\} \qquad (11)$$

where $v_o=i_o/qC_n^o$ depends on the current density i_o. The solution (10) only applies for large profile depths because of the boundary conditions. For lower silver concentrations (diluted melts) the space charge can be neglected, and the profile results from[14,15]

$$c/c_o = \frac{1}{2}\left\{\text{erfc} \, \frac{y-v_a t}{2\sqrt{D_a t}} + \exp \frac{v_a y}{D_a} \cdot \text{erfc} \, \frac{y+v_a t}{2\sqrt{D_a t}}\right\} \qquad (12)$$

where the front velocity v_a obtains from the applied voltage U and the thickness d of the glass substrate

$$v_a = v \ (c_o \to 0) = \mu_a \ U/d \qquad\qquad (13)$$

In the most simple solution the diffusion is neglected, too, and we arrive at

$$c/c_o = \sigma \ (v_a \cdot t - y) \qquad\qquad (14)$$

where σ is the unit step function. The solutions (10), (12), (14) are compared in Figure 4 for the parameters $D_a = 2.6 \cdot 10^{-15} \ m^2/s$, $t=15\text{min}$, $U/d=300V/\text{mm}$. The simple step solution well applies for large depths of the profile. Therefore, this approximation is used to calculate the profiles of strip waveguides for configurations as shown in Figure 3.

The shape of the uniform index distribution then is given by the position vector of the silver ion front

$$\vec{r}_F = x_o \ \vec{e}_x + \int_0^t \mu_a \ \vec{E} \ dt, \quad -x_w/2 \leq x_o \leq x_w/2 \qquad (15)$$

for an exchange time t. The electric field for the configuration of Figure 3 is equal to the static field of the triplate transmission line

$$E = E_x + jE_y = j \ \frac{\pi}{2} \ \frac{U}{dK'(k)} \cdot \left[\frac{\tanh^2 u - 1}{\tanh^2 u - k^2} \right]^{1/2} \qquad (16)$$

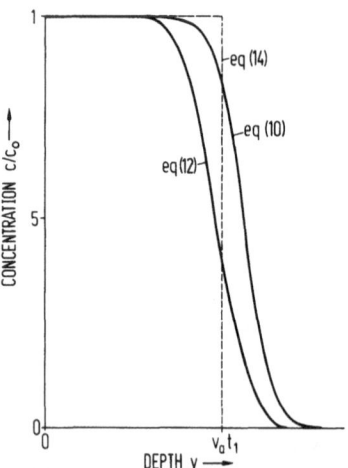

Fig. 4. Theoretical Ag-profiles of ion exchanged film waveguides. $c_o = C_o/C_n^o$ is the normalized source concentration, eq. (10) is plotted for $c_o = 0.3$.

with

$$u = \pi \ (x-j \ y)/2d, \quad k = \tanh(\pi \ x_w/4d) \qquad (17)$$

and the Jacobian elliptic function $K'(k)$. Figure 5a,b shows calculated index profiles for two widths x_w and several applied voltages
U(parameters: T=616°K, t=30min, d=300μm). These results are compared
with experimental index profiles (Figure 5c,d) which are fabricated
for the above parameters and are visualized by Nomarsky interference
contrast (see below).

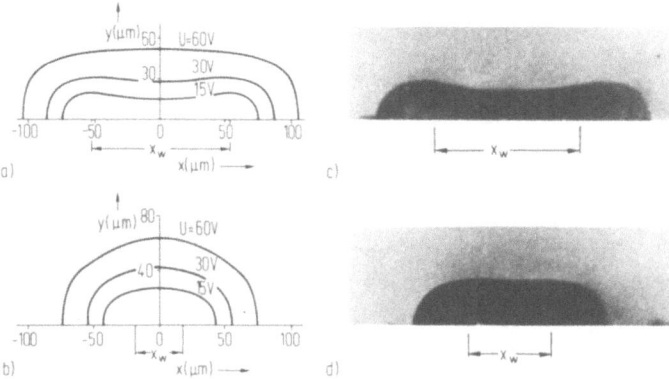

Fig. 5. Theoretical (a,b) index profiles with exchange temperature
T=616°K and an exchange time t=30min for two widths x_w of
the mask window and several applied voltages U, and experimental profiles (c,d) visualized by Nomarsky contrast.

Buried waveguides with reduced losses due to surface scattering and
an improved matching to connecting fibers are obtained by a second
ion exchange with pure $NaNO_3$ melts[16]. This second ion exchange is
usually performed after removing the mask, therefore one-dimensional
solutions are sufficient. Moreover, thermal diffusion may well be
neglected in this case[12], and the distribution of the Ag-ions obtains
from the method of characteristics[17]. We only present the simple case
of $c_o = 1$ (ie. a complete Ag-Na-exchange during the first exchange
process), where the solution is

$$c = \begin{cases} 0 & y < Mv_o t' \\ \dfrac{1}{1-M} \left\{ 1 - \sqrt{Mv_o t'/y} \right\}; & Mv_o t' < y < v_o t'/M \\ 1 & v_o t'/M < y < v_o (t' + t) \end{cases} \qquad (18)$$

for a second ion exchange for a time $t'<Mt/(1-M)$. This solution is
depicted in Figure 6.

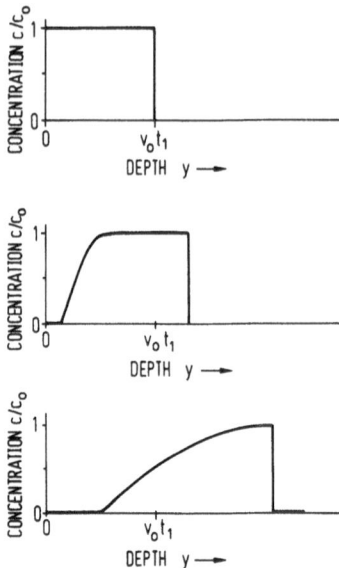

Fig. 6. Ag-profiles of buried film waveguides for exchange times
         $t' = 0$, $t' = t/4$, $t' = t$ for an initial silver exchange time
         t, $c_o=1$, and a mobility ratio $M=0.5$.

B) Experimental index profiles

The optical strip waveguides are fabricated in borosilicate
glasses by a field-assisted migration of silver ions from 10mol%
$AgNO_3$/90mol% $NaNO_3$ melts. The cathode is formed by an evaporated gold
layer (Figure 3), the mask consists of sputtered titanium, SiO_2 or
Si_3N_4 with identical results. The silver ion exchanges are performed
for the following ranges of the parameters:

    temperature T: 500 - 700°K;   time t   : 10min -1h;

    voltage U   : 0 - 50V   ;   thickness d: 200 - 500µm.

Buried waveguides are fabricated in a second step by immersing the
substrate into a pure $NaNO_3$ melt. In this case the exchange times
are usually shorter than for the silver ion exchange, and relatively
high electric field strengths are applied to the sample.

The preparation of the samples is eased when using an evaporated
metal cathode instead of a cathode which is provided by an additional
melt.

The fabrication of optical strip waveguides with a prescribed
index distribution requires the accurate determination of the index

profiles in dependence on the technological parameters. We use the
Nomarsky interference contrast on polished front faces to visualize
the boundaries of the index distributions (see Figure 5). The index
distributions are accurately determined from reflectivity profiling
on beveled surfaces with the aid of an optical multichannel detector
system[18]. For this purpose the surface of a sample is beveled on a
rotating quartz plate to an angle of about $\alpha = 3°$. The roughness of
the bevel is estimated to be smaller than ±20nm from Nomarsky con-
trasts.

The beveled surface of the sample is illuminated monochromatically
at normal incidence. Figure 7a shows the essential parts of the experimental set-up. The reflected light intensity is detected by the
vidicon on the top of the microscope. The vidicon is part of an optical multichannel detector system (PAR, OMA 2).

After adjusting the beveled surface normal to the optical axis
of the microscope the waveguiding surface of the sample (position I
in Figure 7a) is scanned as indicated in Figure 7b. In orientation
(A) the intensity profile is directly related to the depth distribution of the refractive index. When scanning along the x-direction
(orientation B) the lateral intensity profile and the lateral index
distribution are obtained for a given depth y. In addition to these
intensity profiles E1 two more intensity profiles are measured, E2
(position II in Figure 7a) from the homogenous part of the beveled
surface as a reference, and E3 from a non-reflecting optical trap.
The false light E3 which is reflected at surfaces inside the microscope has to be subtracted. Spatial inhomogeneities of the illumin-
ation and of the vidicon sensitivity are eliminated by dividing E1 -
E3 by E2 - E3.

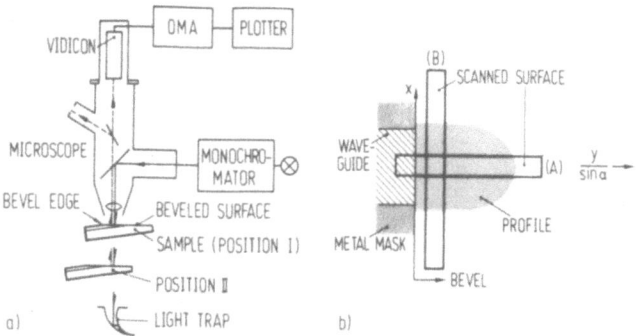

Fig. 7a.   Experimental set-up for reflectivity profiling. 7b. Orien-
           tations (A) and (B) of the scanned surfaces for reflectivity
           profiling. The depth coordinate z is related to the bevel
           angle.

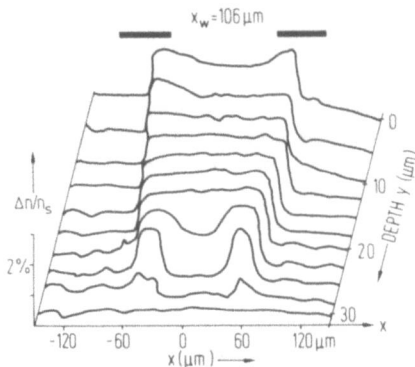

Fig. 8.   Index profile scanned by an optical multichannel analyzer
           on a beveled surface for a silver ion exchange with T=615K,
           t-30min, U=30V, d=300µm.

        The reflectivity R along the bevel is determined in relation to
the reflectivity R_0 of the unchanged glass substrate by

$$R/R_0 = (E1 - E3)/(E2 - E3) \qquad (19)$$

The evaluation of the index profile from the reflectivity profile
according to the formula for homogenous materials

$$n = (1 + \sqrt{R})/(1 - \sqrt{R}) \qquad (20)$$

neglects multibeam interferences within the subsurface profile and
their contribution to the measured reflectivity, since their influence is very small[17]. Figure 8 shows a typical result of a scanned
index profile.

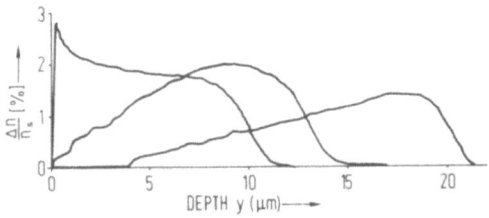

Fig. 9.   Index distributions of buried film waveguides obtained by
           reflectivity profiling.
           Experimental data:
           - initial Ag-exchange   : d=500µm, T=620K,
                                          t=30min, U=50V
           - exchange with NaNO3 melt: T=620K, t'=0min,
                                          t'=10min, t'=30min,
                                          U=50V.

The measured index profiles generally are in very good agreement with calculations according to equation (15), (16), and relevant profile parameters eg. depth, side diffusion, can be determined in dependence on the technological parameters. Figure 9 shows the depth distributions of waveguides buried by a second ion exchange. Again, the results agree with calculated profiles.

4. DESIGN AND REALIZATION OF PLANAR TREES AND STAR COUPLERS

Multimode planar tee couplers are easily analyzed by applying ray theory[6]. Experimentally, rather low insertion losses have been achieved[6]. One should, however, consider the mode dependent coupling.

According to section 2 the insertion loss of tee couplers must be extremely low for data-bus applications, whereas a uniform power distribution is most important for Star couplers.

Figure 10a shows the waveguide configuration of an N-channel transmission Star coupler with waveguides of lengths z_w which are butt-coupled to fibers, coupling horns of lengths z_h, and a mixer of length z_m.

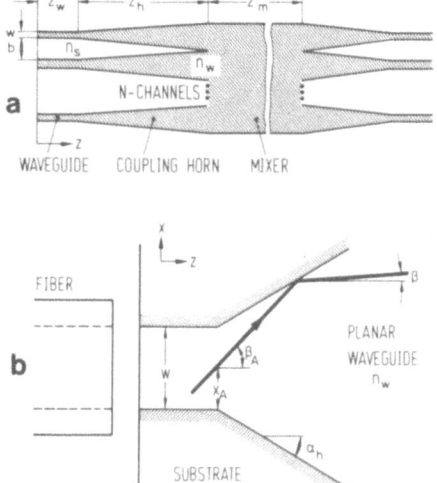

Fig. 10.   (a) Waveguide configuration of an N-channel transmission mixer with coupling horns.   (b) Geometry for a statistical ray tracing.

This planar approach to Star couplers has to compete with efficient bulk optic devices (see for example[19]). Its advantages depend
on low inherent losses and a reduced length z_m of the mixer when
choosing optimum coupling horns which are inserted to eliminate the
packing fraction loss. Since the waveguides are highly multimode,
a numerical ray tracing analysis based on geometric optics is feasible[7]. For this purpose the input rays are characterized by a transverse entrance coordinate x_A and an input angle β_A (Figure 10b), which
are generated by random number generators, and must be statistically
independent. Usually, a large number of rays ($>10^3$) have to be traced.
Since optimum coupling horns cannot be derived analytically, given
Star coupler configurations have to be analyzed. Figure 11 shows the
normalized intensities at the output ports of a 5-channel transmission
Star coupler for several horn taper functions, for an input into the

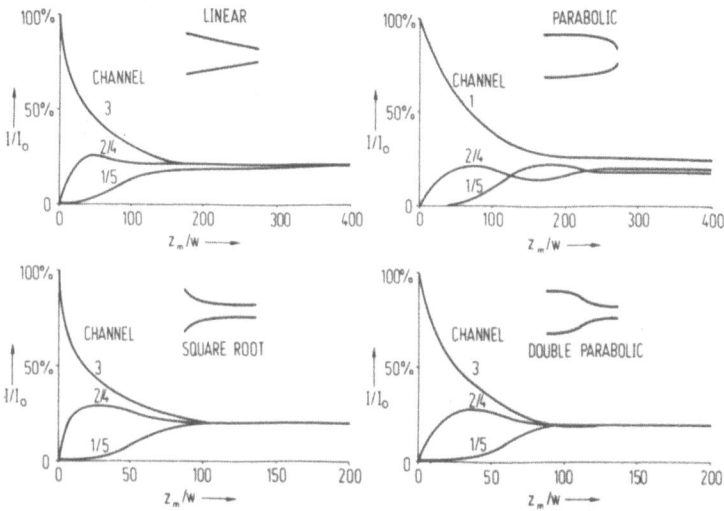

Fig. 11.   Output intensities in dependence on the normalized mixer
            length z_m/w for a 5-channel transmission mixer with
            different horn taper functions.

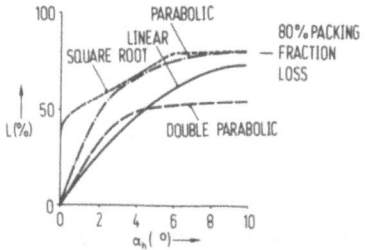

Fig. 12.   Horn loss L in dependence on the average horn angle α_n for
            different horn taper functions and $w=100\mu m$, $b=400\mu m$ (see
            fig. 10a). The packing fraction loss is $L_{max}=w/(w+b)=80\%$.

central horn. Double parabolic horns, in particular, reduce the necessary mixer length for an equal output distribution, when compared to linearly tapered coupling horns[2,7].

Losses only occur within the output horns, Figure 12 shows the horn loss in dependence on the average horn angle α_n for several horn taper functions. One notes, that large horn angles (ie. short coupling horns) are possible for double parabolic horns, if 3dB loss can be tolerated.

Experimentally, a series of 5-channel transmission mixers with different horn configurations, and mixer lengths z_m have been realized by ion exchange processes. Figure 13 shows as an example the influence of the horn configuration on the necessary mixer length z_m for an equal output distribution. The input is at horn 1/5 and the normalized intensity at the output horn 5/1 is shown. The theoretical advantage of inclined coupling horns, and of double parabolic coupling horns for reduced mixer lengths when compared to linear horns[20] is confirmed experimentally. This simple example demonstrates the inherent design flexibility of the planar approach to multimode data distribution.

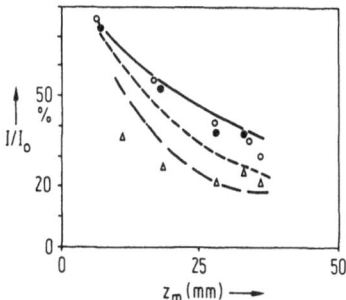

Fig. 13.   Normalized intensity I/I_0 at the output horns 5 or 1 for an input at horn 1 or 5 in dependence on the mixer length z_m for:

      ————● linear coupling horns with w=184μm, b=324μm.
      ------O linear coupling horns inclined by angles $\alpha_n = \tan^{-1}$ $\{n(\omega+b)/z_m\}$ (n=1,2...N, N:odd) with w=221μm, b=301μm, $\alpha_n = 2°$.
      -.—.—..-△ double parabolic coupling horns with w=172μm, b=336μm, mean horn angle $\alpha_n = 2°$.

ACKNOWLEDGEMENTS

The supply of experimental data by D. Ritter and H. J. Lilienhof is gratefully acknowledged.

REFERENCES

1. D. Rosenberger, "Microoptic passive devices for multimode optical
    fiber communication systems", Siemens Forsch. und Entw.-Ber.
    vol. 8, pp. 125-129, 1979.
2. G. L. Tangonan, O. G. Ramer, H. R. Friedrich, C. K. Asawa, D. L.
    Persechini, L. E. Gorre, "Planar multimode devices for fiber
    optics", Opt. Comm. Conference, Amsterdam 1979, pp. 21.5-1 -
    5-4.
3. G. L. Tangonan, L. E. Gorre, and D. L. Persedini, "Planar multi-
    mode couplers for fiber optics", Opt. Commun. vol. 27, pp. 358-
    360, 1978.
4. R. Watanobe and K. Nosu,"Slab waveguide demultiplexer for multi-
    mode optical transmission in the 1.0-1.4µm wavelength region",
    Appl. Opt. vol. 19, pp. 3588-3590, 1980.
5. J. Viljanen and M. Lappihalme, "Planar optical coupling elements
    for multimode fibers with two-step ion migration process",
    Appl. Phys. vol. 24, pp. 61-68, 1981.
6. A. D. De Oliveira and M. G. F. Wilson, "Strip waveguide Y-inter-
    section as efficient coupler for multimode optical communi-
    cation systems", Electron Lett. vol. 17, pp. 100-101, 1981.
7. O. G. Ramer, "Design of planar star couplers for fiber optic
    systems", Appl. Opt. vol. 19, pp. 1294-1296, 1980.
8. M. Hudson and F. Thiel, "The star coupler: A unique intercon-
    nection component for multimode optical waveguide communication
    systems", Appl. Opt. vol. 13, pp. 2540-2545, 1974.
9. M. K. Barnoski, "Data distribution using fiber optics", Appl.
    Opt. vol. 14, pp. 2571-2577, 1975.
10. A. F. Milton and A. B. Lee, "Optical access couplers and a com-
    parison of multiterminal fiber communication systems", Appl.
    Opt. vol. 15, pp. 244-252, 1976.
11. G. Winzer and H. Witte, "Comparison of fiber-optic data bus net-
    works", Siemens Forsch. und Entw.-Ber. vol. 10, pp. 9-15, 1981.
12. M. Abou-el-Leil and A. R. Cooper, "Analysis of field-assisted
    binary ion exchange", J. Amer. Cer. Soc. vol. 62, pp. 390-
    395, 1979.
13. A. R. Cooper and M. Abou-el-leil, "Index variation from field-
    assisted ion exchange", Appl. Opt. vol. 19, pp. 1087-1091,
    1980.
14. T. Kaneko and H. Yamamoto, "On the ionic penetration of silver
    film into glasses under the electric field", Tenth Inter-
    national Congress on Glass, Part I, Ceramic Society of Japan,
    Kyoto 1974, pp. 8-79 - 8-86.
15. G. H. Chartier, P. Jaussaud, A. D. De Oliveira, O. Parriaux,
    "Optical waveguides fabricated by electric-field controlled
    ion exchange in glass", Electron Lett. vol.14, pp. 132-134,
    1978.
16. G. Chartier, P. Collier, A. Guez, P. Jaussaud, Y. Won, "Graded-
    index surface or buried waveguides by ion exchange in glasses",
    Appl. Opt. vol. 19, pp. 1092-1095, 1980.

17. K. S. Spiegler and C. D. Coryell, "Electromigration in a cation
    exchange resin II. Detailed analysis of two-component systems",
    J. Phys. Chem. vol. 56, pp. 106-113, 1952.
18. H. -J. Lilienhof, D. Ritter, E. Voges, "Index profiles of multi-
    mode optical strip waveguides by field-enhanced ion exchange
    in glasses", Opt. Commun. vol. 35, pp. 49-53, 1980.
19. H. -H. Witte and V. Kulick, "Branching elements for optical data
    buses", Appl. Opt. vol. 20, pp. 715-718, 1981.
20. D. Ritter, Dr.-Ing. thesis 1981.

PLANAR TEES AND COUPLERS FOR MULTIMODE DEVICES

G. L. Tangonan

Hughes Research Laboratories
Malibu, California 90265, USA

INTRODUCTION

Multimode fiber-optic systems currently use couplers for data distribution to several user terminals. Couplers are made by modifying the optical fibers (fusing, tapering, lapping and gluing) or with micro-optic components (such as microlenses and beam splitters). These two approaches and the planar approach, which is the subject of this program, are shown in Figure 1. Good couplers have been demonstrated using micro-optic methods, although several key questions have been encountered when attempting to transfer the technology to a manufacturing process.

The cost of mass produced couplers is greatly affected by the reproducibility of the fabrication technique. In the fused fiber approach, the problem of reproducibility is most severe. Controling the electric discharge (or flame torch) conditions and the mechanical devices for tapering the fused region of the fibers continues to require considerable development.

Component placement and packaging in current micro-optic devices requires considerable attention during coupler assembly. Each element must be antireflection (AR) coated to minimize reflection losses and maintain high isolation. The cost of the individual components - SELFOC lenses, beam splitters and precision machined placement devices - may prove to be quite prohibitive.

The planar fabrication approach (shown in Figure 1) attacks directly the problems of reproducibility, one-by-one fabrication, and precise placement of components which are encountered in the two conventional approaches by emphasizing the photolithographic control of

155

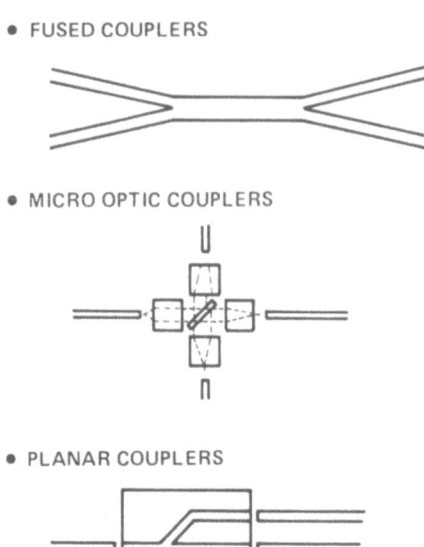

Fig. 1.   Schematic of the three approaches to coupler fabrication

the coupling structure and the batch processing capabilities inherent
in planar processing. A schematic of the fabrication process is shown
in Figure 2. The basic processing steps illustrated are mask deposition, photoetching for pattern definition and ion exchange processing.
These are techniques that are well developed for processing microelec-
tronics, and the benefits of reproducibility, batch processing and low
cost are obvious. In addition, the positioning of components is
largely eliminated by the pattern definition and the use of precision
V-groove holders for interfacing the fibers.

This paper describes the following aspects of planar coupler
development:

* Field-assisted diffusion to provide step index, high-numeri-
  cal-aperture (NA) profiles

* Exchange processes in glass which yield high-NA guides (such
  as Ag into glass)

* Buried channel waveguide coupler formation with low-loss
  guides and efficient coupling

* Utilization of fiber holder technology developed at HRL for
  interfacing fibers and devices (the silicon V-groove approach).

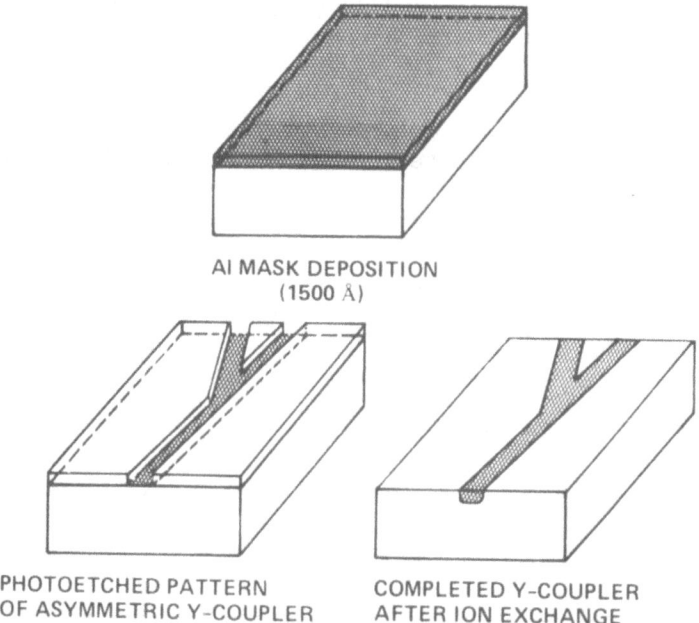

AI MASK DEPOSITION
(1500 Å)

PHOTOETCHED PATTERN
OF ASYMMETRIC Y-COUPLER

COMPLETED Y-COUPLER
AFTER ION EXCHANGE

Fig. 2. Masked ion exchange process for coupler fabrication.

    Our approach to coupler development is based on the recognition
of the essential elements required for the optimization of planar
couplers. The key elements of our approach are:

    *   Guide formation process control

    *   Throughput optimization (fiber to channel to fiber)

    *   Device design (access couplers, star couplers, splitters)

    *   Coupler fabrication and fiber interfacing.

WAVEGUIDE FORMATION

Guide Formation Processes

    Waveguide formation in glass can be accomplished by several tech-
niques, including chemical vapor deposition, ion exchange and diffusion, dip coating, ion implantation and laser heating. Three years
ago, we began investigating these various processes for fabricating
multimode guides. To date, the most successful process has been ion
exchange.

Various ion exchange and diffusion systems have been studied.
The approaches we have studied include:

(1) $Li_2SO_2-K_2SO_4$ eutectic salt melt/soda lime glass

(2) LiCl-KCl eutectic salt melt/soda lime glass

(3) Ag metal field-assisted diffusion (solid phase)/soda lime
     glass

(4) $AgNO_3$ melt/soda lime glass (with and without fields)

(5) $AgNO_3$ melt/borosilicate glass (with and without fields)

(6) Double exchange $NaNO_3$ and $AgNO_3$ borosilicate glass.

The most successful ion exchange systems have been (1), (5), and (6).
Our description of guide formation gives details of these three processes for the sake of conciseness.

The formation of optical waveguides in planar substrates has been
the subject of intensive research in the area of integrated optics.
The thrust of these studies has, however, been directed toward single-
mode device applications. More recently, multimode fiber optic devices have been made that are quite promising. Auracher et al.[1]
reported the first planar branching networks formed in photo-polymer
material of 100µm thickness. Tangonan et al. have described the utilization of ion exchange processes in glass to fabricate couplers[2],
star couplers[3], and even wavelength demultiplexers[4].

The ion exchange process utilized by the Hughes team in these
earlier works was that developed by Chartier et al[5]. In this process,
a eutectic mixture of Li_2SO_4 and K_2SO_4 is heated under an oxygen atmosphere to 580°C. A sodium glass slide is suspended over the melt for
30min to reach thermal equilibrium with the melt. Next, it is dipped
into the melt for 20min, then is again suspended over the melt for
10min to avoid any thermal shock. The process is illustrated in
Figure 3, along with the phase diagram for $LiSO_4-K_2SO_4$. Planar waveguides, 100µm deep, are made by this process. The coupler structures
are formed by masking with a thick (1 to 2µm) Al film. The Al mask
is subsequently removed by dipping the slide in hot 6 M HCl solution.
This process provides an index difference (NA≈0.1) that is too small
to cope with larger numerical aperture fibers (NA>0.2). However, despite this drawback, this process has been quite successful in demon-
strating planar processing of coupler devices.

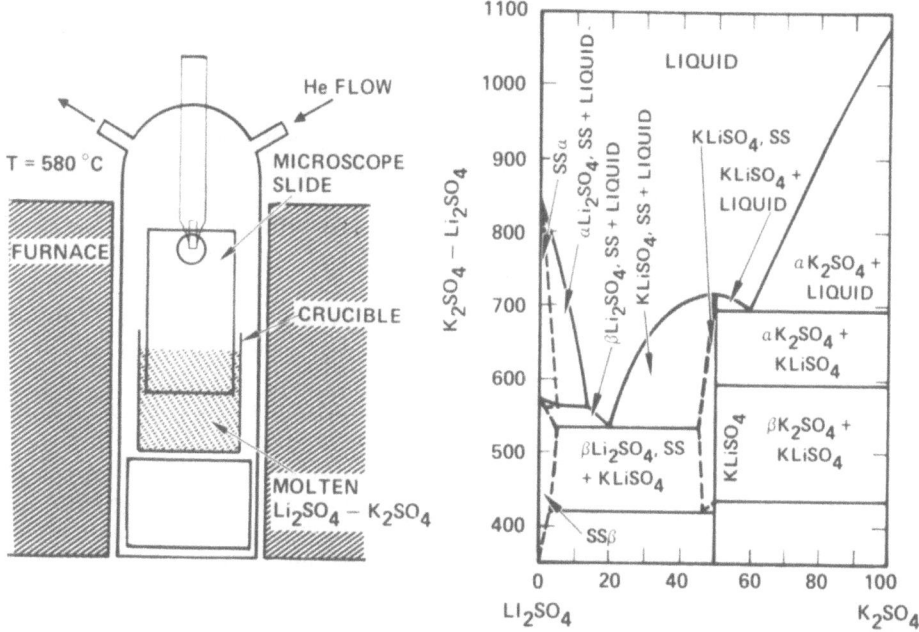

Fig. 3.   Guide formation processes using Li_2SO_4-K_2SO_4 eutectic melt.

Fabrication of Waveguides by Field-Assisted Ion Exchange

To circumvent the problem of low index of refraction change, an
alternative diffusion process was developed: Ag ion exchange under
a field. This process is similar to that described by Chartier et al[6]
and Izawa and NaKagome[7]. When ion exchange is carried out in the
presence of an electric field, the concentration profiles (and thereby
the index of refraction) are considerably modified. The concentration
of the ions during the process is given by:

$$\frac{dC}{dt} = D \frac{d^2C}{dZ^2} - E\mu \frac{dC}{dZ} \quad , \tag{1}$$

where the variables C and t are the concentration and diffusion times,
respectively, and the diffusion parameters D, E, and μ are the diffusion constant, applied electric field, and the ion mobility, respectively. When the initial concentration of Ag in the glass is zero
and an equilibrium surface concentration of C_o is attained, the resultant profile is given by

$$C\,(Z,t) = C_o \; erfc \left[(Z-E\mu t)/2\,(Dt) \right]^{1/2} \quad , \tag{2}$$

which for E = 0, reduces to the result for normal diffusions without
a field.

The schematic in Figure 4 shows the concentration profiles attainable by electric field and normal exchange processes. The implications of the profile differences are:

* The field-assisted profile is predominantly a step index with a graded junction into the substrate.

* The product $E\mu t$ essentially controls the depth of the planar step index guide.

During this conference Dr. E. Voges will describe in detail the important effects determining the index profile of these waveguides.

The experimental apparatus used to demonstrate electric-field-assisted diffusion is shown schematically in Figure 5. A double-crucible arrangement was used to provide a molten salt bath on both sides of the sample. This eliminated Na buildup at the negative electrode side of the substrates. Depending on the polarity of the electrodes, the ion exchange occurs on either side of the sample. Borosilicate glass was used because we found it is readily shaped into the double-crucible configuration, shows little sample warping and is free from the Ag migration problems observed in soda lime glasses. The parameters used for planar guide formation are given in Figure 6. Deep planar waveguides of low loss are obtained using these parameters. From these results, an approximate measure of the ion mobility of Ag at 370°C in the glass is given by

$$\mu = 10 \times 10^{-10} \frac{cm^2}{sec\ V} \ . \tag{3}$$

The index profile obtained for the waveguides was measured using standard interferometric techniques. The result obtained for one of the samples (E) is shown in Figure 7.

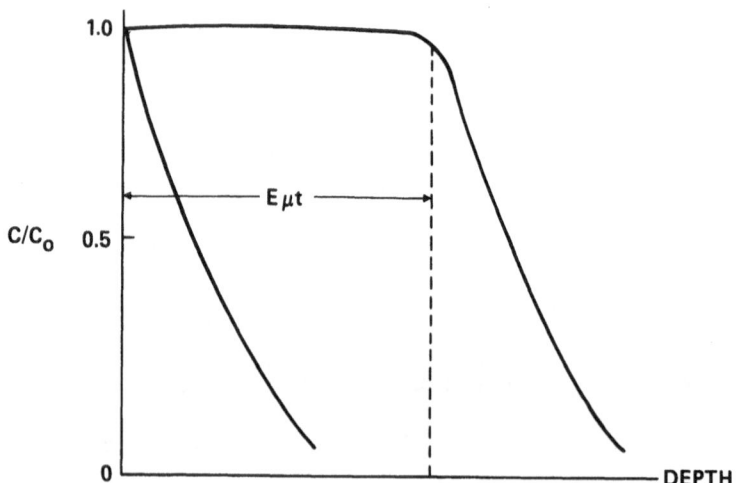

Fig. 4. Diffusion profile obtained with field-assisted diffusion.

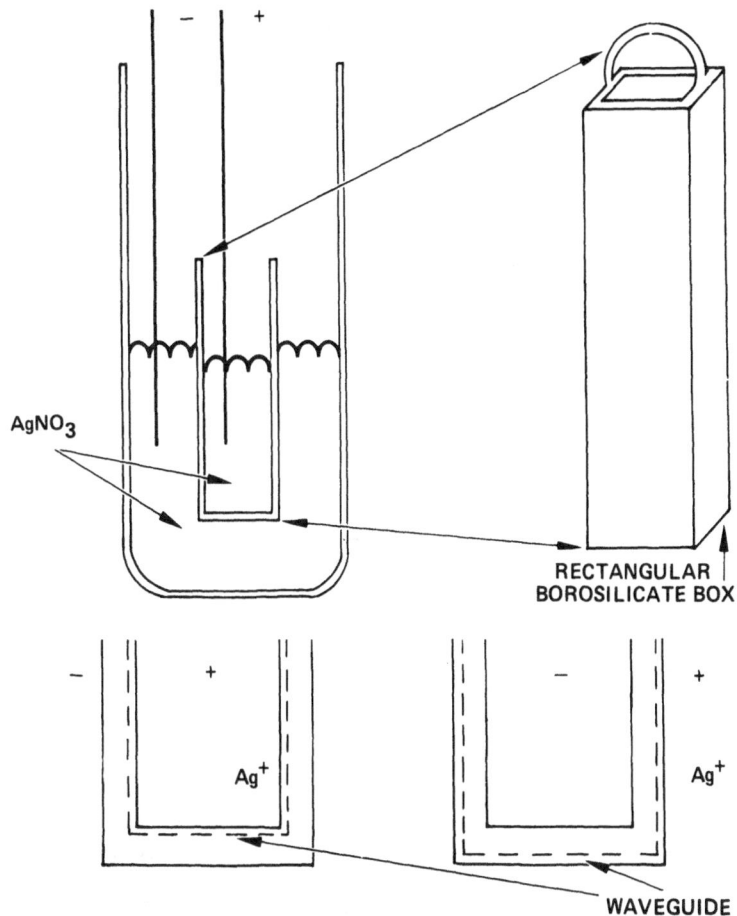

Fig. 5. Double-crucible apparatus for guide formation (a); depending
on the polarity of the electrodes, guide formation occurs on
inner (b) or outer (c) surfaces.

SAMPLE	T, °C	t, HRS	E, V/mm	I, mA	WAVEGUIDE DEPTH, μm
A	370°C	4.5	33	—	52
B	370°C	1.0	100	—	55
C	370°C	2.5	50 – 230	10	25
D	370°C	3.0	50	—	70
E	300°C	4.2	150	—	35

Fig. 6. Waveguide depth for Ag ion exchange versus temperature, time
and applied field.

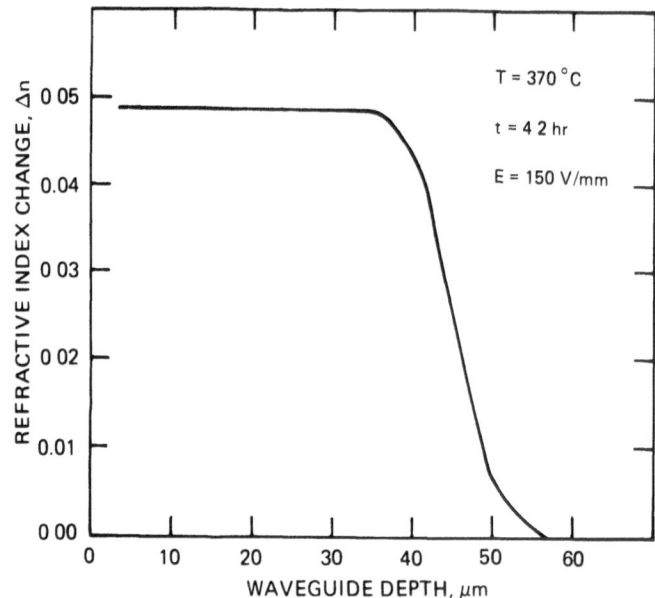

Fig. 7.   Index of refraction profile for sample E.   Note the step-like
index profile measured by interferometric methods.

Buried Waveguide Formation

Buried waveguides have been formed by a second field-assisted
exchange process using $NaNO_3$. In this process, Na is reintroduced in
the surface of the planar guide by the exchange process. A lowering
of the index of refraction at the surface is shown in Figure 8. Waveguiding experiments show that the additional confinement layer at the
surface shows a dramatic difference in the optical waveguide output
profile. This output of two guides, with and without burying, is
shown in Figure 9.

Channel Waveguide Burying

Channel waveguides can be formed by using an aluminum mask for
blocking the exchange process. Channel waveguides formed in this way
can be buried using the same process described above. A comparison
of the output profiles for two waveguides formed with and without
burying is shown in Figure 10. A clear difference is observed between
the two waveguide output profiles. In the laboratory, the buried
waveguides show a remarkable uniformity along the length of the guides
when excited by light. This is in contrast to the case of the unburied guides, in which surface scattering due to cracks and polishing
scratches is seen as bright scattering sites.

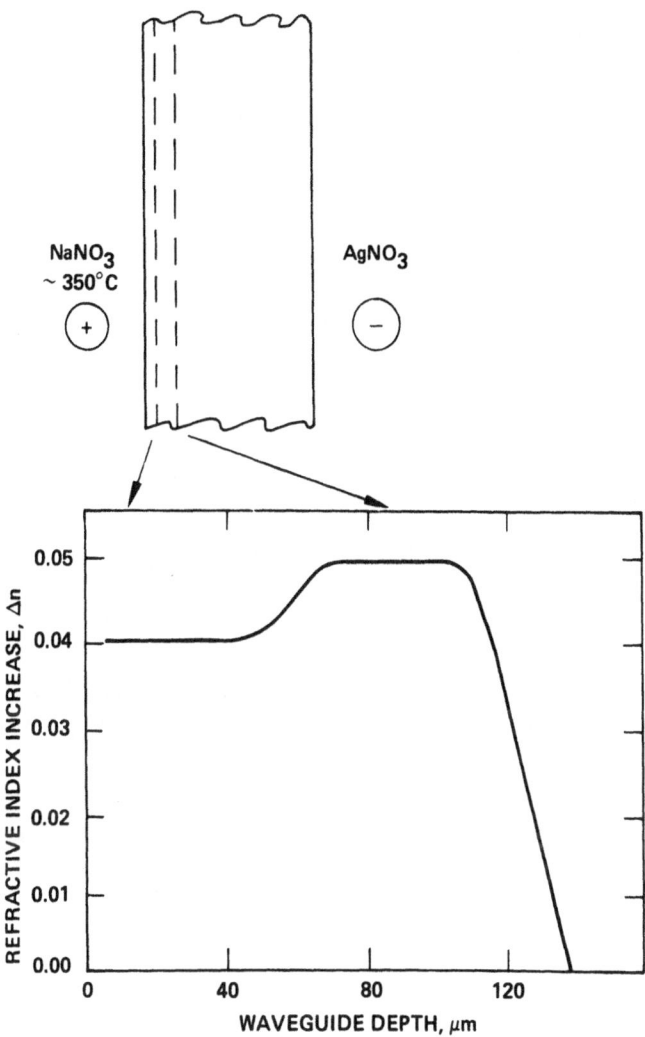

Fig. 8.   Double exchange formation of buried waveguides. The burying
is incomplete, although the index drop of 0.01 is still
substantial.

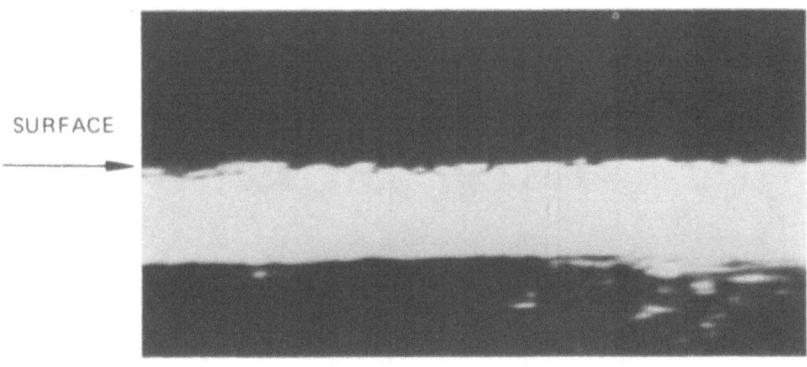

DEPTH = 180 μm
Ag DIFFUSION: 300°C, 4 hr, 30V/mm

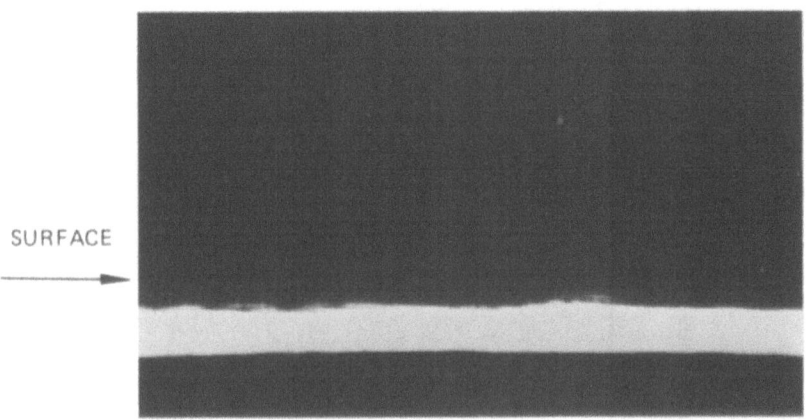

DEPTH = 50 μm
Ag DIFFUSION: 300°C, 1 hr, 30 V/mm
Na BURYING: 400°C, 1 hr, 25 V/mm

Fig. 9.   Buried planar waveguides compared to unburied guides.
(Processing: Ag diffusion (300°C, 1hr, 30V/mm) and Na
diffusion (400°C, 1hr, 25V/mm).

GUIDE — 160 μm x 80 μm
MASK — 25 μm
Ag DIFFUSION — 300°C, 60 min,
150 V/mm

GUIDE — 160 μm x 80 μm
MASK — 25 μm
Ag DIFFUSION — 300°C, 30 min,
150 V/mm
Na BURYING — 400°C, 30 mm,
30 V/mm

Fig. 10. Buried channel waveguides (160μm x 80μ) compared to unburied
guides. (Processing: 25μm mask width, Ag diffusion (300°C,
30min, 150V/mm) and Na burying (400°C, 30min, 30V/mm)).

Problem Areas and Optimization

     We anticipate several problems in the area of materials proces-
sing which should be further studied. The problems which should be
investigated are:

     * Incomplete burying of the waveguides with current techniques

     * Edge effects during masked ion exchange which cause the thick-
        ness variations of the waveguides

     * Optimization of buried channel guides for high throughput
        efficiency

     * Improvements in temperature and field control throughout the
        diffusion process.

The first two of these problem areas require elaboration. Incomplete
burying of the waveguides is deleterious if the effective index barrier
for the buried interface is too low. Presently, the surface index
difference is 0.01. This should be increased to at least 0.015, so
that optical fibers with numerical apertures of >0.20 can be accomo-
dated.

     The second problem area, edge effects caused by diffusion through
a mask, is illustrated in Figure 11. For the same diffusion conditions
the effective diffusion depth varies depending on the position relative to the mask. (The results shown are in fact for the same sample).
This effect depends on the mask width also. The planar guides (no
mask) are remarkably thinner than are guides formed through a mask.
The contrast in results for wide and narrow masks is also quite interesting. For the wide mask (500μm), the tendency for thinner guide
depths is observed at the center of the guide region. Near the mask
edges, an enhanced diffusion is observed. For the narrow mask (35μm),
the expected semicircular profile is observed. Chartier et al[6] recently reported similar observations. Their explanation of the observed shape of the profiles is based on an electromechanical bias
between the silver nitrate bath and the metallic mask. An electric
field is, in their interpretation, generated by this electrochemical
bias.

     Current work at HRL differs significantly from Chartier's inter-
pretation. We have replaced the metallic mask (Al) with a nonconduc-
ting mask (SiO_2) and studied the edge effects. The results in both
experiments are almost identical. However, these results are complicated by the incomplete blocking of the SiO_2 layer. The waveguide
profile for the SiO_2 mask is shown in Figure 12. The same diffusion
conditions were used to closely compare to the results obtained with
the metallic mask. Detailed measurement of the guide shape shows very
close similarities.

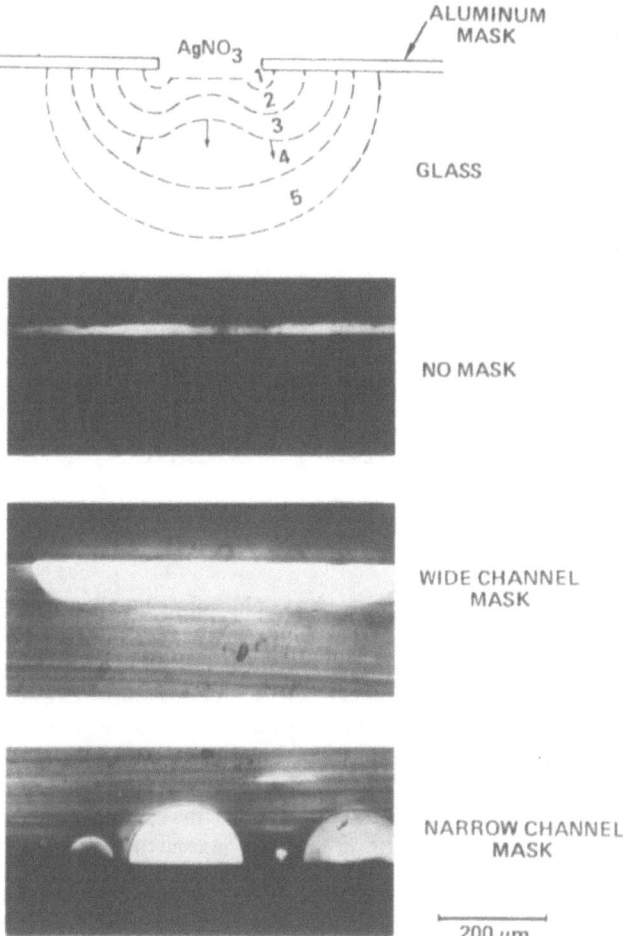

Fig. 11. Effects caused by enhanced diffusion at the mask edge (Al). Note the variation in guide depth with mask width.

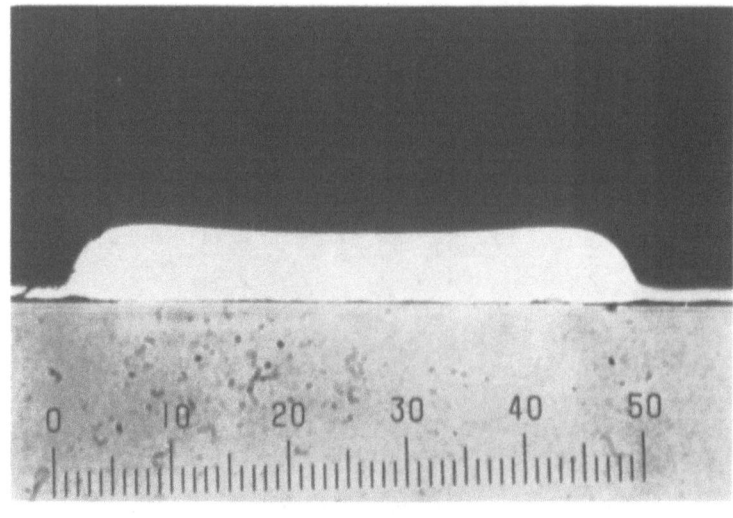

t = 30 min
E = 166 V/M
W = 500 μm

Fig. 12.   SiO$_2$ mask results.   t = 30min, E = 166V/M, W = 500μm.

Our current interpretation is that the observed results are simply
due to current spreading effects at the mask edge. The technical
significance of this problem cannot be minimized. The fabrication
of star couplers, which require very wide masks (several mm) to form
the mixing region, will be complicated by the guide thinning effects.
Fortunately, the high index guides we have used are highly overmoded
(very high NA) when compared to the excitation conditions using conventional fibers (NA ≃ 0.2). Guide thinning effects may not severely
affect star performance, but a closer examination will be necessary.

COUPLER FABRICATION AND OPTIMIZATION

Planar Splitter Development

     The first devices studied with the planar processing were planar
splitters. The results clearly demonstrated the batch processing
potential of planar couplers. They also highlighted the major problem
with the planar approach - coupling the fibers to the channel guides.

     The guides were processed by the Li$_2$SO$_4$-K$_2$SO$_4$ guide formation
techniques described above.

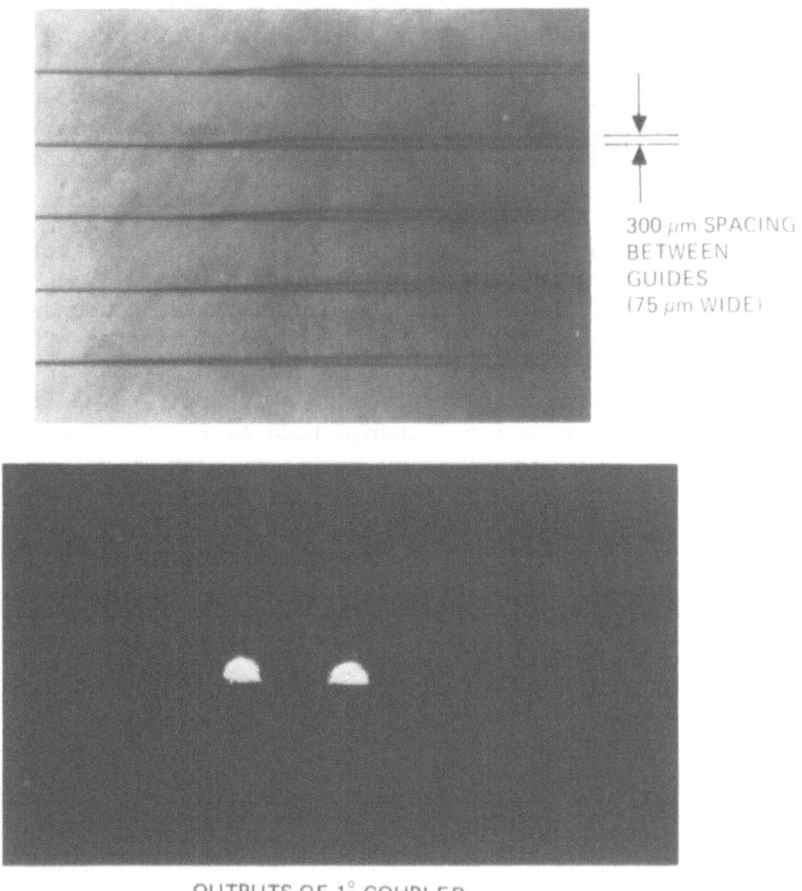

OUTPUTS OF 1° COUPLER
(300 μm CENTER TO CENTER)

Fig. 13.   Planar coupler array formed by ion exchange in a glass slide.

Figure 13 shows a top view of an ion exchange slide with coupling
structures formed using 1, 2, 3, 4, and 5° branches. The coupling
branches are continued parallel to the main branch when 300μm separation is attained. This facilitates the power measurement because
the optical fibers placed at the output ports may be perpendicular
to the slide end face. The output of the 1° branch is also shown in
Figure 13. Using graded-index fibers (Corning) of 65μm core size,
the coupling ratios were measured from the fiber input to fiber output with the planar coupler in between. Mode strippers were used to
ensure that only guided modes were measured. The results are shown
in Figure 14, where, for two outputs P_1 and P_0, we have graphed
$P_1/(P_1 + P_0)$, where P_1 is the branch output, and P_0 is the straight
output. A clear trend of decreasing coupling with increased branching

angle is observed.   The error bars are due mainly to the experimental
difficulties in fiber placement precision and processing variations.
To demonstrate further the versatility of the planar coupler approach,
the tapered coupler structure shown in Figure 15 has been formed.
The coupling is controlled by varying the parameters d_2, θ_1, θ_2, and
θ_3.   For $d_1 = 75$, $d_2 = 50$, $d_3 = 100\mu m$, $\theta_1 = 2^o$, $\theta_2 = 0.66^o$, and $\theta_3 = 1^o$,
a coupler with a 20:1 tap ratio was measured.

The ion exchange process seems well suited for fiber optic coupler
applications; in fact, the guide losses are 0.1 to 0.5dB/cm depending
on processing variations.   The dominant scattering mechanism is due
to surface cracks, which have been largely eliminated by preprocessing
the slide.   The crack density is reduced by a process involving a 5min
etching of the slide in 5% aqueous HF.   In addition, a fire polishing
of the slides prior to ion exchange reduces the final crack density.

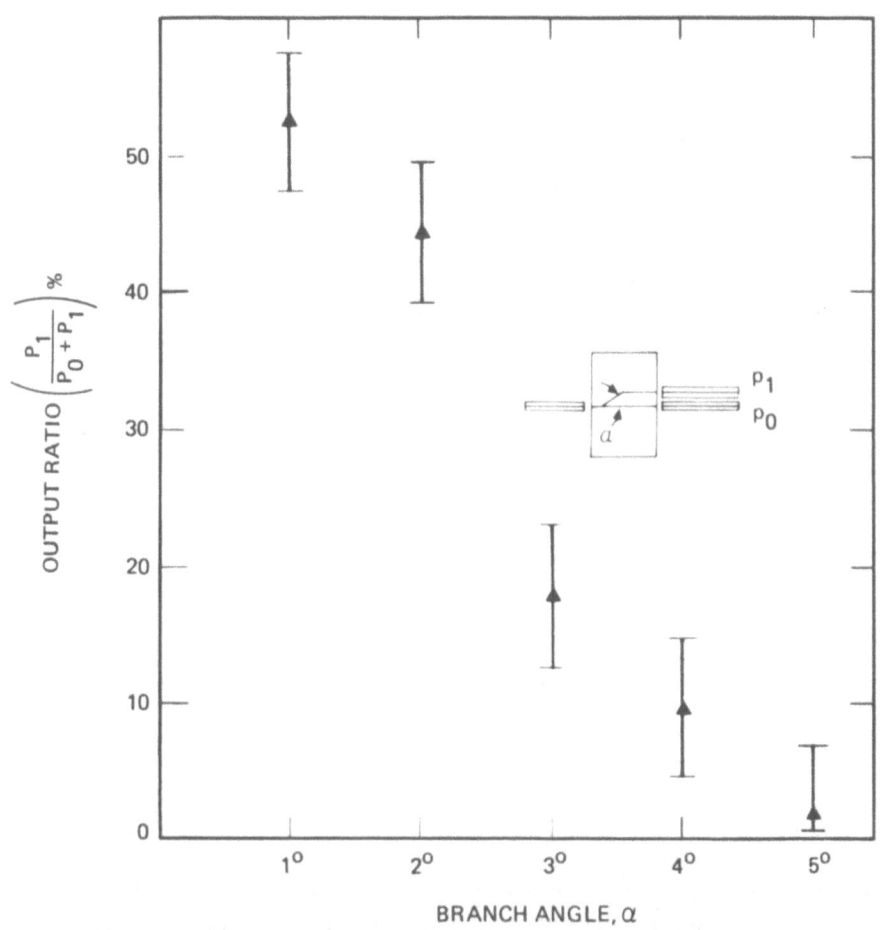

Fig. 14.   Coupling ratio results for different branch angles.

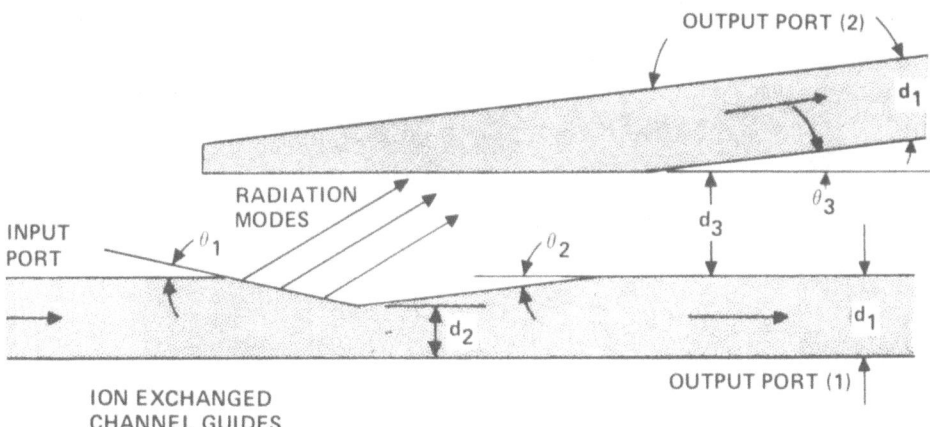

Fig. 15.   Tapered coupler of planar variety formed by ion exchange.
Coupling is due to radiation excitation by the taper.

Planar Star Coupler Development

We have designed planar star devices using the ray-tracing
computer program. In contrast to conventional star designs in which
a simple mixing region is formed, our planar design utilizes horns to
form a transition from the channel guides to the planar mixing region.
This is shown in Figure 16 for a reflective star.

An 8-port star coupler device was fabricated using the ion exchange process. The output of a transmission star when excited at one
of the ports by a graded-index fiber (Corning) is shown in Figure 17.

The measured throughput matrices of a transmission star coupler
with 8 input fibers and 8 output fibers cemented in place are given
here. Plates containing the stars were immersed in the Li_2SO_4-K_2SO_4
melt for the following temperatures and time intervals: 605°C, 5min;
605°C, 10min; 605°C, 15min; 585°C, 20min; 585°C, 40min. Two stars were
located on each plate. The plate immersed for 20min at 585°C appeared
to yield the best stars. A fairly clean near-field output pattern
is observed when light from a fiber is inputted into any of the entrance ports. With light from a fiber (~200μW) inputted into any of the
entrance ports, the maximum sum total power intercepted by a fiber was
24μW (-9.2dB).

The highest throughput was down ~-9dB relative to the power in
the input fiber. We can partition the loss as follows:

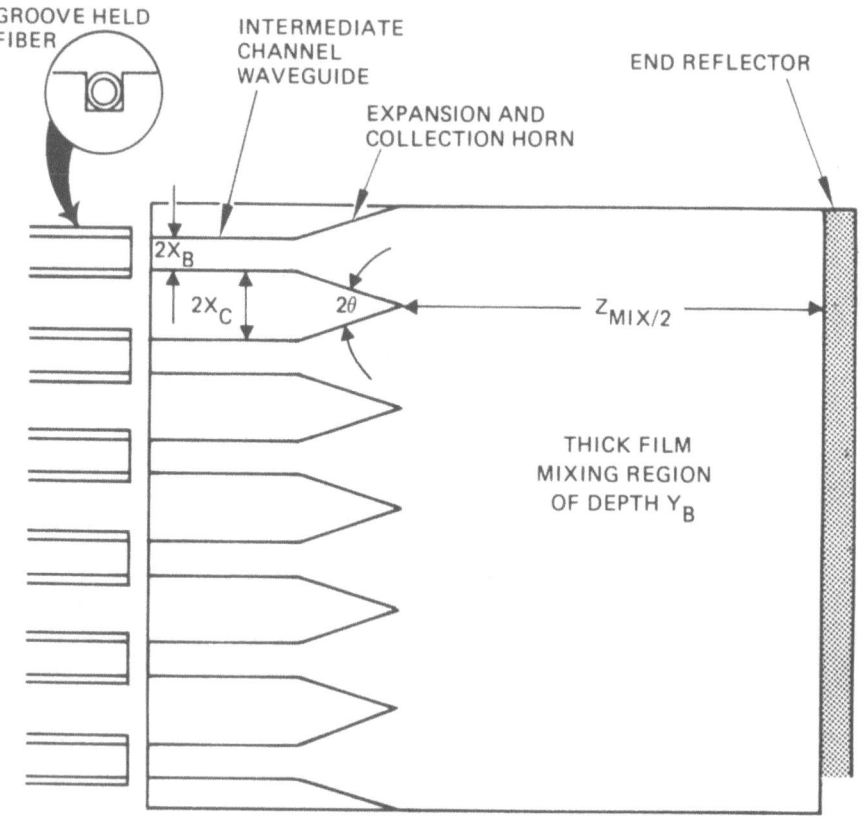

Fig. 16.   Reflection star-coupler design using a horn transition to
           the mixing region.

Measured		Measured		Estimated		Estimated
Insertion	=	Coupling	+	Propagation	+	Internal
Loss		Loss		Loss		Loss

or

$$9dB = 4.2dB + (0.1dB/cm)\ (8cm) + (4.0dB)$$

The measured coupling loss of 4.2dB was determined from the experiments with parallel channels described previously. The remainder of
the insertion loss is partitioned to propagation loss and the internal
loss of the star structure. If the propagation loss is taken to be
very small (0.1dB/cm), an estimated internal loss of 4.0dB remains.
The latter loss may be compared with the packing fraction loss of
3.8dB for a linear array of fiber.

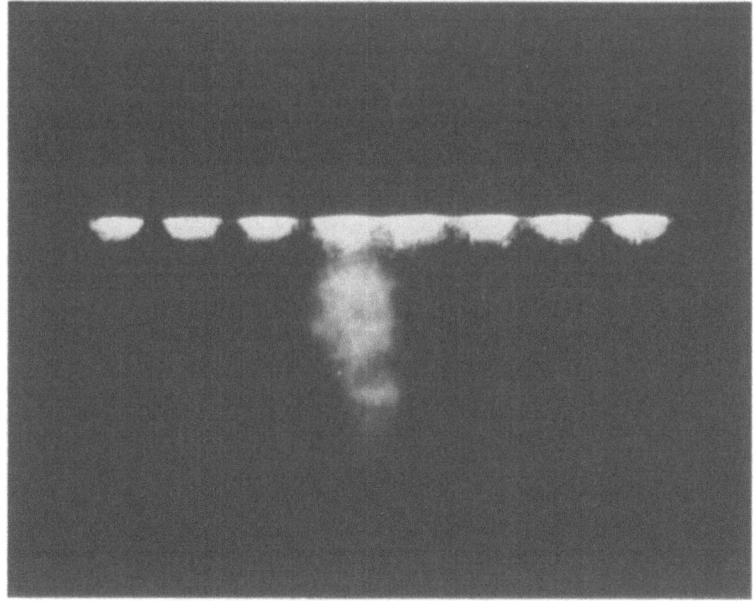

CENTER TO CENTER DISTANCE = 241 μm

Fig. 17. Output of an 8-port star coupler.

Fiber Interfacing

Photographs of the fiber holders are given in Figures 18 and 19.
Conventional silicon selective etching was used to fabricate these
holders. The experience gained by these experiments adds to the con-
fidence level we have in this approach to fiber interfacing. The precision control of photolithography is now applicable to both guide
formation and fiber holder arrays. We are certain that optimum coupling conditions can be attained by this approach.

Y Couplers Formed by Ag Ion Exchange and Double Exchange (Ag/Na)

We have fabricated optical couplers with the Ag ion exchange and
double exchange with Ag and Na. The coupler mask is similar to those
described earlier in this section, the only difference is the mask
width of 30μm. The new results arise from an attempt to use the high
throughput parameters for forming the coupling structures. The parameters were

    Unburied - 300°C, 15min, 150V/mm
    Buried   - 300°C, 7.5min, 150V/mm; Na burying
             - 400°C, 10min, 150V/mm.

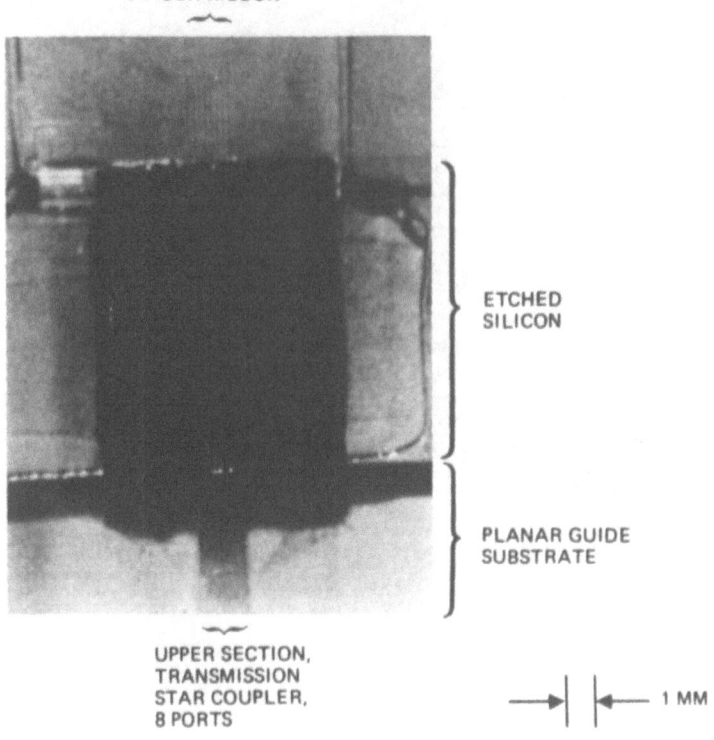

Fig. 18.   Top view of 8 fiber ribbon epoxied to an 8 port trans-
           mission star coupler.

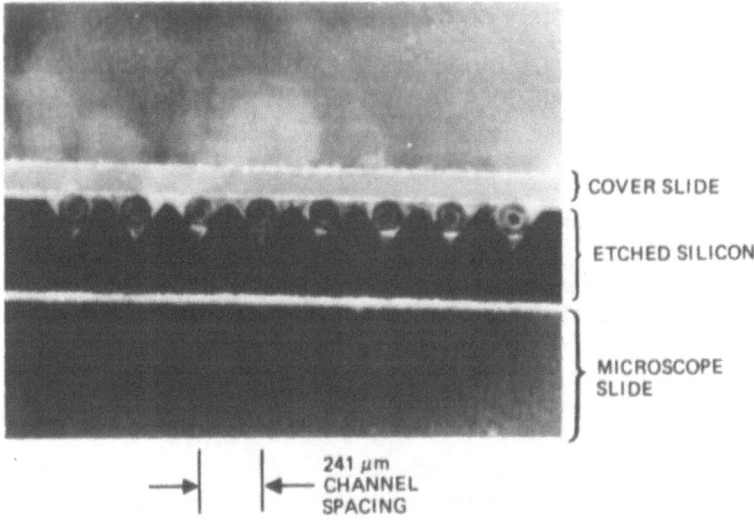

Fig. 19.   End view of 8 fiber ribbon epoxied into etched silicon
           spacers.

The ends of the 1° Y-coupler structures for both the unburied
and buried structures are shown in Figure 20. The difference between
the index profile is evident from the guiding of the microscope illu-
mination. The mask utilized had 1, 2, 3, 4, and 5° angle couplers.
The total sample length was 1.2cm.

COUPLER PERFORMANCE EVALUATION

The evaluation of couplers was carried out to determine the internal losses caused by the branching structures separate from the
fiber to channel interface losses. To accomplish this we determined
in a separate experiment the throughput from fiber-to-channel-to-fiber
(F/C/F) for straight channels. From this measurement an adequate estimate of the coupling losses without the structured guide patterns
are determined. Subsequently the excess loss of couplers was determined with fibers at the input and output (with index matching liquid).
From these measurements the interval structural losses of the couplers
were estimated.

The samples studied were processed under similar conditions.
Guide formation was accomplished with T = 300°C for $AgNO_3$, t = 7.5min,
and E = 150V/mm. For the burying with $NaNO_3$ the parameters are T =
400°C, t = 10min, and E = 150V/mm. For the couplers the mask width
was 30µm, for the channel guides width ranging from 5 to 25µm were
studied. The output of the buried Y coupler (1°) is shown in Figure
21.

The straight channel measurements showed that the average value
of F/C/F throughput loss was -1.4dB. Our best result was -0.9dB. The
best results were obtained with 85µm step index fiber (NA = 0.20).
It was important in these measurements to ensure that the excitation
was reproducible. We accomplished this by monitoring the input spot
on the fiber end face. In general, meridional ray excitation yielded
the best throughput results. Skew ray excitation, which in a step
index guide can be achieved by moving the input spot across the input
face with input NA equal to the fiber NA, causes a degradation of the
throughput. By moving the input spot across the fiber end face, we
were able to determine the weighted average loss as -1.9dB. The
throughput loss is constant until spot positions farther than $r_{CLAD}/2$
are attained. This measurement allows an understanding of the losses
when an LED is used to excite the fiber. In this case the throughput
losses were -2.8dB. The value of -1.8dB is reasonably close, considering the difficulties in stripping the leaky modes caused by the
LED excitation.

The performance of branch couplers, fabricated as previously described, was evaluated. The insertion loss and coupling ratio were
measured for different waveguide branches with 1, 2, 3, 4, and 5°
angle couplers. The total length was 1.2cm.

UNBURIED COUPLER

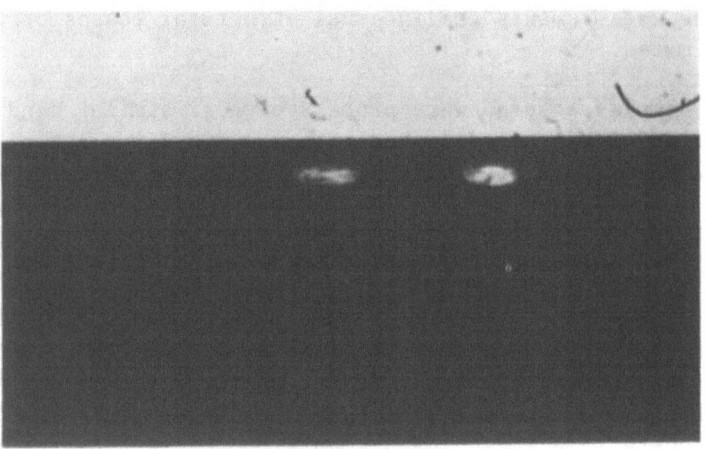

BURIED COUPLER

Fig. 20.   End view of the 1° Y-coupler guides:
(a) unburied and (b) buried.

The insertion loss of each coupler was measured using different diameter fibers. The fibers were excited by meridional rays only for these measurements. The input fiber and the output fiber for the straight channel were adjusted for optimum throughput (lowest loss) of the straight channel. The output fiber for the branch was carefully moved to optimize the throughput of the branch channel with the input fiber fixed. With the fibers in position, the total throughput and tap ratio were measured directly.

The results of these measurements for a series of samples are summarized in Figures 22 and 23. These results indicate a few simple trends:

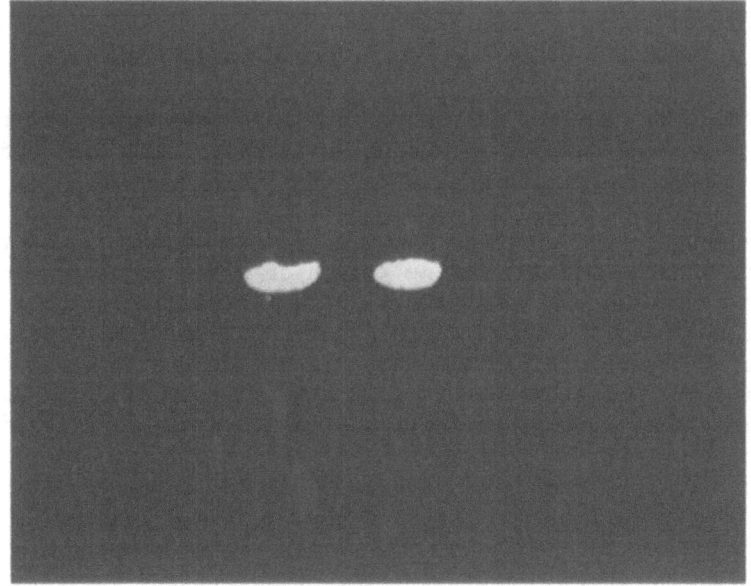

BURIED Y COUPLER (1°)

Fig. 21. Output of the buried Y coupler (1°) when excited by a fiber.

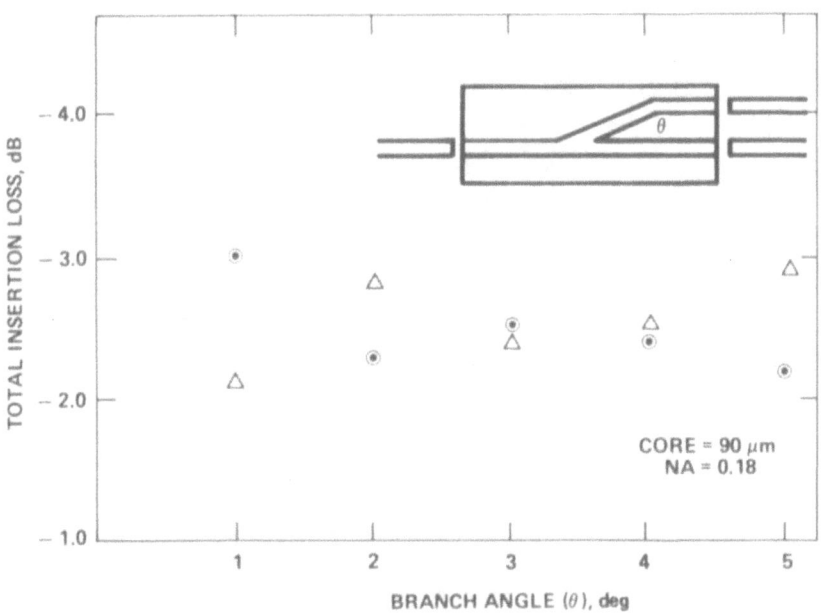

Fig. 22. Total insertion loss of two sets of taps.

(1)   The 90μm core fiber produces the best insertion loss results.

(2)   The insertion loss is in the 2 to 3dB range.

(3)   The variance in the tap ratio is large ±1dB, if the approach
      of optimizing throughput is taken. Care can lead to exactly
      3dB taps with some sacrifice in throughput.

A measurement of the effects of launch conditions on the coupler
performance was made by moving the input spot across the input face
to the center and R/4, R/2, 3R/4. The measured results of the coupler
insertion loss is summarized in Figure 24. The insertion loss increase at the center position is a result of the index hole at the
center of the fiber which was used for the measurement. As a general
statement the skew ray excitation leads to a higher insertion loss.
The coupling ratio is not as sensitive to excitation as would be expected. This is presumably due to the mode mixing occurring in structure, as a result of the graded index profile of the guide boundaries.

Based on these measurements, we estimate the internal loss of
the couplers to be 1dB. It is difficult to assess the causes of this
loss at this time.

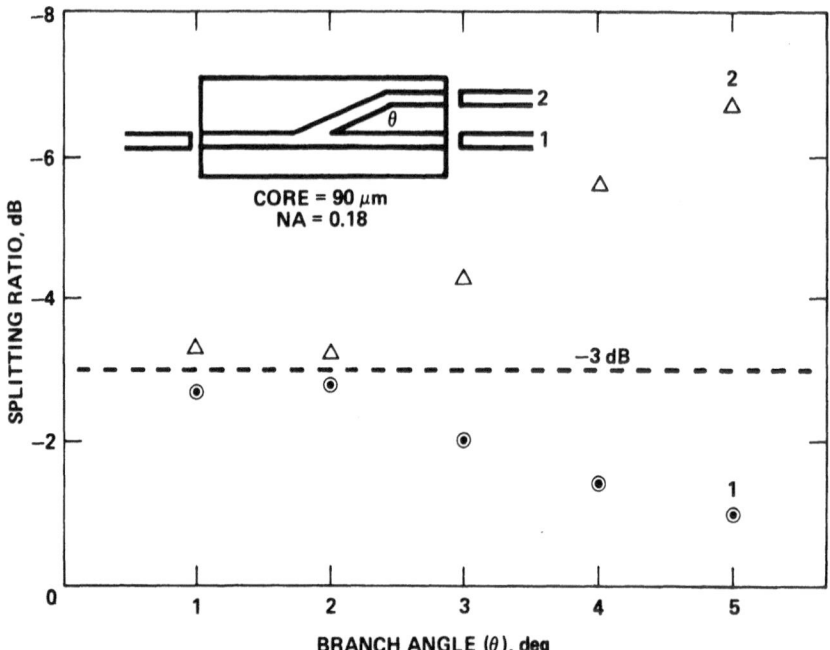

Fig. 23.   Splitting ratio for planar tap.

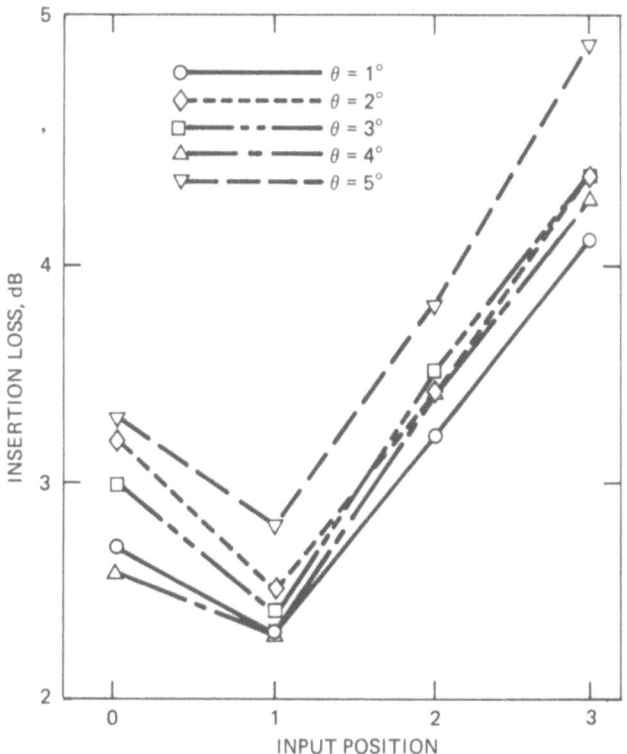

Fig. 24. Coupler insertion loss as a function of input fiber
excitation condition.

REFERENCES

1. F. Auracher, Opt. Commun. 17:129, 1976.
2. G. L. Tangonan, L. E. Gorre and D. L. Persechini, Opt. Commun.
    27:358, 1978.
3. G. L. Tangonan, O. G. Ramer, L. E. Gorre, H. R. Friedrich, C. K.
    Asawa, M. K. Barnoski and D. L. Persechini, Topical Meeting
    on Optical Fiber Communications, Washington, D. C., paper
    WD3, March 6-8, 1979.
4. G. L. Tangonan, O. G. Ramer, H. R. Friedrich, C. K. Asawa, D. L.
    Persechini and L. E. Gorre, Optical Communications Conference,
    Amsterdam, paper 215, 1979.
5. G. Chartier, P. Jaussaud, A. D. de Oliveira and O. Parriaux,
    Electron. Lett. 14 132, 1978.
6. G. Chartier, P. Collier, A. Guez, P. Jaussaud and Y. Won, Appl.
    Opt. 19:1092, 1980.
7. T. Izawa and H. NaKagome, Appl. Phys. Lett. 21:584, 1972.

INTEGRATED OPTICS SWITCHES AND MODULATORS

R. V. Schmidt

Hughes Research Laboratories
3011 Malibu Canyon Road
Malibu, CA 90265

INTRODUCTION

Integrated optics has been an active field of research for about 10 years. During this time, considerable progress has been made toward realizing high performance optical devices using guided-wave techniques[1]. The electro-optic modulator and switch are two devices that have received considerable attention.

The use of guided-wave techniques has led to switches and modulators with increased compactness, lower drive voltages and drive power compared to bulk electro-optical devices. In fact, switches using single-mode optical waveguides have been demonstrated with electrical drive requirements compatible with electrical integrated circuits. Modulators with bandwidths greater than 10GHz have also been demonstrated.

The commonly studied integrated optic switch is a four-port device with two optical inputs and two optical outputs that have two distinct switching states. It is the analog of the electrical double-pull double-throw reversing switching. In the straight-through state a signal entering in the top or bottom port, as illustrated in Figure 1, exists in the top or bottom port, respectively. In the other state, called the crossover state, a signal entering in the top port exists in the bottom port, while a signal entering in the bottom port exists in the top port. This switch may be used either as an amplitude modulator or as an element for switching the path of optical signals. A modulator, of course, requires only two optical ports. The basic electro-optic modulator is a phase modulator by an amount proportional to the applied electric field. For amplitude modulation, electro-optic phase modulation is converted to intensity changes typically through the use of an interferometer or a directional coupler.

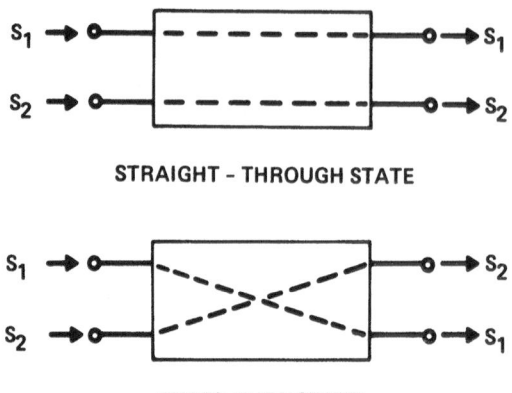

Fig. 1. Diagram showing the straight-through state (top) and cross-over state (bottom) of a four-port directional coupler switch.

In this paper we describe the general properties of electro-optic modulators and switches. Emphasis will be placed on the electro-optic phase modulator because of its simplicity and importance to other devices. The travelling wave modulator and lumped element modulator will be defined and compared. The operation of waveguide directional couplers will be reviewed and the results used to describe the operation of directional coupler switches and modulators. Much of the discussion on directional coupler switches comes from reference 2. Interferometric modulators will be covered by first describing the operation of 3-port waveguide power splitters and combiners. Both Y-junctions and three-coupled waveguide power splitters will be treated. Then the operation of an interferometric modulator for both digital and analog modulation will be described.

GENERAL BACKGROUND

Efficient electro-optic switches and modulators utilize single-mode strip optical waveguides. Figure 2 illustrates the cross-section of a strip waveguide. Typically, three refractive indices define the waveguide: the substrate index n_s, the waveguide strip index n_g, and the cladding or cover layer index n_c. The dimensions of the guide and the value of refractive indices determine the number of guided modes and their propagation constants β. The propagation constant can be written as

$$\beta = 2\pi N_o/\lambda \quad , \tag{1}$$

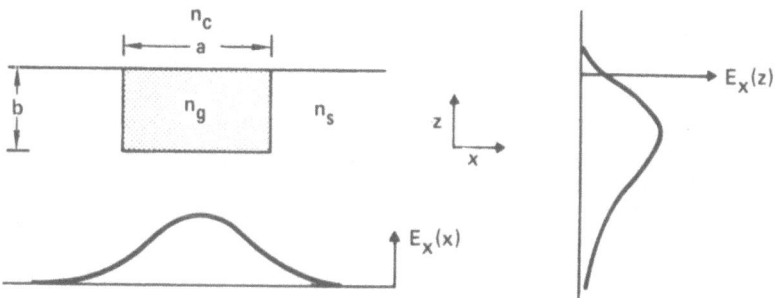

Fig. 2.   Illustration of single-mode waveguide cross section and
          typical modal field distributions.

where λ is the free-space optical wavelength, and N_0 is the so-called
optical effective index of the guided mode. The effective index is
determined by waveguide dispersion, and in general must be calculated
using the numerical or approximate techniques. The value of N lies
between n_s and n_g, being larger (closer to n_g) for more tightly confined modes. Typically, strip waveguides are designed with n_g just
slightly larger than n_s, so that a single-mode guide has the dimensions of several microns.

Two material systems have received particular attention for implementation of integrated optics switches and modulators. They are
titanium-diffused LiNbO3 and the GaAs/GaAlAs material system.

Lithium niobate is a ferroelectric crystal with excellent electro-
optic properties. A strip waveguide is easily fabricated in LiNbO3
by first photolithographically delineating the waveguide circuit pattern in Ti metal. The metal is then diffused into the LiNbO3 in an
oxygen atmosphere at $\sim$1000°C for several hours. For λ = 0.83μm, a
4-μm Ti strip 400-Å thick diffused for 6h creates a single-mode waveguide with a depth of $\sim$3μm. The refractive index of the crystal in
the volume where the metal is diffused is increased proportionally
to the Ti concentration. The guided mode has an intensity distribution similar to that shown in Figure 2, although the diffused waveguide has graded index boundaries. There is little optical energy
at the crystal surface because of the large index difference between
LiNbO3 (n_s = 2.2) and air (n_c = 1). The index difference between the
diffused region and the substrate is of the order 10^{-2}. Because the
diffusion temperature is below the crystal's Curie temperature there
is no need to repole the crystal after diffusion. The loss in Ti-
diffused waveguides is typically less than 1 dB/cm at 0.6323μm.

The semiconductor GaAs/GaAlAs material system is interesting for
integrated optics devices because of the ability to create excellent
optical sources, optical detectors, and electronic devices with it.

This provides the potential of realizing true optical integrated circuits. To date, this potential has not been realized because of the required fabrication complexity. Strip waveguides have been formed by a variety of techniques including diffusion, proton bombardment, and epitaxial growth techniques. The propagation-loss of GaAs waveguides is typically a few dB/cm at 1.06µm wavelength. One potentially useful feature of the GaAs waveguide system is that velocities of microwaves and light waves are almost identical. This will be shown to be a substantial advantage for traveling wave electro-optic interactions.

The lack of diffraction effects and the small feature size of single-mode waveguides are the primary reasons for the improved performance of integrated electro-optic devices over their bulk counterparts. The advantages can be seen by considering the guided-wave phase modulator. Phase modulation is a basic electro-optic function which is used in nearly every electro-optic device. It is thus worthwhile to review the theoretical operation of a waveguide phase modulator because the results are applicable to a large variety of waveguide devices.

The waveguide phase modulator consists of a strip waveguide which supports a single mode of effective index N_0. Electrodes of length L are fabricated so that an applied voltage creates a predominantly transverse or normal electric field across the waveguide. Two possible electrode configurations are illustrated in Figure 3. In one case the elctrodes are placed symmetrically on either side of the waveguide. A transverse electric field is created with this configuration. In the other case, one electrode is placed on top of the waveguide, and a predominantly normal electric field is created across the waveguide.

A voltage V applied across electrodes separated by a distance d creates an electric field $E \approx V/d$ within the waveguide. The applied field electro-optically induces a change in effective index:

$$\Delta N_o = \frac{1}{2} \eta N_o^3 rE \tag{2}$$

The coefficient η whose value is near unity, accounts for incomplete overlap of the optical and applied electrical fields. The symbol r is the relevant electro-optic coefficient which is determined by crystal orientation and the direction of applied field.

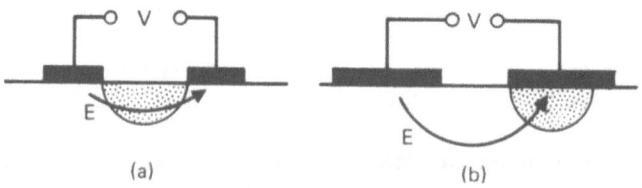

(a)   (b)

Fig. 3.   Electro-optic phase modulator electrode configurations for
          (a) transverse applied fields, and (b) normal applied fields.

For a static applied electric field the refractive index change produces a phase change in the waveguided light, propagating the length L of the electrodes by an amount,

$$\Delta\phi = \Delta\beta L = 2\pi\Delta N_o/\lambda \quad . \tag{3}$$

Using equation (3), the electro-optically induced phase change produced by a voltage V is

$$\Delta\phi = \eta\pi N_o^3 \, r \left(\frac{L}{d\lambda}\right) V \quad , \tag{4}$$

where d is the separation between the electrodes.

Of particular importance in characterizing an electro-optic phase modulator's performance in the bandwidth and the voltage required to produce a π-phase change. Two modulator configurations will be considered: the traveling wave modulator and the lumped element modulator. For the traveling wave modulator the device bandwidth is limited by transit time difference between the microwaves and light waves. The lumped element modulator bandwidth is determined by the termination resistor and electrode capacitance. Transit time effects are not important.

The traveling wave phase modulator fabricated on a substrate of relative dielectric permitivity ϵ_r is illustrated in Figure 4. It consists of a strip waveguide surrounded on either side by two electrodes which form a planar strip line of characteristic impedance Z_o. The strip line is driven by a transmission line of impedance Z_o and is terminated in a matched load, $R_o = Z_o$. Several different types of planar strip lines can be used. They include the coplanar strip, asymmetric coplanar strip, and the coplanar waveguide which are illustrated in Figure 5. The choice of gap, d, and strip width, W, determines the characteristic impedance of the transmission line. Figure 5 provides the characteristic impedance for the strip lines formed on $LiNbO_3$, calculated with a conformal mapping tecgnique[3]. It has been shown that the asymmetric coplanar parallel strip line yields a more uniform frequency response than the balanced strip lines[4]. Because the dimensions of the strips are only a few microns wide, the strip lines can have significant attenuation which can limit modulator performance.

The bandwidth of the traveling wave modulator can be analyzed easily. The optical field traveling down the modulator is given by

$$\mathcal{E}_o(x,t) = E_o \cos\left[\omega_o\left(t - \frac{N_o}{c}x\right)\right] \quad , \tag{5}$$

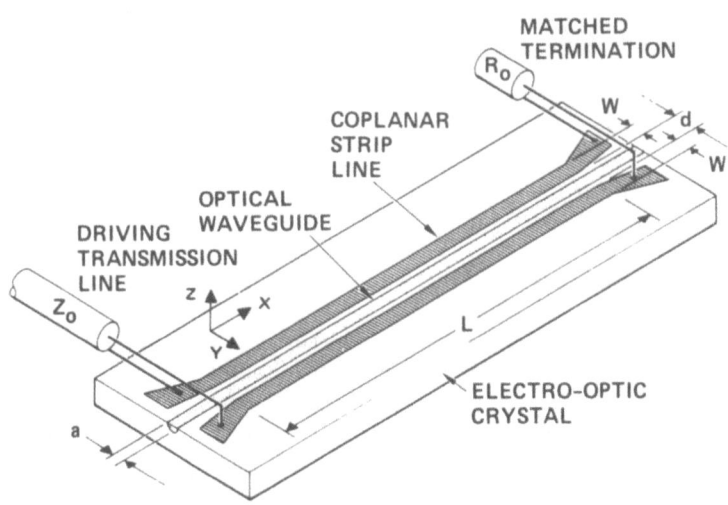

Fig. 4.   Illustration of a traveling-wave electro-optic waveguide
          phase modulator.

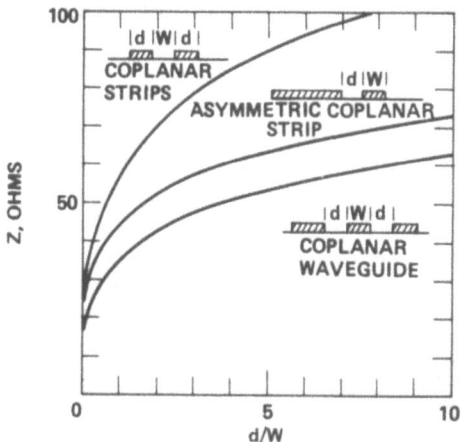

Fig. 5.   The characteristic impedance of three different strip line
          configurations on LiNbO$_3$ as a function of the gap, d, to
          strip width, W, ratio.

where ω_0 is the optical radian frequency, N_0 is the optical effective
index, and c is the velocity of light in free space. The microwave
field is given by

$$\mathcal{E}_m(x,t) = E_m \cos\left[\omega_m\left(t - \frac{N_m}{c}x\right)\right] e^{-\alpha x} \quad , \tag{6}$$

where ω_m is microwave radian frequency, $N_m = \sqrt{(\epsilon_r + 1)/2}$ is the microwave effective index, and α is the strip line attenuation. For planar
strip lines the frequency dependence of the attenuation is approximately

$$\alpha(f) = \alpha_0 f^{1/2} \tag{7}$$

where f is the microwave frequency, and α_0 is a coefficient which
depends on geometrical and electrical parameters of the strip line[5].

An optical disturbance starting at the microwave input of the
modulator at time $t = t_0$ travels to a position $x = (t - t_0)c/N_0$ by
time t. Thus, at time $t = t_0 + (N_0/c)x$ the optical disturbance ob-
serves the microwave field

$$\mathcal{E}_m(x,\omega_m) = E_m e^{-\alpha x} \cos\left\{\omega_m\left[t_0 + \left(\frac{N_0 - N_m}{c}\right)x\right]\right\} \quad . \tag{8}$$

After the optical disturbance transverses the modulator it is phase
shifted by an amount

$$\Delta\phi(\omega_m) = \frac{2\pi}{\lambda} \int_0^L \Delta N_0(x,\omega_m) \, dx \quad , \tag{9}$$

where

$$\Delta N_0(x,\omega_m) = \frac{1}{2} N_0^3 \, r\mathcal{E}_m(x,\omega_m) \quad . \tag{10}$$

Performing the integral one obtains the frequency response of the
phase modulator,

$$\Delta\phi(\omega_m) = \frac{\pi N_0^3 rL}{\lambda} E_m \sqrt{\frac{(e^{-\alpha L}-1)^2 + 4e^{-\alpha L}\sin^2 u}{(\alpha L)^2 + (2u)^2}} \sin\{\omega_m t_0 + \theta\}, \tag{11}$$

where

$$\theta = \tan^{-1}\left(\frac{e^{-\alpha L}\sin 2u}{e^{-\alpha L}\cos 2u - 1}\right) - \tan^{-1}\left(\frac{\alpha L}{2u}\right) , \tag{12}$$

and

$$u = \frac{\omega_m}{2} \frac{N_o - N_m}{c} L \quad . \tag{13}$$

When the attenuation of the strip line can be neglected, Equation (11) simplifies to

$$\Delta\phi(\omega_m) = \frac{\pi N_o^3 rL}{\lambda} E_m \frac{\sin u}{u} \cos(\omega_o t + u) \quad . \tag{14}$$

The modulators' 3 dB bandwidth, Δf_{m3dB} (ie., when the power in the optical side bands are reduced by one-half), occurs when the sinc function is equal to $1\sqrt{2}$, which corresponds to $u = 1.4$ and

$$\Delta f_{m3dB} = \frac{1.4c}{\pi |N_o - N_m| L} \quad . \tag{15}$$

Thus, the traveling wave modulator bandwidth is limited by the difference in velocity between the microwaves and the light waves. For LiNbO$_3$, $N_o = 2.2$ and $N_m = 4.2$, and the modulator has a bandwidth of 6.6GHz/cm. However, because of the small dimensions of the microwave strip lines, the strip line attenuation contributes to reducing the bandwidth. In Figure 6 the response of a LiNbO$_3$ phase modulator, $|\Delta\phi(\omega)/\Delta\phi(o)|^2$, is plotted using Equation (11) for several different values of α_o and a 0.61cm modulator length. Reported experimental devices[5] have exhibited an attenuation characterized by $\alpha_o \approx 1.5$dB/cm GHz$^{1/2}$. A modulator with a lossless strip line 0.61cm long would be expected to have a 10.8GHz bandwidth. With an attenuation of $\alpha_o = 1.7$dB/cm GHz$^{1/2}$ the bandwidth is reduced to 8GHz.

For a modulator constructed in the GaAs/GaAlAs material system in which the substrate is also covered with GaAs so that $N_m = \sqrt{\epsilon_r}$, the microwave velocity and the light wave velocity are almost identical. In this case only electrode attenuation will limit the frequency response. To illustrate, the frequency responses of a LiNbO$_3$ modulator and a GaAs modulator are presented in Figures 7 and 8 respectively, for modulators of different lengths and the same strip line attenuation $\alpha_o = 1.7$dB/cm GHz$^{1/2}$. The bandwidth of the LiNbO$_3$ devices is limited by both velocity mismatch and strip line attenuation. The GaAs devices have considerably greater bandwidth which is limited by strip line attenuation. In both cases improved bandwidth can be obtained by reducing strip line attenuation. In a practical device the bandwidth can also be reduced by impedance mismatches between the strip line and the terminating load or the driving transmission line[5].

In the case of the lumped element modulator transit time considerations are not significant. An illustration of a lumped element modulator is provided in Figure 9. It consists of a coplanar electrode structure with capacitance

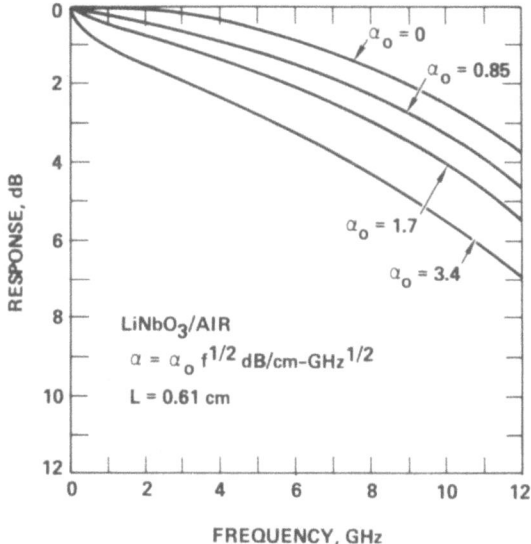

Fig. 6.   Plot of a LiNbO3 phase response for different values of
strip line attenuation.

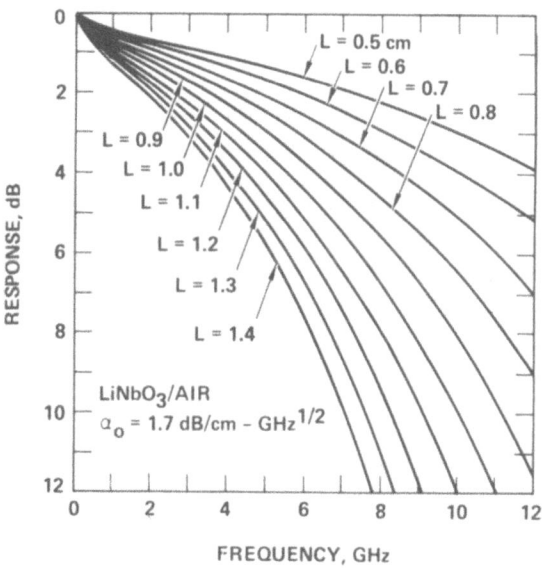

Fig. 7.   Plot of LiNbO$_3$ phase modulator response for devices of
different length and strip line attenuation
$\alpha = 1.7$dB/cm GHz$^{1/2}$.

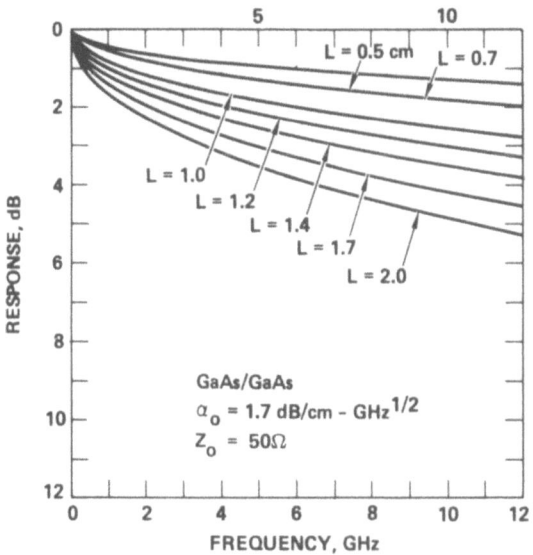

Fig. 8.   Plot of GaAs phase modulator response for devices of
          different length and strip line attenuation
          $\alpha_o = 1.7 \text{dB/cm GHz}^{1/2}$.

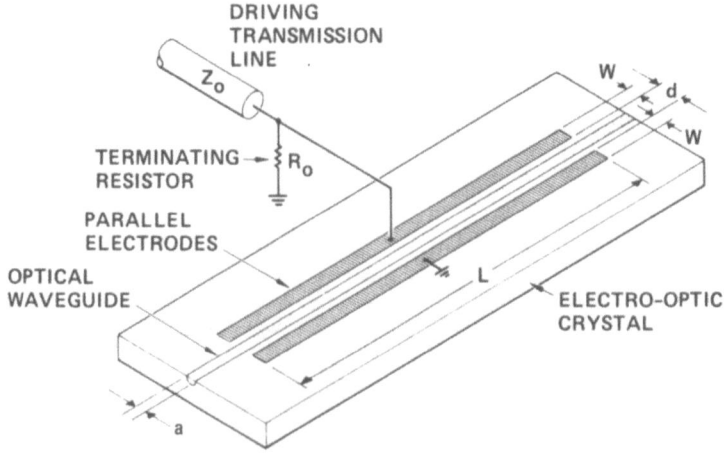

Fig. 9.   Illustration of a lumped element electro-optic phase
          modulator.

$$C = \frac{\varepsilon_o L}{\pi} (1 + \varepsilon_r) \ln \left| 4 \frac{W}{d} \right| \quad , \tag{16}$$

which is driven by a transmission line terminated in its characteristic impedance, $Z_o = R_o$. Neglecting parasitics, the speed of the phase modulator is limited by the capacitor charging time $\tau_{RC} = R_o C$. The phase retardation of a light wave traversing the phase modulator is simply

$$\Delta\phi(\omega_m) = \frac{\pi N_o^3 rL}{\lambda} E_m \frac{\cos\left[\omega_m t - \tan^{-1}\left(\frac{\omega_m RC}{2}\right)\right]}{\sqrt{1 + \left(\frac{\omega_m RC}{2}\right)^2}} \quad , \tag{17}$$

where $E_m = \eta\, V/d$ is the electric field produced by the applied voltage V. The 3dB bandwidth is

$$\Delta f_{m_{3dB}} = \frac{1}{\pi RC} \quad . \tag{18}$$

It is interesting to compare traveling wave modulators and lumped-element modulators in a manner first described by Izutsu and coworkers[6]. The comparison assumes that the same electrode structure is used for both devices. If C_o is the capacitance per unit length of the electrode structure in free space ($\varepsilon_r = 1$), then the modulator electrode capacitance per unit length is

$$C_\ell = N_m^2 C_o \quad , \tag{19}$$

where N_m, the microwave effective index, was defined with Equation (6). Accordingly, the microwave phase velocity on the electrode is

$$v_m = c/N_m \quad , \tag{20}$$

and the characteristic impedance is given by

$$Z_o = \frac{1}{v_m C_\ell} = \frac{1}{cN_m C_o} \quad . \tag{21}$$

For both modulators the voltage required for a π-phase shift is given by

$$V_\pi = \left(\frac{\lambda}{\pi N_o^3 r}\right) \frac{d}{L} \quad . \tag{22}$$

A figure of merit for comparing modulators is the power per unit band-width required for a π-phase shift. For both modulators the power required is

$$P_\pi = \frac{1}{2} \frac{V_\pi^2}{R_o} .$$

(23)

Then, using Equations (21) and (22) and the expressions in Equations (15) and (18), one obtains

$$\frac{P_\pi}{\Delta f} = \frac{\pi}{2} N_m^2 \, C_o \, \frac{\lambda^2}{\eta N_o^3 \, r} \frac{d^2}{L}$$

(24)

for the lumped element modulator, and

$$\frac{P_\pi}{\Delta f} = \frac{\pi}{2} N_m^2 C_o \left(\frac{\lambda}{\eta N_o^3 \, r}\right)^2 \frac{d^2}{L} \frac{\left|\frac{N_o}{N_m} - 1\right|}{1.4}$$

(25)

for the traveling wave modulator. A comparison of Equation (24) and (25) shows that the $\Delta P_\pi/\Delta f$ figure of a lumped-element modulator is a factor of $1.4/|1 - N_o/N_m|$ larger than that of a traveling-wave modu-lator. For $LiNbO_3$, the lumped-element modulator requires a factor of 2.8 more power for a given bandwidth. Expressed another way, the traveling wave modulator has a bandwidth of 6.6GHz/cm, while a lumped-element modulator driven with the same power has a 2.3GHz/cm bandwidth. Of course, this discussion presumes the same electrode configurations for both devices and neglects electrode losses and parasitics. In practice, optimized devices of either type would not have similar elec-trodes, and losses and parasitics would have significant influence on device performance.

The advantage of strip waveguide phase modulators over bulk de-vices is apparent from Equations (24) and (25). Low drive power re-quires a small d/L ratio. In bulk devices where d is the width of the crystal, diffraction limits the interaction length L over which the beam may be focused to approximately d^2/λ. So if d is made small, the interaction length is restricted. For widths comparable to single-mode guides, the maximum interaction length is limited to around 100μm. Because a waveguide does not allow diffraction or spreading of light, the waveguide modulator does not have this restriction, and its length is determined by speed requirements or fabrication convenience.

OPTICAL DIRECTIONAL COUPLERS

An integrated optics directional coupler is formed by fabricating two parallel waveguides in proximity so that the light in one wave-guide can couple to the other waveguide via the evanescent fields. In Figure 10 a waveguide directional coupler circuit pattern is

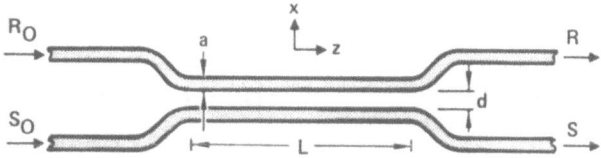

Fig. 10. Top view of a waveguide directional coupler where L is the coupler length, a is the waveguide width, and d the separation between the waveguides in the interaction region.

illustrated. Two waveguides with propagation constants β_R and β_S are brought within a distance d of one another for a length L. Over this length the waveguides are coupled so that optical energy can transfer between the two guides. If the waveguides have the same propagation constants and energy is incident in only one guide, it will transfer completely to the other guide in a distance $\ell = \pi/2\kappa$, called the transfer length, where κ is the coupling coefficient which describes the strength of the interguide coupling. Typically, the coupling coefficient is related to separation between the guides, d, by an exponential relation of the form

$$\kappa = \kappa_o \exp(-d/\lambda) \qquad (26)$$

where κ_o and λ are coefficients that depend upon the various waveguide parameters[7]. In Figure 11, the coupling coefficient and the transfer length are given for various separations between guides. These measurements were taken on single-mode Ti-diffused waveguides at three different wavelengths. The straight lines are best fits of the exponential relation[7]. Transfer lengths as short as 200μm, and as long as 1cm have been experimentally observed in Ti-diffused waveguides.

Electrically controlled optical switching can be achieved with a directional coupler because the degree-of-light transfer between the waveguides depends upon the difference in propagation constants, $\Delta\beta = \beta_R - \beta_S$, which can be controlled via the electro-optic effect. If the light in the two guides is characterized by complex amplitudes R and S, which vary slowly in the propagation direction, the interaction between the two guides is described by the coupled-wave equations[8],

$$R' - j\delta R = -j\kappa S \quad , \qquad (27)$$

and

$$S' + j\delta S = -j\kappa R \quad , \qquad (28)$$

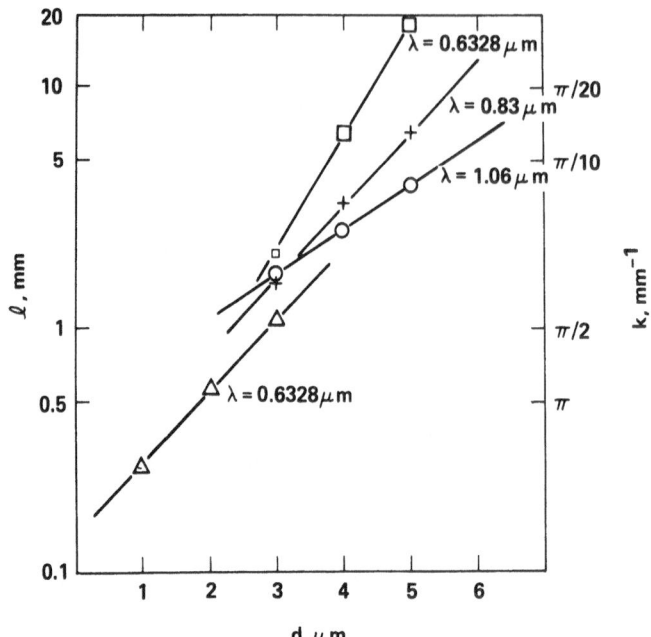

Fig. 11. Coupling coefficient of Ti-diffused LiNbO₃ waveguides at
1.06-, 0.83-, and 0.633-μm wavelengths as a function of
waveguide separation. For each wavelength the waveguides
were fabricated to be single mode using the following fabrication parameters: (□)λ = 0.6328μm, τ(Ti metal thickness) = 300Å, a = 3μm; (Δ)λ = 0.6328μm, τ = 300Å, a = 2μm;
(+)λ = 0.83μm, τ = 400Å, a = 4μm; and (0) λ = 1.06μm, τ =
460Å, a = 5μm.

where the primes denote differentiation with respect to the propa-
gation direction, $\delta = \Delta\beta/2$, and κ is the coupling coefficient. For
arbitrary input amplitudes R_o and S_o, as shown in Figure 10, the solution of (27) and (28) can be expressed in matrix form (29):

$$\begin{bmatrix} R \\ S \end{bmatrix} = \begin{bmatrix} A & -jB \\ -jB* & A* \end{bmatrix} \begin{bmatrix} R_o \\ S_o \end{bmatrix} , \qquad (29)$$

where the asterisk denotes a complex conjugate. The matric coefficients are

$$A = \cos\left(\kappa L \sqrt{1 + (\delta/\kappa)^2}\right) + j(\delta/\kappa) \sin\left(\kappa L \sqrt{1 + (\delta/\kappa)^2}\right) \quad (30)$$

$$B = \sin\left(\kappa L \sqrt{1 + (\delta/\kappa)^2}\right) \Big/ \sqrt{1 + (\delta/\kappa)^2} . \qquad (31)$$

If light is launched in the R guide as z = 0, the power in the two
guides at the coupler output z = L is given by

$$SS^* = \frac{\sin^2\left(\kappa L \sqrt{1 + (\delta/\kappa)^2}\right)}{1 + (\delta/\kappa)^2} R_o R_o^* \quad , \tag{32}$$

and

$$RR^* = 1 - SS^* \quad . \tag{33}$$

Equation (32) illustrates several directional coupler properties that
are important to the design of optical switches. Energy can transfer
completely from one guide to the other only if $\Delta\beta = 0$ and

$$\kappa L = (2m + 1)\pi/2 \quad , \tag{34}$$

where m is an integer. If the guides do not have the same propagation
constants it is impossible for energy to be completely transferred
across the coupler and some optical energy always remains in the
input waveguide. However, in this phase mismatched case ($\Delta\beta \neq 0$),
energy exits entirely in the input waveguide only if

$$(\kappa L)^2 + (\delta L)^2 = (m\pi)^2 \tag{35}$$

is satisfied.

DIRECTIONAL COUPLER SWITCH/MODULATOR

We are now ready to explain the operation of directional coupler
switches. The first type of directional coupler switch that was experimentally demonstrated is illustrated in Figure 12. It consists
of a directional coupler with electrodes placed over the waveguides
in the coupler region. The applied electric fields normal to crystal
surface are oppositely directed in each guide so that the linear
electro-optic effect produces the phase mismatch between the guides
by increasing the refractive index in one guide and decreasing it in
the other. The switch is designed so that it is in the crossover
state with no voltage applied to the electrodes. This requires fabricating the coupler length to be equal to one transfer length, satis-
fying (34). The straight-through state is obtained by applying an
appropriate voltage to the electrodes to satisfy the phase mismatch
condition (35). In Figure 12 the optical power distribution in the
two guides is shown as a function of position for both switching
states. The results are shown for a coupler length equal to one transfer length $L/\ell = 1$.

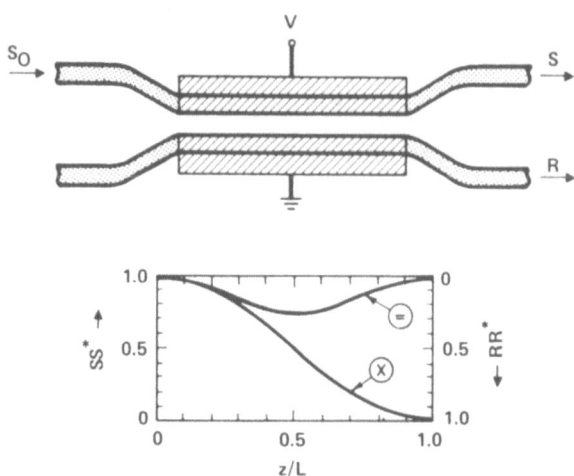

Fig. 12.   Top view of a directional coupler switch with uniform elec-
           trodes placed over the waveguides in the coupling region.
           Also shown are the intensity distributions along the coupler
           in crossover Ⓧ and straight-through ⊜ switching states.

The condition for the crossover and straight-through states can
be graphically represented in a switching diagram or so-called crossbar diagram shown in Figure 13.   The coordinates are the device length,
normalized by the transfer length, L/ℓ, and the normalized phase mismatch, $\Delta\beta L/\pi = 2\delta L/\pi$.   The requirements for a perfect crossover state
as specified by (34) are denoted by isolated Ⓧ points on the L/ℓ
axis.   The locus of straight-through states, denoted by ⊜, are
circles prescribed by (35).   For a one transfer length coupler ($L/\ell =$
1) the required phase mismatch to switch is shown by the dashed line
and is

$$\Delta\beta L = \sqrt{3}\ \pi\quad . \tag{36}$$

Because the electrode configuration is identical to that of the phase
modulator previously discussed, the results obtained for phase modulator performance apply also to the directional coupler switch, except
that the drive voltage and switching energy requirements are increased
by a factor of $\sqrt{3}$ and 3 respectively.

Several switches of this type have been demonstrated[10-14].   It
has proven difficult, however, to reproducibly build this directional
coupler switch with a good crossover state because of the necessity
to fabricate in order that $L = \ell$.   It is not possible to significantly
alter the value of ℓ by electro-optically changing N.   Furthermore,

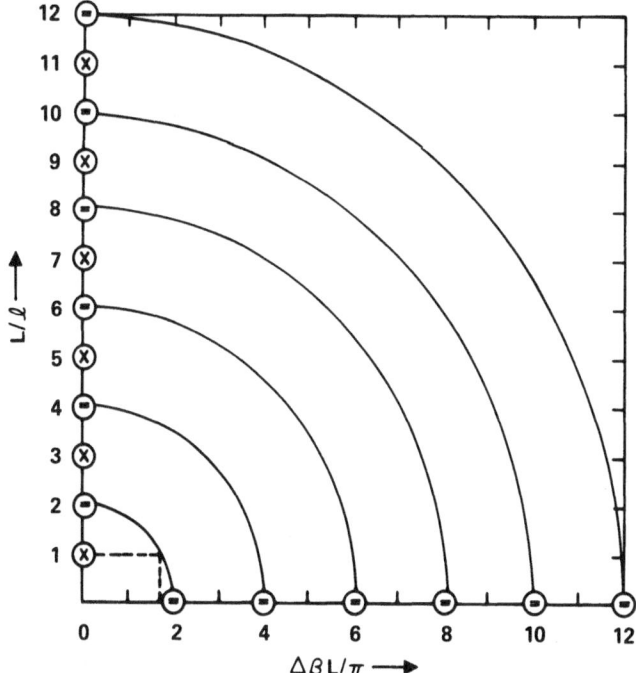

Fig. 13. The crossbar switching diagram for a switch with uniform
          electrodes. The isolated points marking the conditions
          required for complete energy crossover are marked by ⊗ ,
          and arcs indicating the conditions required for the straight-
          through state are marked by ⊜ .

as indicated in (32) and (33), for this directional coupler switch,
it is impossible to obtain the cross-state via a phase mismatch which
can be achieved electro-optically. As a result, the cross-state crosstalk is limited by fabrication tolerances. From (34) we note that to
achieve a cross-state with better than -20-dB crosstalk requires fabrication with $L/\ell = 1$ to within ∿ ±6percent.

     A directional coupler switch that allows electrical adjustment
to achieve both switch states with low cross-talk, thus eliminating
severe fabrication requirements, is illustrated in Figure 14. The
switch, which uses a technique called stepped Δβ reversal, requires
two sets of electrodes[15]. The electrodes divide the coupler into two
equal-length sections in which phase mismatch, electrical adjustment
of both switching states is obtained. Electrical adjustment of the
straight-through state can also be obtained as before by applying a
uniform phase mismatch along the entire length of the coupler.

     Electrical adjustability of the crossover state due to stepped
Δβ reversal can be understood by considering how the energy is exchanged between the guides. Consider the switch illustrated in

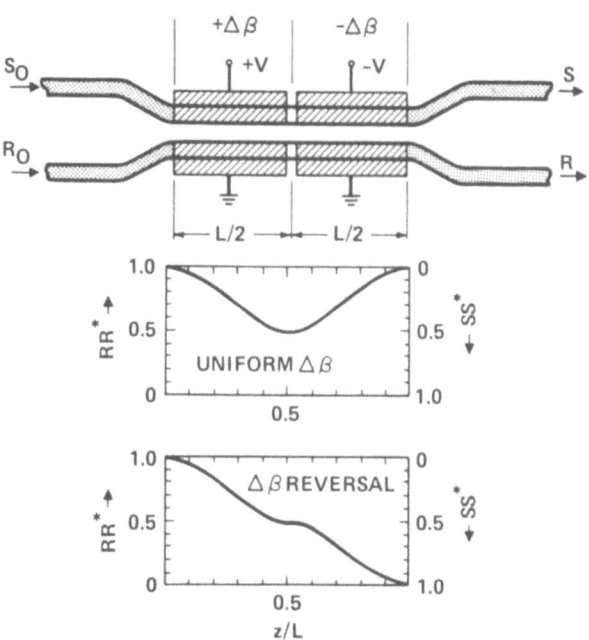

Fig. 14. Top view of stepped $\Delta\beta$ reversal switch requiring two sets
          of electrodes. Also shown are the intensity distributions
          along the coupler for a coupler of length $L/\ell = \sqrt{2}$.

Figure 14, where R and S waves travel in the top and bottom waveguides,
respectively. For simplicity, it is convenient to examine the special
case where the coupler length is $L/\ell = \sqrt{2}$. In this case the light
is equally divided between the R and S waves in the middle of the
coupler when the switch is in the straight-through state with uniform
phase mismatch along the entire coupler length. In Figure 14 this
case is illustrated when light is incident in the R waveguide. At
this midpoint along the coupler, where the amplitudes are equal, the
phases of the R and S waves, with respect to the electro-optically
induced slow and fast waveguides, determine that the light completely
returns to the R wave at the end of the coupler. When the switch is
used in the stepped $\Delta\beta$ configuration, the fast and the slow waveguides
are effectively interchanged at the midpoint of the coupler by virtue
of the change in sign of applied voltage. In this case light starting
in the R waveguides will exit in the S waveguide, as illustrated in
Figure 14. Similarly, for other L/ℓ ratios from 1 to 3, the crossover
states can be obtained by using stepped $\Delta\beta$ reversal. However, in
general the voltage required to obtain the crossover state is different
from the voltage required to obtain the straight-through state. A
more detailed analysis of these switches can be found in References
9 and 15.

In Figure 15 the switching diagram of a stepped $\Delta\beta$ reversal switch is shown. The solid lines are the locus of states obtainable with equal and opposite voltages, while the dashed lines indicate the state which can be obtained with uniform phase mismatch along the coupler. A coupler whose length is between one and two transfer lengths will provide the lowest switching voltage requirements for both switching states. The fact that both switching states can be obtained by electrical adjustment greatly reduces the precision required to fabricate the switches. Switches with reversed $\Delta\beta$ electrodes have been demonstrated in Ti-diffused LiNbO$_3$[15] and in GaAs waveguide structures[16,17] with reported crosstalk levels of -26dB. Recently, crosstalk as low as -35dB has been observed[18].

The drive voltage requirements for stepped $\Delta\beta$ reversal switches cannot be as simply expressed as those for the previous cases because there is a range of acceptable coupler lengths. However, both switching states can be obtained for

$$\Delta\beta L < 2\pi \tag{37}$$

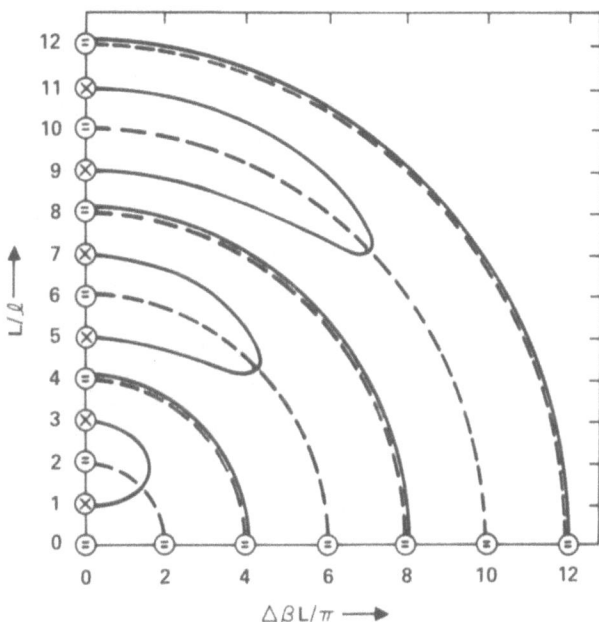

Fig. 15. The crossbar diagram of a stepped $\Delta\beta$ switch. The solid curves are switching states available with equal but opposite polarity voltages applied to the electrodes. The dashed curves are the straight-through states available with the same polarity voltages applied to the electrodes. The solid and dashed curves were slightly separated for clarity, but are actually identical.

From this condition it is clear that the drive voltage can be reduced
by increasing L. However, (37) is only valid for $\ell < L < 2\ell$; larger
L/ℓ ratios required larger phase mismatch (see Figure 15). Thus, an
increase in L requires a similar increase in ℓ, which can be achieved
most readily by increasing the separation between the waveguides, d.
This increase in d, however, results in a lower electric field for
the same applied voltage (24) which does not allow a voltage reduction
proportional to the increase in L. This difficulty can be overcome
by using multiple sections of alternating $\Delta\beta$ reversal. For N sections
of alternating $\Delta\beta$ reversal, (37) is valid for a coupler length between
N - 1 and N + 1 transfer lengths, and electrically adjustable crossover and straight-through states can be obtained with full advantage
of the increased length[9]. Figure 16 shows the switching diagram for
a switch with six sections of alternating $\Delta\beta$ reversal. The dashed
line gives the operating line of a switch 5.75 transfer lengths long.
As the voltage is increased, the switch first reaches the crossover
state, and as the voltage is further increased the straight-through
state is obtained. A detailed analysis of the properties of alternating $\Delta\beta$ switches can be found elsewhere[9].

The actual switching characteristics versus voltage for a six-
section switch formed in Ti-diffused $LiNbO_3$[19] is also shown in Figure
16. A Z-cut crystal was used with the TE waveguide mode, which utilizes the r_{13} electro-optic coefficient. The coupler was 10.5mm long,
corresponding to a 5.75 transfer length operating line for 0.6328μm
wavelength operation. The guides were 3μm wide, separated by 3μm.
The switch required 3V for the crossover state and an additional 3V
to switch to the straight-through state. This device can be used
either as a switch or as a modulator. As a digital modulator it would
be dc biased in the crossover state and switched to the straight-
through state with a modulating voltage. Alternatively, for analog
modulation, the switch would be biased in the linear region around
4.5V. As a digital modulator this device exhibited a 1 Gbit/sec modulation rate capability[20]. A single electrode pair modulator, which
was designed to be very small with 1μm wide waveguides and electrode
separation, has been reported with a 100 psec rise time[21].

A traveling wave directional coupler switch has also been demonstrated by Kubota and co-workers[5]. A traveling wave directional coupler switch is illustrated in Figure 17. The experimental device used
asymmetric coplanar strip lines rather than the coplanar strip lines
illustrated. The optical waveguides were fabricated by Ti diffusion
into a $LiNbO_3$ Z-cut plate. The waveguides were single mode at 1.32μm
wavelength. The characteristic impedance for aluminum electrodes
14μm wide, 16mm long, and 3μm thick with a 6μm gap was about 40 Ω.
The electrodes had a loss of approximately $1dB\text{-}cm^{-1}\text{-}GHz^{-1/2}$. The
switch was fabricated in the crossover state by simultaneously fabricating several slightly different directional couplers on the same
chip and selecting the one which was nearest to the cross-over state[22].

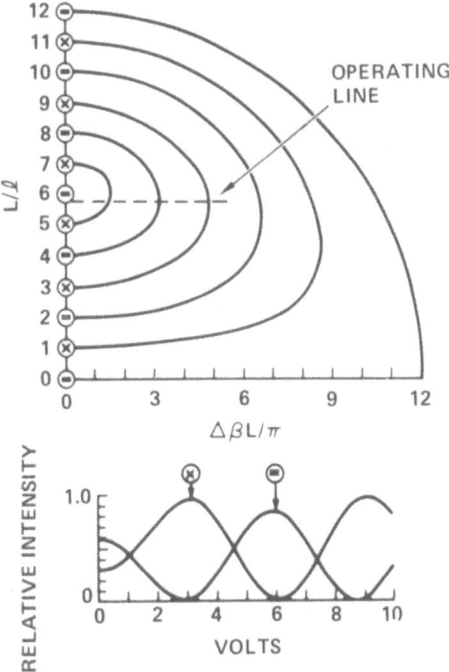

Fig. 16. The crossbar diagram of a six-section alternating $\Delta\beta$ switch.
The dashed line is the operating line of a switch 5.75 trans-
fer lengths long. The bottom curves are the output switching characteristics of a six-section switch as a function
of applied voltage.

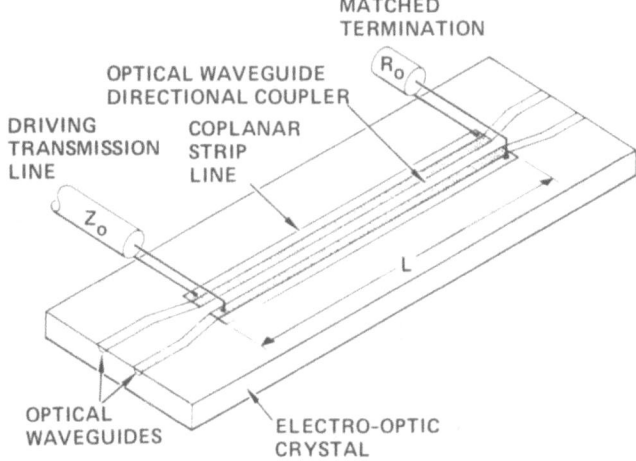

Fig. 17. Illustration of traveling wave directional coupler switch.

Extinction ratios as large as -17dB were obtained. The device required 4V to switch and had a rise time of less than 400 psec.

THREE PORT WAVEGUIDE JUNCTIONS

The three port optical waveguide junction is the basic element used to split and recombine light in an integrated optical interfero- meter. In this section a brief description will be provided of the three-waveguide directional coupler junction and the Y-junction.

Three coupled waveguides can be used to form a 3-port power splitting junction. In Figure 18, an implementation using three ident- ical waveguides is illustrated. All three waveguides have the same propagation constant β, and the two outer waveguides are spaced the same distance from the center waveguide. Thus, the coupling from the center waveguide to the outer waveguides is identical. The strength of coupling is characterized by the coupling coefficient $\kappa = \pi/2\ell$, where ℓ is the length required for complete transfer of energy between two identical waveguides with the same separation. Coupled mode analy- sis can be used to characterize the operation of three coupled wave- guides. A transfer matrix analysis has been performed[23] for the three coupled waveguides illustrated in Figure 19. The input amplitudes are $A_i(0)$, and the output amplitudes of a coupler of length L are $A_i(L)$. The coupled mode analysis yields the transfer matrix,

$$
\begin{bmatrix} A_2(L) \\ A_1(L) \\ A_3(L) \end{bmatrix} =
\begin{bmatrix}
\frac{1}{2}(1+\cos\sqrt{2}\kappa L) & \frac{i}{\sqrt{2}}\sin\sqrt{2}\kappa L & -\frac{1}{2}(1-\cos\sqrt{2}\kappa L) \\
\frac{i}{\sqrt{2}}\sin\sqrt{2}\kappa L & \cos\sqrt{2}\kappa L & \frac{i}{\sqrt{2}}\sin\sqrt{2}\kappa L \\
-\frac{1}{2}(1-\cos\sqrt{2}\kappa L) & \frac{i}{\sqrt{2}}\sin\sqrt{2}\kappa L & \frac{1}{2}(1+\cos\sqrt{2}\kappa L)
\end{bmatrix}
\begin{bmatrix} A_2(0) \\ A_1(0) \\ A_3(0) \end{bmatrix}
\tag{38}
$$

Examination of Equation (38) reveals that an input $A_i(0)$ in the center waveguide is split equally between output amplitudes $A_2(L)$ and $A_3(L)$ when $L = \ell/\sqrt{2}$. Under this condition a scattering matrix for the junc- tion illustrated in Figure 18 can be found. If a_i are the input ampli- tudes, and b_i are the output amplitudes, the scattering matrix becomes

$$
\begin{bmatrix} b_1 \\ b_2 \\ b_3 \end{bmatrix} =
\begin{bmatrix}
0 & \frac{i}{\sqrt{2}} & \frac{i}{\sqrt{2}} \\
\frac{i}{\sqrt{2}} & 0 & 0 \\
\frac{i}{\sqrt{2}} & 0 & 0
\end{bmatrix}
\begin{bmatrix} a_1 \\ a_2 \\ a_3 \end{bmatrix}
\tag{39}
$$

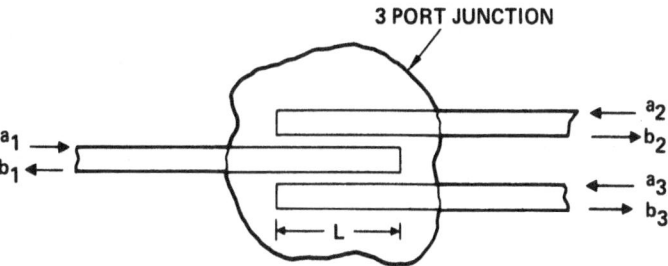

Fig. 18. Drawing of three coupled waveguides forming a power splitter.

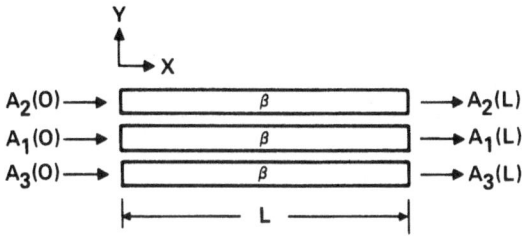

Fig. 19. Schematic of three coupled identical waveguides used to
calculate the transfer matrix of inputs $A_i(0)$ to outputs
$A_i(L)$, i = 1, 2, 3.

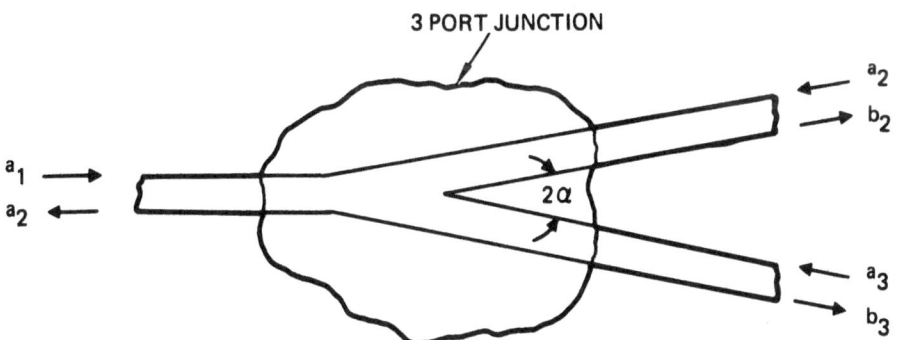

Fig. 20. Diagram of a waveguide Y junction forming power splitter.

Investigation of the scattering matrix reveals the important properties of a 3-port junction which are utilized in an interfero-metric modulator. First, an input in port 1 is equally split between ports 2 and 3 with no loss. However, the junction cannot be used to combine the inputs of port 2 and 3 with no loss unless the inputs have identical phase.

To construct the three waveguide directional coupler junction it is necessary to fabricate waveguides with identical propagation constants. It is also necessary to fabricate the waveguides with the exact coupling coefficient to effect equal and lossless coupling in a length L. However, if the exact coupling coefficient is not obtained the splitting ratio will still be unchanged; but excess loss would be incurred because light would still be in the center waveguide when it is terminated.

The Y-junction coupler is illustrated in Figure 20. When used as a power splitter, an input at port 1 is equally divided between ports 2 and 3. The symmetry of the junction assures equal splitting. The losses in the junction have been shown[24] to be a sensitive func-tion of the splitting angle 2α. A simple zero-order analysis of a Y-junction is presented to illustrate its operation and to deduce the loss sensitivity on 2α.

To analyze the Y-junction, it is convenient to model the junction as illustrated in Figure 21. The junction is started by a waveguide of width W, which then spreads to a uniform-waveguide of width 2W via an adibatic transition region. It is assumed that no losses occur during the transition region or during the joining of the transition region to the uniform waveguides of widths W and 2W. The waveguide fork with apex angle 2α is composed of two waveguides of width W which separate uniformly about the symmetry axis of the junction. Two sources of loss are apparent in this model. One is the mode mismatch loss which occurs between guided modes in the uniform guide of width 2W and fork waveguide modes. The other is loss associated with the directional changes between the wide waveguide and the fork arrange-ments.

The loss may be calculated approximately without too much diffi-culty. To do this, it is necessary to approximate the fundamental TE guided modes as follows:

1. Input guide of width 2W

$$E_x^{2W}(x) = \begin{cases} E_o^{2W} e^{j(\beta z - \omega t)} & -W \le x \le W \\ 0 & |x| \ge W \end{cases}$$

$$H_y^{2W}(x) = \frac{\beta}{\omega\mu} E_x^{2W}(x) \quad . \tag{40}$$

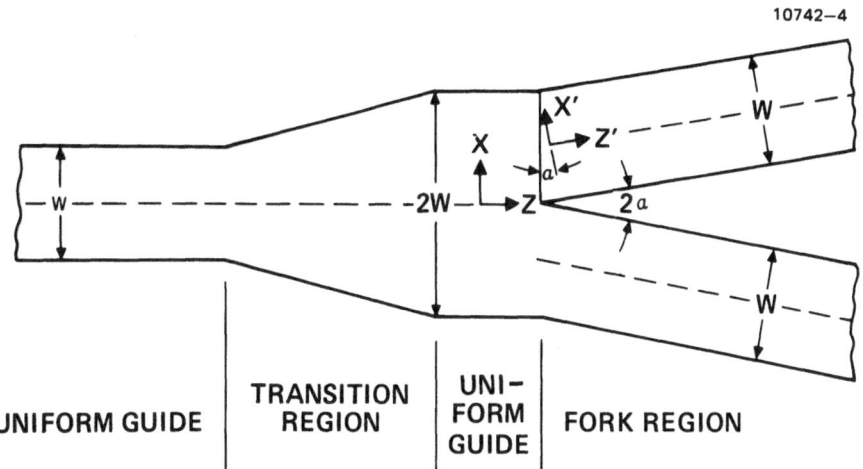

10742—4

TRANSITION | UNI-
UNIFORM GUIDE | REGION | FORM GUIDE | FORK REGION

Fig. 21. Idealized waveguide representation of a waveguide Y junction which allows approximate analysis of junction performance.

2.   In the fork guide of width W

$$E^W(x') = \begin{cases} E_o^W e^{j(\beta z' - \omega t)} & -\frac{W}{2} \leq x' \leq \frac{W}{2} \\ 0 & |x'| \geq \frac{W}{2} \end{cases}$$

$$H_y^W(x') = \frac{\beta}{\omega\mu} E_o^W(x') \qquad . \tag{41}$$

We have assumed identical propagation constants in both guides.
The power orthogonality/normalization condition

$$P = 2 \int \vec{E}_i(\xi) \times H_j(\xi) \, d\xi = \delta_{ij} \tag{42}$$

for the ith and j^{th} modes gives the wave amplitudes,

$$E_o^{2W} = \sqrt{\frac{\mu\omega}{4\beta W}} \tag{43}$$

and

$$E_o^W = \sqrt{\frac{\mu\omega}{2\beta W}} \qquad . \tag{44}$$

If it is assumed that the field distribution E(x') incident on the
forked waveguide can be expanded by the complete set of guided
modes $E_v^W(x')$ and radiation modes $E_p^W(x')$, we can write

$$E(x') = \sum_v a_v E^W(x') + \int_p q_p E_p^W(x') \, dp \quad . \tag{45}$$

Using the orthogonality condition Equation (42), we can express the amplitude coefficient of an excited fundamental mode as

$$a_o = 2 \int E(x') \times H_o^W(x')^* \, dx' \quad . \tag{46}$$

Since we obtain a 3 dB power split, no coupling between waveguides occurs after the fork. Letting

$$E(x') = E_x^{2W}(x') e^{j\beta x' \alpha} \quad , \tag{47}$$

which incorporates the effects of the direction changes through the wavefront phase shift $e^{j\beta x' \alpha}$, we have

$$a_o = 2 \int_{-\frac{W}{2}}^{\frac{W}{2}} \sqrt{\frac{\mu\omega}{2\beta W} \frac{\beta}{\omega\mu}} \sqrt{\frac{\mu\omega}{4\beta W}} e^{j\beta x' \alpha} \, dx' \quad ; \tag{48}$$

and performing the integration, we have

$$a_o = \frac{1}{\sqrt{2}} \frac{\sin\left(\frac{\beta W \alpha}{2}\right)}{\left(\frac{\beta W \alpha}{2}\right)} \quad . \tag{49}$$

The $1/\sqrt{2}$ factor represents the mismatch loss, and the $\text{sinc}(\beta W \alpha/2)$ represents the directional change loss. By symmetry, the same result is obtained for the lower waveguide of the fork. Similarly, if light enters the junction from one of the fork waveguides, the transmission to the 2W wide waveguide will be identical to that specified by Equation (49) by reciprocity considerations. Thus, the scattering matrix of the Y-junction can be written,

$$\begin{bmatrix} b_1 \\ b_2 \\ b_3 \end{bmatrix} = \begin{bmatrix} 0 & a_o & a_o \\ a_o & 0 & 0 \\ a_o & 0 & 0 \end{bmatrix} \begin{bmatrix} a_1 \\ a_2 \\ a_3 \end{bmatrix} \quad . \tag{50}$$

In the limit of small angles, $a_0 \rightarrow 1/\sqrt{2}$ and the scattering matrix reduces to the previous form, Equation (39). Thus the Y-junction and the directional coupler junctions are formally identical in their operating characteristics. It should be noted that the directional loss term in Equation (49) should also be included in the directional coupler analysis. This follows because the waveguides of the directional coupler must be bent to separate the waveguides in a practical implementation.

In a Y-junction fabricated as illustrated in Figure 20 there exists an optimum fork angle which minimizes the direction change loss. In the region where the input waveguide is expanding, the wavefront of the mode must be curved to account for the lateral spread in energy. When the waveguide forks are normal to the wavefronts in the expanding waveguide, the direction change loss is minimized. Anderson has provided a complete analysis of Y-junctions[24] which properly treats mode mismatching and wavefront curvature.

INTERFEROMETRIC MODULATORS

Modulators based on Mach-Zehnder interferometers have been demonstrated using both three-coupled-waveguide power splitters and Y-junctions. In Figure 22 an illustration of a Mach-Zehnder interferometer modulator using a traveling wave interaction is presented. With no drive voltage applied to the electrodes, the interferometer is balanced; that is, the optical path length through both arms of the interferometer is identical. Thus, neglecting losses, light entering the interferometer is split in half by the first Y-junction and recombined at the second Y-junction to its original intensity. When an electrical drive signal is placed on the modulator so that light traversing the two arms of the interferometer becomes π out of phase, there is destructive interference at the output Y-junction. Light is scattered into the substrate and the guided wave transmission is reduced to zero. If V_π is the applied voltage required to provide a π-phase shift in the interferometer, it can simply be shown that the transmitted intensity of the modulator is

$$I(V) = I_0 \cos^2\left[\frac{\pi}{2} \, (V/V_\pi)\right] \quad , \qquad (51)$$

where I_0 is the incident intensity and V is the applied voltage.

In configuration of Figure 22 the interferometric modulator is suitable for use with digital signals, which simply require transmitted light to be turned on or off. The modulator can also be used for analog modulation if a dc bias voltage is applied to optically bias the modulator at the $\pi/4$ point. In this condition the intensity through the modulator is reduced to one-half. It is also the condition required for maximum linearity between the electrical drive

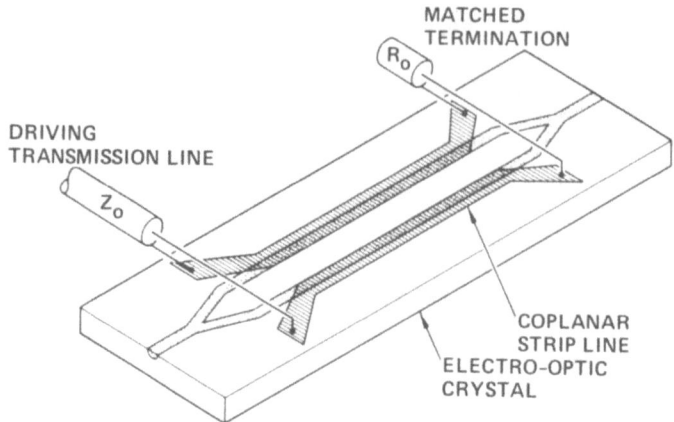

Fig. 22. Illustration of a Mach-Zehnder interferometer traveling
          wave modulator.

signal and the transmitted optical signal. In a practical device it
may be preferred to build the optical bias into the waveguides by
having waveguides of different width. This would eliminate the
requirement for a dc electrical bias. It has been found experimentally[25] that optical bias drift problems can occur with a dc electrical
bias. Even a slight drift in the optical will degrade the linearity
of an analog modulator.

Monokata[26] and co-workers first reported a lumped element interferometric amplitude modulator using three coupled waveguides to
perform the power splitting and recombination. The device was constructed in Ti diffused $LiNbO_3$ and had V_π = 3.5V at 0.633μm. Recently[27] a traveling wave interferometric modulator using Y-junctions has
been reported with a $1\sqrt{2}$ bandwidth of 11.2GHz. The power per unit
bandwidth figure of merit of this device is 11.6μW/MHz.

REFERENCES

1. H. Kogelnik, "An introduction to integrated optics", IEEE Trans
        Microwave Theory Tech. MTT-23:2-16, Jan. 1975.
2. R. V. Schmidt and Rod. C. Alferness, "Directional coupler switches,
        modulators and filters using alternating Δβ techniques, "IEEE
        Trans. Circuits and Systems CAS-26:1099-1108, Dec. 1979.
3. K. C. Gupta, Ramesh Garg, and I. J. Bahl, "Microstrip Lines and
        Slotlines", Artech House, Inc., Dednam, Massachusetts, 1979.
4. M. Izutsu, et al., "10 GHz bandwidth travelling-wave $LiNbO_3$
        optical waveguide modulator", IEEE J. Quant. Electron. QE-14:
        394-395, June 1978.

5. K. Kubota, et al., "Traveling wave optical modulator using a directional coupler LiNbO$_3$ waveguide", IEEE j. Quant. Electron. QE-16:754-760, July 1980.

6. M. Izutsu, et al., "Broad-band traveling-wave modulator using a LiNbO$_3$ optical waveguide", IEEE J. Quant. Electron. QE-13:287-290, April 1977.

7. R. C. Alferness, R. V. Schmidt and E. M. Turner, "Characteristics of Ti-diffused LiNbO$_3$ optical directional couplers", Appl. Opt. 18, Nov. 1979.

8. S. E. Miller, "Coupled-wave theory and waveguide applications", Bell Syst. Tech. J. 33:661-719, May, 1954.

9. H. Kogelnik and R. V. Schmidt, "Switched directional couplers with alternating $\Delta\beta$, IEEE J. Quant. Electron. QE-12:396-401, July, 1976.

10. K. Tada and K. Mirose, "A new light modulator using perturbation of synchronism between two coupled guides", Appl. Phys. Lett. 25:561-562, Nov. 1974.

11. M. Papuchon, et al., "Electrically switched optical directional coupler: COBRA, "Appl. Phys. Lett. 27:289-291, Sept. 1975.

12. J. C. Campbell et al., "GaAs electro-optic directional coupler switch", Appl. Phys. Lett. 27:203-205, Aug. 1975.

13. F. J. Leonberger, J. P. Donnelly and C. O. Bozler, "GaAs p$^+$n$^-$n$^+$ directional coupler switch", Appl. Phys. Lett. 29:652-654, 1976.

14. H. Kawaguchi, "Directional-coupler switch with Schottky barriers", Electron. Lett. 14:387-388, June 1978.

15. R. V. Schmidt and H. Kogelnik, "Electro-optically switched coupler with stepped $\Delta\beta$ reversal using Ti-diffused LiNbO$_3$ waveguides", Appl. Phys. Lett. 28:503-506, May 1976.

16. F. J. Leonberger and C. O. Bozler, "GaAs directional coupler switch with stepped $\Delta\beta$ reversal", Appl. Phys.Lett. 27:202-205, 1977.

17. J. C. Shelton, F. K. Reinhart and R. A. Logan, "GaAs-Al$_x$Ga$_{1-x}$ As rib waveguide switches with MOS electro-optic control for monolithic integrated optics", Appl. Opt. 17:2548-2555, Aug. 1978.

18. M. Papuchon, private communication.

19. R. V. Schmidt and P. S. Cross, "Efficient optical waveguide switch/ amplitude modulator, "Opt. Lett. 2:45-47, Feb. 1978.

20. P. S. Cross and R. V. Schmidt, "A 1 G bit/second integrated optical modulator", IEEE J. Quant. Electron. QE-15: 1415-1418, Dec. 1979.

21. R. C. Alferness, N. P. Economou and L. L. Buhl, "Fast compact optical waveguide switch modulator", Appl. Phys. Lett. 38:214-217, Feb.15th, 1981.

22. Juichi Noda, private communication.

23. H. Ogiwara, "Optical waveguide 3x3 switch: theory of tuning and control", Appl. Opt. 18:510-515, Feb. 15th, 1979.

24. Iain Anderson, "Transmission performance of Y-junctions in planar dielectric waveguides", IEEE J. Microwave, Optics and Acoust. 2:7-12, Jan. 1978.

25. R. V. Schmidt, P. S. Cross and A. M. Glass, "Optically induced crosstalk in LiNbO$_3$ waveguide switches", J. Appl. Phys. 51: 90-93, Jan. 1980.

26. M. Minokata, T. Yamada and S. Uehara, "Optical intensity modulation using a pair of gate couplers and electro-optic phase shifters, Proc. 1977 IOOC, pp. 145-148, July 18-2)th, 1977, Tokyo, Japan.

27. T. Sueta and M. Izutsu, "High-speed guided-wave optical components", Proc. 1981 IOOC, p. 82, April 27-29, 1981, San Francisco, California.

PLANAR ELECTROOPTIC AND ACOUSTOOPTIC BRAGG-DEFLECTORS

F. Auracher

Research Laboratories of Siemens AG
D-8000 Muenchen 83

INTRODUCTION

Planar i.e. waveguide Bragg-deflectors are becoming increasingly important for applications of low laser power, eg. in signal processing devices. There are several advantages of planar devices compared to bulk devices. Figure 1 compares the two versions of a light deflector or modulator. In the bulk version the laser beam is focused into the interaction volume. For a given modulating field strength, which is limited by dielectric break down, nonlinear or thermal effects, a certain interaction length L is required to achieve a given diffraction efficiency or modulation depth. Due to diffraction of the freely propagating beam in a bulk device (Figure 1a) the beam spreads to a diameter $D = (\lambda L/\pi)^{1/2}$ yielding a minimum interaction volume $V = D^2 L \approx \lambda L^2/\pi$. If we confine instead the light beam in a slab waveguide of depth $d \approx \lambda$ *, we only have diffraction spreading in the lateral dimension of the beam and the necessary interaction volume is reduced by a factor $d/D \approx \lambda_0/D$. For the same interaction medium the required drive power is consequently reduced by a factor of $(\lambda \pi/L)^{1/2}$, ie. typically several orders of magnitude.

Of course, one has to make sure that the penetration depth of the modulating field matches the waveguide depth. In the case of the electrooptic Bragg-deflector this is achieved by the proper choice of electrode finger spacing; in the acoustooptic device we use a surface acoustic wave (SAW) of suitable frequency, so that the acoustic penetration depth matches the depth of the optical wave-guide. In addition to the low drive power there are further advantages of waveguide devices, namely small size, easy and reproducible

---

*λ_0 is the light-wavelength in vacuo, λ_0/n in the medium.

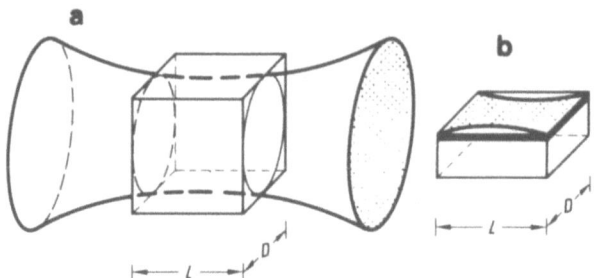

Fig. 1.   Comparison of bulk device (a) and planar-waveguide
          device (b).

fabrication of complicated, high resolution structures for transducer electrodes and the possibility of integration. Disadvantages
of planar devices are the difficult coupling of the laser beam into
the waveguide, the high optical intensity due to the confinement of
the light in the waveguide which can lead to nonlinear effects or
optical damage, the presumably poor beam quality due to nonuniform
coupling efficiency or waveguide properties and the limited choice
of materials due to the necessity to fabricate an optical waveguide
in it.

1. Bragg-effect

When a planar light wave encounters a refractive index grating
with amplitude Δn and periodicity Λ:

$$n(z) = n_o + \Delta n \cos \frac{2\pi z}{\Lambda} \qquad (1)$$

it is generally diffracted into several orders. If the thickness L
of the grating is large enough, however, these diffracted beams
interfere and cancel each other. There is then only one angle of
incidence - the so called Bragg-angle - for which the diffracted
light beams superimpose to form a single Bragg-deflected beam.
The Bragg-angle is given by

$$\sin\theta_B = \frac{\lambda_o}{2n\Lambda} \qquad (2)$$

The required value of the thickness parameter Q for Bragg reflection
is

$$Q = \frac{2\pi\lambda_o L}{n\Lambda^2} \geq 4\pi . \qquad (3)$$

The deflection efficiency η defined as the ratio of the intensity of the deflected beam to that of the undeflected beam is given by

$$\eta = \frac{I_1}{I_o} = \frac{1}{1 + (\frac{Kn}{k\Delta n}\Delta\theta)^2} \sin^2 (qL) \qquad (4)$$

with $q = \left[q_o^2 + (\frac{K\Delta\theta}{2})^2 \right]^{1/2}$

and $q_o = \frac{\pi}{\lambda_o \cos \theta_B} \Delta n$.

$\overline{K}$ and $\overline{k}$ are the wave vectors of the grating ($K = \frac{2\pi}{\Lambda}$) and the light wave, respectively and $\Delta\theta$ is the deviation from the Bragg angle.

2. Optical waveguide

      In planar Bragg-deflectors the light is guided in slab wave-guides. These waveguides are usually fabricated by a diffusion process whereby the refractive index is increased near the surface of the substrate. Presently LiNbO$_3$ is the most favored material for wide band planar Bragg-deflectors because low-loss waveguides can be fabricated easily by indiffusion of metals (mostly Titanium) or by outdiffusion of Li$_2$O. Moreover, this material shows a strong electrooptic and piezoelectric effect and very low sound attenuation even in the GHz frequency range.

      The diffused waveguides are typically of a Gaussian profile with nearly parabolic refractive index profile near the surface. The guided light modes have then depth profiles of the E field resembling Hermite-Gauss functions. Specifically, a single moded waveguide has a normalized intensity profile given by

$$u^2_{opt}(y) \approx \begin{cases} 0 & y < 0 \\ \\ \frac{4}{\sqrt{\pi} \, \delta^2_H} \; y^2 \, \exp \left[- (\frac{y}{\delta H})^2 \right] & y > 0 \end{cases} \qquad (5)$$

A measured depth profile together with the best fitting theoretical curve for a Li$_2$O-outdiffused waveguide in LiNbO$_3$ is shown in Figure 2. The optical depth profile $u^2_{opt}(y)$ has to be known for calculations of the deflection efficiency and, in the case of the acoustooptic deflector, also its frequency dependence.

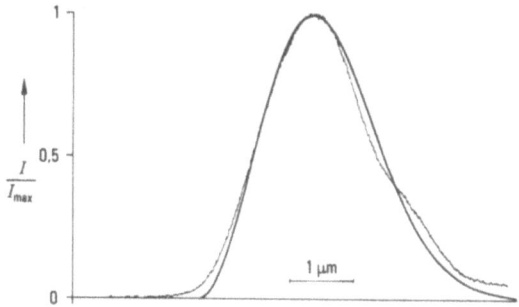

Fig. 2. Measured depth profile of optical intensity in
out-diffused LiNbO$_3$ waveguide at λ_0 = 0.633μm
(thin line) and best fit to theoretical model
(heavy line).

3. Electrooptic Bragg deflector

Figure 3 shows a line drawing of a planar electrooptic Bragg-
deflector. The device consists of an electrooptic crystal
(usually a y-cut LiNbO$_3$ crystal) with a thin slab waveguide at the
surface. Interdigitated electrodes are applied to the surface with
their fingers normal to the c-axis of the crystal. Polarized light
is coupled into the waveguide via a prism coupler.

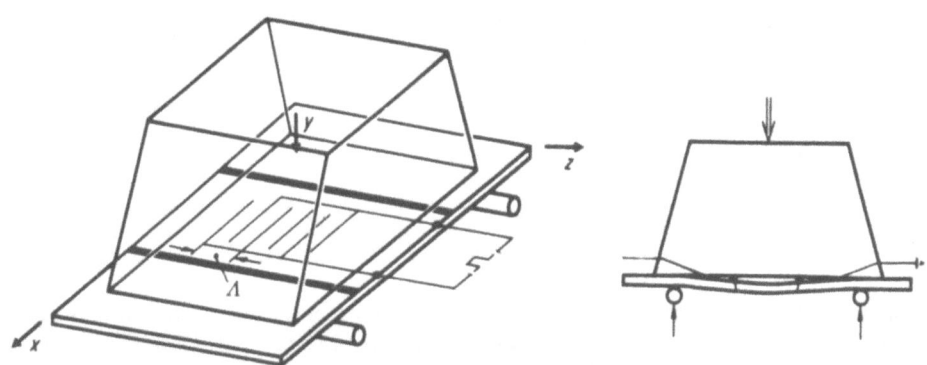

Fig. 3. Schematic arrangement of planar electrooptic
Bragg-deflector.

3.1 <u>Theory of operation</u>. When a voltage V is applied to the
electrodes a refractive index grating with periodicity Λ determined
by the finger separation is induced due to the linear electrooptic
effect. The amplitude Δn of this grating is given by[1]

$$\Delta n = -n^3 r \gamma(\Lambda) \frac{V}{1.854\Lambda} \quad , \tag{6}$$

where n is the refractive index of the waveguide and r the appropriate electrooptic coefficient. The overlap integral

$$\gamma(\Lambda) = \int_0^\infty u_{opt}^2(y) \, \exp \left(- \frac{2\pi}{\Lambda} \sqrt{\frac{\epsilon_{r,11}}{\epsilon_{r,1}}} \, y \right) \, dy \tag{7}$$

takes into account the spatial overlap of the normalized depth
profile of the optical intensity with the depth profile of the
induced refractive index grating[1]; $\epsilon_{r,11}$ and $\epsilon_{r,1}$ are the relative
dielectric constants of the waveguide parallel and perpendicular
to the c-axis, respectively. The required voltage for maximum
deflection efficiency is then given by[1]

$$V_o = \frac{1.854\lambda_o \Lambda}{2 \, n^3 r \gamma(\Lambda)} \, \frac{1}{L} \, . \tag{8}$$

The drive voltage can be minimized for a given interaction length
L by choosing the minimum value of $\frac{\Lambda}{\gamma(\Lambda)}$.

3. 2. <u>Experimental results</u>. Several devices have been reported
in the literature with low drive voltages and also high bandwidths
[1-4]. We report here on recent very efficient devices fabricated in
LiNbO3[1]. The deflectors use y-cut LiNbO3 crystals as substrates.
The waveguides were produced by outdiffusion of Li_2O for 0.5h at
980°C in a flowing O_2 atmosphere. The waveguides support a single
TE mode at $\lambda_o = 0.633\mu m$. The electrodes were produced by evaporating $0.2\mu m$ aluminum onto the photoresist pattern and using the
lift-off technique. The number of fingers was chosen to give an
electrode structure 0.5mm in width so that only slight focusing of
the laser beam is required. Various electrode lengths L and finger
spacings were tried in order to determine the optimum design parameters, which are presented in Table 1.

A single LiNbO3 prism coupler with the c-axis normal to its
base is used for both input and output coupling. The prism angle
was chosen such that the input and output beams run collinear. For
TE polarization the appropriate refractive indices are n_o in the
prism and n_e in the waveguide. Because $n_o > n_e$, input coupling is

Table 1. Design parameters and performance data of
electrooptic waveguide deflectors.

	Symbol	Unit	Deflector No. 1	2	3
finger gap/width	a	μm	3	4	6
length of fingers	L	mm	1,6	3	6,6
no. of finger pairs	N		42	31	21
capacitance	C	pF	42	58	88
cut-off frequency (theory)	f_c	MHz	151	110	72
drive voltage for maximum deflection	V_o	V	8,3	4,1	2,6
deflection angle	$2\theta_B$	deg.	3,03	2,26	1,51
deflection efficiency	η	%	96	93	94

possible. Two spacers in the form of aluminum strips evaporation-
deposited on both sides of the electrode structure define the gap
between the prism and the substrate. The prism presses two steel
rolls, 12mm apart, supportively against the substrate (Figure 3).
The bending of the flexible substrate results in a tapered gap
responsible for high input coupling efficiency. With this coupling
scheme and antireflection coated prism entrance and exit faces a
total optical insertion loss of only 1.5dB was obtained. The
deflection efficiency of all devices was about 95% and the cross
talk in the deflected beam position below -30dB. Figure 4 shows
the measured intensities of the deflected and undeflected beams as
a function of the drive voltage for deflector no.2. The expected
sinusoidal energy exchange and high deflection efficiency are
clearly documented.

Taking advantage of planar technology, we integrated four
electrode structures on a single substrate, thus permitting individual control of four beam positions. The neighboring electrode
structures were tilted relative to each other by 0.1°. The finger
spacing was chosen to satisfy the Bragg condition; the widths of
the electrode structures were chosen to yield equal intensities in
all deflected spots when a Gaussian beam of 1.2mm width is coupled
into the deflector. Figure 5 shows a far-field scan with all four
beams turned on. The residual intensity in the undeflected beam
is about 15%. The larger widths of the two center beams are due to
the narrower electrode structures; the intensities of the individual
beams (represented by the areas under the peaks in the scan) are

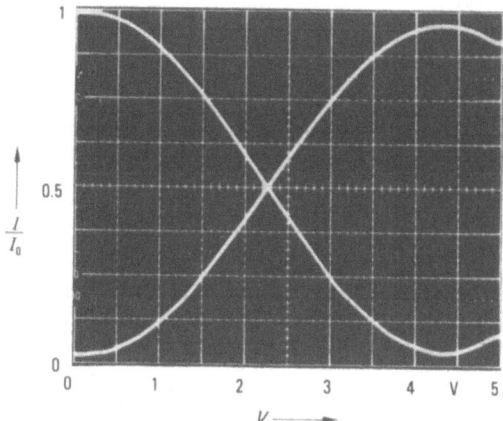

Fig. 4.   Measured deflected and undeflected beam intensity as a
function of applied voltage for electrooptic Bragg-
deflector according to[1].

however almost identical, which is as desired. On this device we
measured the near field intensity profile of the guided light finding a close agreement with the assumed profile (5) for δ_H = 1.55μm.
Using this value we calculated the values V_0L from (8). The obtained
data are shown as dots in Figure 6 together with the theoretical
curve. Excellent agreement between theory and experiment is evident.
The above example of an electrooptic Bragg deflector shows all
the mentioned advantages of waveguide devices. The integration
of several (four) different electrode structures does not add a
single fabrication step yet increases the performance of the device
with respect to the number of deflected beam positions considerably.

There are, however, many applications where one needs a continuously deflected beam or a high number of deflected beam positions; this is the realm of acoustooptic deflectors.

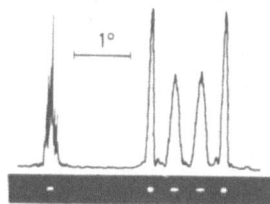

Fig. 5.   Scan of far-field of multiple-beam Bragg-deflector
when all four electrodes are energized as
reported in[1].

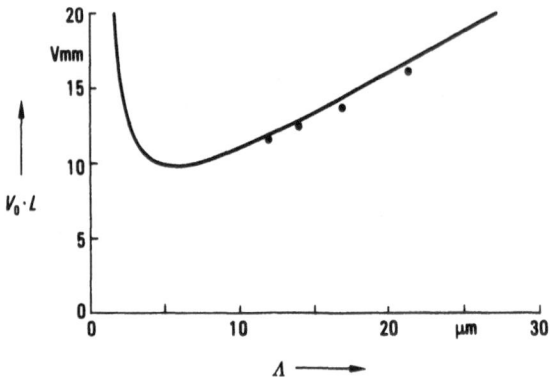

Fig. 6.   Calculated dependence of $V_o L$ on the grating period
          and experimental data for a planar electrooptic
          Bragg-deflector at λ_o = 0.633μm according to[1].

4. Acoustooptic Bragg-deflector

There is presently a strong interest in high bandwidth
acoustooptic Bragg cells (Bragg-deflectors) for signal processing
devices eg. spectrum analyzers or correlators for microwave signals
[5-7]. The basic principle of a planar acoustooptic Bragg-cell is
shown in Figure 7. In its most common form a LiNbO$_3$ substrate is
used again. Usually a y-cut crystal with a Ti-diffused waveguide
is used with an interdigital finger type transducer for the generation of a SAW. By applying an rf signal of frequency f to the
transducer, a SAW of wavelength $\Lambda = \frac{v_s}{f}$ is generated, propagating in
the z-direction with velocity v_s. The periodic distortions of the
medium cause the desired refractive index grating. The penetration-
depth of this grating is of the order of the acoustic wavelength.
Bragg deflection occurs again for the right incidence angle Θ_B.
In contrast to the electrooptic Bragg-deflector, we can vary the
grating periodicity and therefore also the deflection angle.

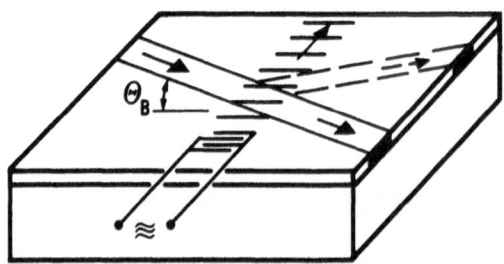

Fig. 7.   Schematic arrangement of planar acoustooptic
          Bragg-deflector.

4.1 <u>Theory of acoustooptic interaction</u>. Let us first, for simplicity, assume a bulk deflector with an isotropic interaction
medium with an elastooptic constant p, refractive index n and density ϕ. The acoustooptically induced refractive index grating has
an amplitude

$$\Delta n = - \frac{1}{2} n^3 \, p \, S \, , \tag{9}$$

where S is the strain amplitude. The corresponding intensity of
the sound is then

$$I_S = \frac{1}{2} \delta \, v_S^3 \, S^2 \, . \tag{10}$$

For complete deflection an intensity

$$I_{SO} = \frac{1}{2M_2} \left(\frac{\lambda_0}{L} \right)^2 \tag{11}$$

is required where $M_2 = \dfrac{n^6 p^2}{v_S^3}$ is the well known acoustooptic figure

of merit. In anisotropic materials such as, LiNbO$_3$ we have to
replace p and S by the appropriate tensor elements p_{ijkl} and S_{kl},
of course. In the planar device we have to include in addition an
overlap integral taking into account the normalized depth profiles
$u^2_{opt}(y)$ of the guided light intensity and $u_n(y/\Lambda)$ of the induced
refractive index grating. The overlap integral is again given by

$$\gamma(\Lambda) = \int_0^\infty u^2_{opt}(y) \, u_n(y/\Lambda) dy \, . \tag{12}$$

As the depth profile $u_n(y/\Lambda)$ scales with Λ the overlap integral
and therefore the acoustooptic interaction in planar devices is
frequency dependent. This frequency dependence can be compensated,
however, by a proper transducer design.

     In the case of LiNbO$_3$ which also shows a strong piezoelectric
and electrooptic effect, the refractive index grating is actually
caused by two effects: The elastooptic effect causes a grating of
amplitude $\Delta n_1 = - \frac{1}{2} n^3 pS$ with a depth profile $u_s(y/\Lambda)$. Due to the
piezoelectric effect a space charge wave accompanies the elastic
wave. The associated electric field E_p creates via the electrooptic
effect an additional index grating of amplitude $\Delta n_2 = - \frac{1}{2} n^3 r E_p$
and depth profile $u_p(y/\Lambda)$; r is the appropriate electrooptic
coefficient. Thus the resulting refractive index grating has an
amplitude

$$\Delta n = (pS + rE_p)(\frac{1}{2}n^3) \tag{13}$$

and a depth profile

$$u_n(y/\Lambda) = \frac{pS}{pS+rE_p} u_s(y/\Lambda) + \frac{rE_p}{pS+rE_p} u_p(y/\Lambda) .$$ (14)

Of course, due to the anisotropic properties of the crystal, the appropriate tensor elements p_{ijkl}, S_{kl} and r_{ijk} have to be used. A detailed analysis requires numerical calculation. For the standard crystal orientation some results are given in[5] and[8-10].

4.2 Number of resolvable spots. The main advantage of the acousto-optic Bragg-deflector over the electrooptic one is the high number of resolvable spots. This quantity is defined as $N = \Delta\Phi/\delta\Phi$, where $\Delta\Phi$ is the change in deflection angle and $\delta\Phi$ the beam spreading angle due to diffraction. One simply obtains

$$N = \frac{\tau_s \Delta f}{\alpha} ,$$ (15)

where τ_s is the acoustic transit time through the optical aperture, Δf the bandwidth and $\alpha \approx 1$ a correction factor taking into account the influence of the beam shape on $\delta\Phi$. As the usable optical aperture is limited by the crystal size, the acoustic attenuation or by the width of the optical beam that can be coupled uniformly into the waveguide a high number of resolvable spots always requires a large bandwidth Δf.

4.3 Bandwidth of Bragg-deflector. The frequency response of the planar Bragg-deflector is governed by the frequency response of the transducer, that of the Bragg-interaction itself and that of the overlap integral. The frequency response of the transducer can be found from its equivalent circuit which is simply the transducer capacitance

$$C_T = N C_s$$

in parallel with the frequency dependent radiation conductance

$$G_a(f) = G_o \frac{\sin^2 x}{x^2} ,$$ (16)

where $x = N\pi (f-f_o)/f_o$ and $G_o = 8k^2 C_s f_o N^2$. In these expressions N is the number of finger pairs, C_s the capacitance of a single finger pair and k^2 the coupling constant. Clearly the radiated acoustic power increases with N^2 whereas the relative bandwidth decreases with $1/N$. To obtain high bandwidth transducers one has to choose a small value of N causing a high insertion loss of the transducer. Increasing the length of the fingers raises the small signal diffraction efficiency proportional to L^2. One cannot, however, increase L arbitrarily, because C_T increases with L and it becomes difficult to match the signal source to the transducer. The standard design of wide band transducers has to be altered for Bragg-cells because of the frequency dependence of the Bragg interaction itself. Strictly speaking the Bragg condition is only satisfied for a single frequency. Deviation from that frequency causes a roll-off in deflection efficiency because the incidence angle θ deviates then from the Bragg angle θ_B. The roll-off can be calculated from (4). The resulting relative bandwidth of the Bragg interaction is given by[5]

$$\left(\frac{\Delta f}{f_o}\right)_{Bragg} \approx \frac{3.6\pi}{Q_o} \, , \qquad (17)$$

where Q_o is the Q-parameter at the center frequency f_o. Therefore, for efficient transducers with large L-(large Q-) values one has to find means to overcome this limitation. Several designs have been suggested and successfully implemented. All of them use the principle of beam steerings, ie. the direction of the sound wave is varied with frequency in such a way that the Bragg condition is always satisfied. The most elegant ways to achieve this are the use of multiple tilted transducers[11] shown in Figure 8a and of a single chirped transducer with tilted fingers[12,13] shown in Figure 8b. Both methods clearly show the advantage of using a planar structure because such complicated transducer structures can be fabricated easily and reproducibly in a single planar process. Moreover, the overall frequency response can be corrected by proper weighting and phasing of the individual transducers or by apodizing the finger overlaps of the chirped transducer. Figure 8c clearly demonstrates the variable overlap w of the fingers. With the multiple transducer design a 3 dB-bandwidth of 680 MHz has been achieved using four transducers[5,13]. A bandwidth of over 400 MHz has been reported for the chirped transducer[12,14]. Bandwidths up to 1 GHz appear possible with these designs. The performance of various reported planar Bragg cells is summarized in Table 2.

5. Comparison of electrooptic and acoustooptic Bragg-deflectors

Having looked at both types of deflectors it is interesting to look at their relative merits. Both deflectors are limited to low laser power - typically 1mW in the visible wavelength range - due to the high light intensity in the waveguides. Both deflector types exhibit fast response ($\gtrsim$ 500 MHz bandwidth). The deflection efficiency of the electrooptic deflector is very high (>90%); acoustooptic deflectors can achieve high deflection efficiency ($\approx$ 50%) for moderate bandwidths, this is limited by available drive powers. The most interesting high bandwidth ($\gtrsim$ 500 MHz) acoustooptic devices are presently limited to a few percent deflection efficiency, however.

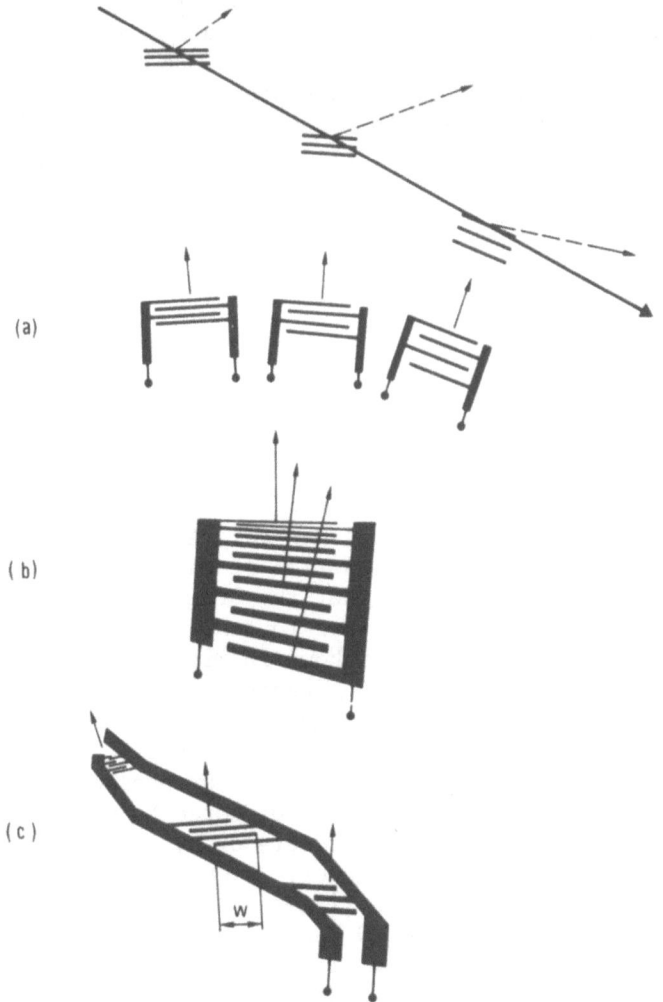

Fig. 8. Various transducer structures for wide-band planar Bragg-
         cells. (a) multiple tilted transducer, (b) simple tilted
         chirp transducer, (c) advanced tilted chirp transducer.

Table 2. Performance of planar Bragg-cells.

Bandwidth	Center frequency	Diffraction efficiency	Ref.
425 MHz	600 MHz	{ 5% for λ_o = 0.633μm 1,6% for λ_o = 0.83 μm }	6
680 MHz	650 MHz	1% for λ_o = 0.633μm	5
470 MHz	600 MHz	8% for λ_o = 0.633μm	14

The most distinct difference is the high number of resolvable spots (several hundred) of the acoustooptic deflector whereas the electro-optic version has only two in its basic design. The acoustooptic deflector, however, requires a carrier frequency whereas the electro-optic deflector operates in the base band and needs only very low drive power.

CONCLUSION

We have reviewed the principles and performance of waveguide Bragg-deflectors. Both, the electrooptic and acoustooptic devices have achieved a high performance and it can be expected that they will be implemented in systems in the near future.

<u>References</u>

1. F. Auracher, R. Keil and K.H. Zeitler, New Electrooptic Bragg Deflectors with Low-Insertion Loss and Multiple-Beam Capability, <u>Siemens Forsch. - u.Entw. - Ber.</u> 10:44-47 (1981).
2. J.M. Hammer and W. Phillips, Low-loss single-mode optical waveguides and efficient high-speed modulators of $LiNb_xTa_{1-x}O_3$ on $LiTaO_3$, <u>Appl.Phys.Lett.</u> 24:545-547 (1974).
3. J. Noda, N. Uchida and S. Saku, Electro-optic diffraction modulator using out-diffused waveguide layer in $LiNbO_3$, <u>Appl.Phys.Lett.</u> 25:131-133 (1974).
4. G.L. Tangonan et al., Electrooptic diffraction modulation in Ti-diffused $LiTaO_3$, <u>Appl.Opt.</u> 17:3259-3263 (1978).
5. Chen S. Tsai, Guided-Wave Acoustooptic Bragg Modulators for Wide-Band Integrated Optic Communications and Signal Processing, <u>IEEE Transact. on Circuits and Systems</u>, CAS-26: 1072-1098 (1979).
6. M.K. Barnoski et al., Integrated-Optic Spectrum Analyzer, <u>IEEE Transactions on Circuits and Systems</u>, CAS-26:1113-1124 (1979).
7. M.W. Casseday et al., Wide-Band Signal Processing Using the Two-Beam Surface Acoustic Wave Acoustooptic Time Integrating Correlator, <u>IEEE Transact. on Sonics and Ultrasonics</u>, SU-28: 205-212 (1981).
8. J.M. White, P.F. Heidrich and E.G. Lean, Thin-Film Acoustooptic Interaction in $LiNbO_3$, <u>Electr.Lett.</u> 10:510-511 (1974).
9. W.R. Smith Jr., Design of Bragg Cells for SAW/Integrated Optic Signal Processing Devices, Proc. of the 1979 IEEE Ultrasonics Symposium, 98-101.
10. E.A. Kolosovskii, D.V. Petrov and A.V. Tsarev, Influence of the parameters of a diffused waveguide on the frequency dependence of the acoustooptic interaction efficiency, <u>Sov.J. Quant.Electr.</u> 10:998-1000 (1980).

11. C.S. Tsai, M.A. Alhaider, Le T. Nguyen and B. Kim, Wideband
     guided-wave acoustooptic Bragg diffraction and devices using
     multiple tilted surface acoustic waves, Proc.IEEE. 64:318-328
     (1976).
12. T.R. Joseph and Bor-Uei Chen, Broadband Chirp Transducers for
     Integrated Optics Spectrum Analyzers, Proc. of the 1979 IEEE
     Ultrasonics Symposium, 28-33.
13. Chin C. Lee, Kuan-Yang Liao, Chin L. Chang and Chen S. Tsai,
     Wide-Band Guided Wave Acoustooptic Bragg-Deflector Using a
     Tilted-Finger Chirp Transducer, IEEE J.Quant.Electr., QE-15:
     1166-1170 (1979).
14. Kuan Y. Liao, Chin L. Chang, Chin C. Lee and Chen S. Tsai,
     Progress on Guided-Wave Acoustooptic Bragg-Deflector Using
     a Tilted-Finger Chirp Transducer, Proc. of the 1979 IEEE
     Ultrasonics Symposium, 24-27.

ULTIMATE LIMITS IN INTEGRATED OPTICS

H. Kogelnik

Bell Laboratories
Holmdel, NJ 07733 USA

INTRODUCTION

Before exploring the boundaries of this intriguing new dis-
cipline, we have to remind ourselves that the field of integrated
optics is still in its infancy, still in its research stage, and
still searching for its proper role. At this early stage, when
many discoveries are yet to be made, it may appear somewhat futile
to attempt to forecast the limitations of this infant technology,
and it is certainly premature to try detailed comparisons with
established technologies such as silicon integrated circuits. Yet,
there are some crude patterns emerging that indicate limits in the
size, speed, and power consumption of integrated optical devices
and circuits.

The scope and principles of integrated optics have been
thoroughly reviewed and summarized in recent review articles[1-10]
and textbooks[11,12]. The reader is referred to these for detailed
explanations. The devices of interest include junctions and
directional couplers, thin-film prisms and lenses, polarizers and
polarization converters, filters and wavelength multiplexers,
switches and modulators, lasers and amplifiers, detectors and
bistable elements.

Comparing integrated optics to other technologies one should
bear in mind that there are classes of functions that can be performed by optical devices only. Wavelength multiplexing of optical
channels is one example. Here, there is no direct competition with
silicon IC's. On the other hand there are other classes of functions, eg. optical logic and optical computing elements, where comparisons would be appropriate. However, these latter kinds of

devices are, at present, in an even earlier stage of exploration and
are thus evading detailed comparison.

The arguments in the following discussion are order of magnitude
arguments and they are presented from two interconnected points of
view. First, we shall focus on three principal device characteristics: size, speed and power consumption. Then we shall discuss
specific features of two classes of devices: filters and switches
and modulators. For further detail on limits in integrated optics
see Reference 13.

SIZE

In integrated optics there are two limitations that prevent a
reduction of component size. One is a limit of the confinement of
light due to the finite wavelength of light, the other is the minimum interaction length required to produce useful effects such as
efficient switching.

Confinement: The 1-μm Barrier

The near-infrared region from about 0.8 to 1.6μm represents
the spectral region of principal interest to integrated optics[1-12].
In this region one uses dielectric waveguides, planar relatives of
optical fibers, to confine and guide the light. The simple example
of a film guide is shown in Figure 1, where n_f, n_s and n_c are the
refractive indices of the film, substrate and cover materials,
respectively. The film performs as a waveguide if

$$n_f > n_s, n_c. \tag{1}$$

Table I lists the refractive indices of some typical integrated
optics materials[1-12]. The waveguide will support a single guided
mode when the film height h is very small and approximately

$$h \approx (\lambda/2)(n_f^2-n_s^2)^{-1/2}. \tag{2}$$

However, the light is not exclusively confined to the film region.
Evanescent fields penetrate into the substrate and cover regions.
If we add the corresponding penetration depths $1/\gamma_s$ and $1/\gamma_c$ to the
guide height h, we obtain a measure for the degree of confinement
of the light. This is the effective guide thickness h_{eff}

$$h_{eff} = h + 1/\gamma_s + 1/\gamma_c. \tag{3}$$

Figure 2 shows a normalized plot[14,15] for this confinement as a
function of wavelength and film thickness, where $k = 2\pi/\lambda$. The

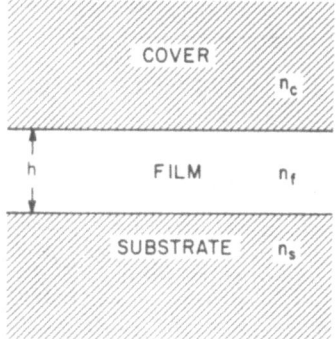

Fig. 1.   Schematic of a dielectric film waveguide.

limit on confinement is seen to be

$$(h_{eff})_{min} \approx \lambda / \sqrt{n_f^2 - n_s^2} \, . \tag{4}$$

Similar arguments hold for strip waveguides. Practical refractive index values combine with the above mentioned wavelength range to impose a barrier to confinement of about 1μm.

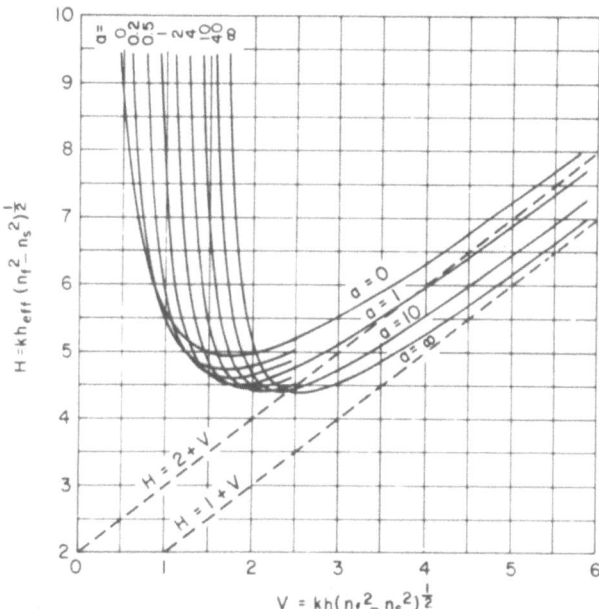

Fig. 2.   Normalized effective guide thickness H as a function of the normalized frequency V for film waveguide with various degrees of asymmetry.

Table 1.

Wavelength μm	Material	Refractive Index	Material	Refractive Index
.9	GaAs	3.59	$Al_{.3}Ga_{.7}As$	3.40
1.7	InP	3.15	$In_{.47}Ga_{.53}As$	3.48
.6	$LiNbO_3$	2.214	$Ti:LiNbO_3$	2.234

Note that the effective width h_{eff} is a very critical measure of confinement. This is appropriate for considerations of packing density or the avoidance of crosstalk. The concentrations of light obtainable for the effective application of gain or electric fields are better by about a factor of four.

Interaction Length

The physical effects of interest to integrated optics are usually small over distances of the order of a wavelength. Examples for this are the gain of a junction laser or a phase change induced via the electrooptic effect. In order to construct efficient devices, these effects have to be accumulated over relatively long interaction distances which represent the device length L. As we will discuss in later sections, these lengths vary somewhat from one device type to the other, and some are subject to power trade-offs. Typical device lengths range from 0.5 to 10mm.

SPEED

Speed is one of the key attributes of optical technologies and the ultimately available bandwidths appear virtually unlimited. Limitations are usually introduced when optical components have to be interfaced with electronics. We consider here only those limitations due to optics as they occur in the form of propagation delay, pulse broadening in transmission lines, and the generation of short pulses. Most of those limitations appear to be no burden. And, on the contrary, we are faced with the challenge to exploit all the speed that is offered to us.

Propagation Delays

The propagation constant β of a dielectric waveguide or optical

fiber is usually somewhat dispersive, ie. it is dependent on the
optical carrier frequency ω. The group velocity in the guide

$$v_g = 1/\beta' \stackrel{\sim}{\sim} c/n \tag{5}$$

is essentially that of the speed of light in the material. With
refractive index values n as given above, this means group velocities of about c/3, a third the speed of light in vacuum. The
propagation delay τ_d due to a connection or a device of length L
is, therefore, limited to

$$\tau_d = L/v_g \stackrel{\sim}{\sim} nL/c \stackrel{\sim}{\sim} 3L/c \tag{6}$$

which is about 10ps/mm.

Pulse Broadening

The pulsewidth of light pulses propagating in an optical
transmission line broadens when different parts of the pulse
spectrum travel at different group velocities. This is called
group velocity dispersion and is measured by β. For a coherent
pulse with a Gaussian pulse shape the broadening follows the law

$$\tau^2 = \tau_o^2 + (8\beta''L/\tau_o)^2 = \tau_o^2 + (DL\Delta\lambda)^2 \tag{7}$$

where τ and τ_o are the full widths of the output and input pulses
measured between 1/e amplitude points, L is the length of the line,
and

$$\Delta\lambda = (\lambda/\omega) \quad = 8\lambda/\omega\tau_o \tag{8}$$

is the full spectral width of the pulse. Measurements indicate a
"zero-dispersion point" near 1.3µm for fibers made of fused-silica.
The ultimate limitations of dielectric fiber guides near this zero-
dispersion point have been tested by recent experiments[16] using
color center lasers that were tunable to the zero-dispersion region.
In these experiments, pulses of 5ps width were transmitted near the
zero-dispersion point over kilometer lengths of fiber without
measurable pulse broadening. Theory indicates[17] that higher order
dispersion terms (eg. β''') will ultimately limit pulse transmission
to minimum pulsewidths τ_{min} of the order of

$$\tau_{min} \cdot L^{-1/3} \stackrel{\sim}{\sim} 1ps \cdot km^{-1/3} \tag{9}$$

where the limiting pulsewidth scales with the cube root of the
fiber length.

Ultrashort Pulse Generation

One of the principal limitations for the generation of ultrashort pulses is the bandwidth over which gain is available for the
amplification of light. In semiconductor junction lasers, so far
the only laser devices compatible with integrated optics, the gain
bandwidth exceeds 100Å. The corresponding limiting pulsewidths
(see Eq.(8)) are well below lps. Experiments in pulse generations
employing mode-locking techniques in junction lasers with external
cavities are now approaching the pulsewidth mark of lps.

POWER REQUIREMENTS

The power requirements and power dissipation of integrated
optics devices vary appreciably from one device type to the other.
Filters consume virtually no power at all, switches in switching
networks may not have to be switched that often and consume medium
power levels, while junction lasers clearly have the highest power
consumption. The various power tradeoffs will be discussed under
the sections dealing with devices. There are just about no power
tradeoffs in semiconductor junction lasers. For CW operation at
room temperature, modern laser devices[18] require threshold power
densities of $1V \times 1kA/cm^2 = 1kW/cm^2$ and drive powers of about
10 to 100mW per device. There has been considerable progress[18] in
recent years in improving laser life and fabrication yield. But
further advances will be needed before one can contemplate the
integration of many junction lasers on a single chip. In addition,
the close packing of junction lasers will require the resolution of
a power dissipation problem of the order of $1kW/cm^2$ due to the high
threshold powers. This is more than two orders of magnitude higher
than the power dissipation of modern silicon IC's, and represents
a formidable problem.

FILTERS

Filter devices are the building blocks of optical wavelength
multiplexing circuits which are one of the goals of integrated
optics. We shall discuss here two filter types that have been
explored theoretically and experimentally. They are the corrugated
waveguide filters and the directional coupler filters. One of their
principal characteristics is the bandwidth of their filter response
which is subject to limitations described below.

Corrugated Waveguide Filters

These filter devices are made by machining ultrafine corrugations into the surface of a film waveguide as sketched in Figure 3,

which also shows the measured filter response of such a device[19].
To adjust the peak reflectance to a given wavelength λ, one has to
machine a corrugation period Λ according to the Bragg condition

$$\Lambda = \lambda/2N \tag{10}$$

where $N \sim n$ is the effective guide index defined by

$$\beta = 2\pi N/\lambda. \tag{11}$$

Depending on the choice of materials and wavelength, values for Λ
lie in the range from 1000 to 3000Å. The effective index N depends
on wavelength and film thickness. As the corrugation consists of
thickness variations, it can be viewed as inducing periodic index
variations of amplitude ΔN which depend on the corrugation height.
Related to ΔN is another important device parameter, the coupling
length ℓ, which is approximately given by

$$\ell = \lambda/2\Delta N. \tag{12}$$

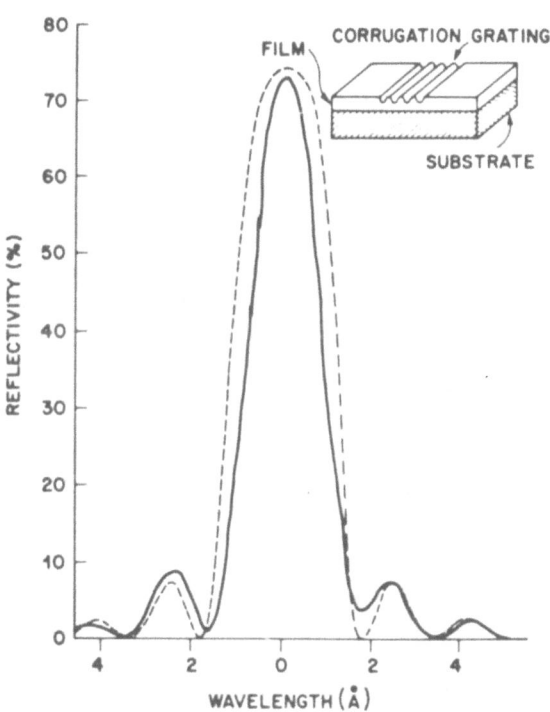

Fig. 3.   Sketch of corrugated waveguide filter and its
           measured filter response (from Reference 19).

To obtain an efficient device, we need a device length

$$L \gtrsim \ell. \tag{13}$$

Lengthening the device much beyound ℓ has very little effect on device performance. The 3dB bandwidth of the filter is, approximately[20-23],

$$\Delta\lambda/\lambda \simeq \Lambda/\ell \simeq \Delta N/N \tag{14}$$

For a uniform corrugation, this imposes both maximum and minimum limits on bandwidth. The smallest bandwidths are achieved by making ℓ and L as large as possible. This approach is limited by fabrication tolerances, and particularly by the achievable uniformity of both the waveguide and the corrugations. Maximum lengths are of the order of 1cm. This means that the lower limit for bandwidth is about 0.1Å, and experiments approaching this limit have been reported[19]. The maximum achievable bandwidth of a uniform corrugation is dictated by the maximum ΔN that can be induced by the corrugation. For practical waveguides $\Delta N_{max}/N$ is of the order of 10^{-2} to 10^{-3}, resulting in a limit of about 10Å. Values close to this have been achieved experimentally[20]. Techniques such as chirping, ie. fabricating corrugations of non-uniform periodicity, have been proposed for broadening the bandwidth of these filters up to about 200Å[22-24].

Directional Coupler Filters

Coupler filters[25] are a device option with relatively broad bandwidth. As indicated in Figure 4, these devices consist of a directional coupler with two guides of different effective index N_1 and N_2. The dispersion $dN/d\lambda = N'$ of the two guides is designed to be unequal in order to achieve intersecting dispersion characteristics. The bandwidth of this device is given by

$$\Delta\lambda/\lambda = 1/L(N_1' - N_2'). \tag{15}$$

There are no obvious limitations on achieving broad bandwidth in this device, and efficient coupler filters with $\Delta\lambda = 200$ Angstroms have been reported[25]. The lowest bandwidths achievable with practical guide dispersion have been estimated at about 50Å.*

SWITCHES AND MODULATORS

The class of electrooptic devices[27-29] of interest in integrated optics serves essentially two functions: the first is to modulate an information carrying signal onto the lightwave carrier, and the

---

*Note added in proof: Filter bandwidths between 5Å and 50Å have been demonstrated recently in mode-converter filters[26].

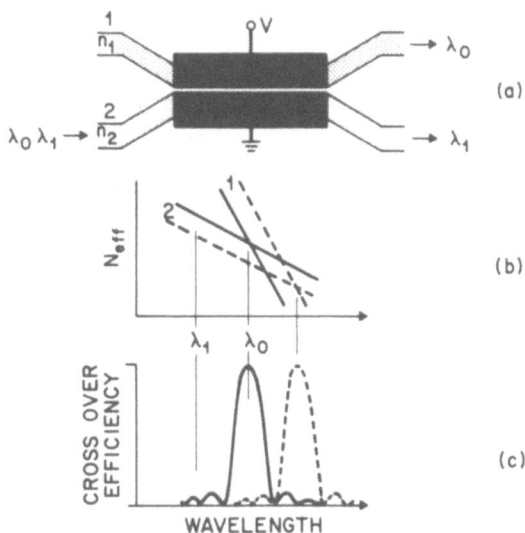

Fig. 4. Sketch of directional coupler filter and its
characteristics (from Reference 25).

second is to switch the path of the light from one guide to another
in response to an electrical signal. In probing the limits, our
order of magnitude arguments are about the same for a set of devices,
including phase modulators, switched directional couplers, and
switched $\Delta\beta$ couplers. The reason is that the geometry of these
structures, including the electrodes, are basically the same, and
that the required drive voltage is approximately equal to the half-
wave retardation voltage V_π in all these cases. As an illustration,
Figure 5 shows the example of a switched $\Delta\beta$ coupler. Two guides are
indicated, which are typically about 3µm wide, and spaced about 3µm
apart. The electrodes, shown shaded, serve to apply the drive
voltage in six sections of reversed polarity. This device functions
both as an amplitude modulator and as an electrically controlled
switch[29,30].

The application of an electric field E to a guide induces a
change ΔN of the effective index of magnitude

$$\Delta N = \frac{1}{2} N^3 rE \tag{16}$$

where r is the electrooptic coefficient of the material, and good
overlap between the optical and electric fields is assumed.

The electrooptic coefficients of some preferred electrooptic
materials are listed in Table II[31].

Table II

Material	Wavelength μm	r-coefficients, 10^{-12} m/v		
LiNbO$_3$	.63	r_{33}=32.2	r_{13}=10	r_{22}=6.8
LiTaO$_3$	.63	r_{33}=30.3	r_{13}=7	r_{21}=20
GaAs	.9	r_{41}=1.2	–	

Referring to Figure 5, we note that the (coplanar) electrode capacitance C is almost independent of the gap spacing d and proportional to the device length L. It is approximately given by

$$C = \epsilon L \tag{17}$$

where ϵ is the low frequency dielectric constant ($\epsilon_{relative} \sim 35$ for LiNbO3). The half-wave voltage V_π is

$$V_\pi = (\lambda / N^3 r)(d/L). \tag{18}$$

Drive voltages have been lowered to $3V^{30}$ by designing devices with lengths of about 10mm. With this we can calculate the electrical

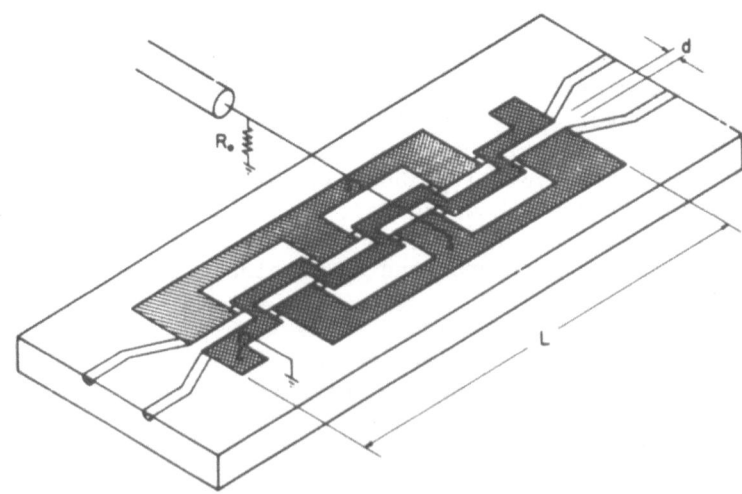

Fig. 5.   Sketch of switched Δβ coupler showing electrode arrangement for six sections of alternating Δβ (after Reference 30).

energy E_{SW} needed for one switch of the optical path (or to modulate 1 bit of information) as[29]

$$E_{SW} = \frac{1}{2} CV_\pi^2 = (\varepsilon\lambda^2/2N^6 r^2) \cdot (d^2/L).\qquad(19)$$

Pushing the electrode-gap and guide-width dimensions to the limits of about 1μm, one obtains an ultimate figure of merit of about

$$E_{SW} \cdot L = 1 \text{ pJ} \cdot \text{cm}\qquad(20)$$

which will be difficult to obtain in practice. Note that there is a direct tradeoff between energy and device length, with less energy required for the longer devices and associated longer transit delay.

CONCLUSIONS

Many of the functions envisaged for integrated optics circuits and devices cannot be provided by other technologies such as the silicon or Josephson technologies. The characteristics and limitations of integrated optics are also rather different. It appears that the available speed is almost limitless, and it will be a challenge to exploit this speed. Because of limits on confinement and interaction lengths, integrated optics devices may look rather large in comparison to future devices of the silicon and Josephson technologies.

As the field is very young there is a stream of new ideas, and there are several areas that are, as yet, little explored. This includes the area of optical bistability, as well as the promising possibility of integrating optical and electronic devices (eg. p-i-n-detectors or lasers and FET's) on the same semiconductor substrate, a new area not discussed in the main text. Among other proposed possibilities that have as yet seen little exploration are operation at low temperatures to reduce power dissipation in lasers, the use of short pulses at long repetition rates to exploit the high peak powers for the enhancement of nonlinear interactions and the shortening of interaction lengths, optical subpicosecond gates[31], and serial processing to exploit the high speed of optics.

In the above discussion, we were principally concerned with the limitation of discrete devices and discrete structures. Some of these have obvious implications for optical integrated circuits. We have not addressed the formidable technological task of putting many multifunctional devices together on a single chip[32,33]. While this interconnection problem is a highly important task in integrated optics, the ultimate limitations imposed by the interconnections themselves appear to reside in the realm of fabrication technology rather than that of physics.

REFERENCES

1. W. S. Chang, W. M. Muller, and F. J. Rosenbaum, "Integrated Optics in Laser Applications", Vol. 2, Academic Press, New York (1974).

2. H. F. Taylor and A. Yariv, "Guided Wave Optics", Proc. IEEE 62:1044 (1974).

3. P. K. Tien, "Integrated Optics", Sci. Amer. 230:28 (1974).

4. P. K. Cheo, "Thin-Film Waveguide Devices", Appl. Phys. 6:1 (1975).

5. H. Kogelnik, "An Introduction to Integrated Optics", IEEE Trans. Microwave Theory Tech. MTT-23:2 (1975).

6. D. Ostrowsky, "L'Optique Integrée", La Recherche 6:740 (1975).

7. Y. Suematsu, "The Progress of Integrated Optics in Japan", IEEE Trans. Microwave Theory Tech. MTT-23:16 (1975).

8. E. M.Conwell, "Integrated Optics", Phys. Today 29:48 (1976).

9. P. K. Tien, "Integrated Optics and New Wave Phenomena in Optical Waveguides", Rev. Mod. Phys. 49:361 (1977).

10. H. Kogelnik, "Review of Integrated Optics", Fiber Integrated Opt. 1:227 (1978).

11. M. K. Barnoski, Ed., "Introduction to Integrated Optics", Plenum Press, New York (1973).

12. T. Tamir, Ed., "Integrated Optics", Springer, Berlin, Germany (1975).

13. H. Kogelnik, "Limits in Integrated Optics", Proc. of IEEE 69:232 (1981).

14. H. Kogelnik and V. Ramaswamy, "Scaling Rules for Thin-Film Optical Waveguides", Appl. Opt. 13:1857 (1974).

15. H. Kogelnik, "Theory of Dielectric Waveguides", in: "Integrated Optics", T. Tamir, Ed., Springer, Berlin, Germany (1975).

16. D. M. Bloom, L. F. Mollenauer, C. Lin, D. W. Taylor and A. M. DelGaudio, "Direct Demonstration of Distortionless Picosecond-Pulse Propagation in Kilometer-Length Optical Fibers", Opt. Lett. 4:297 (1979).

17. H. G. Unger, "Optical Pulse Distortion in Glass Fibers at the Wavelength of Minimum Dispersion", Arch. Elek. Ubertragung 31:518 (1977).

18. R. W. Dixon, "Current Directions in GaAs Laser Device Development", Bell Syst. Tech. J. 59:669 (1980).

19. H. A. Haus and P. T. Ho, "Effect of Noise on Active Modelocking of a Diode Laser", IEEE J. Quantum. Electron. QE-15:1258 (1979).

20. D. C. Flanders, H. Kogelnik, R. V. Schmidt, and C. V. Shank, "Grating Filters for Thin-Film Optical Waveguides", Appl. Phys. Lett. 24:194 (1974).

21. R. V. Schmidt, D. C. Flanders, C. V. Shank, and R. D. Standley, Appl. Phys. Lett. 25:651 (1974).

22. M. Matsuhura, K. O. Hill, and A. Watanabe, J. Opt. Soc. Amer. 65:804 (1975).

23. P. S. Cross and H. Kogelnik, "Sidelobe Suppression in Corrugated-Waveguide Filters", Opt. Lett. 1:43 (1977).

24. C. S. Hong, J. B. Shellan, A. C. Livanos, A. Yariv, and A. Katsir, "Broadband Grating Filters for Thin-Film Optical Waveguides", Appl. Phys. Lett. 31:276 (1977).

25. R. C. Alferness and R. V. Schmidt, "Tunable Optical Waveguide Directional Coupler Filter", Appl. Phys. Lett. 33:161 (1978).

26. R. C. Alferness and L. L. Buhl, Opt. Lett. 5:473 (1980).

27. I. P. Kaminow, "Optical Waveguide Modulators", IEEE Trans. Microwave Theory Tech. MTT-23:57 (1975).

28. J. M. Hammer, "Modulation and Switching of Light in Dielectric Waveguides", in: "Integrated Optics", T. Tamir, Ed., Springer, Berlin, Germany (1975).

29. R. V. Schmidt and R. C. Alferness, "Directional Coupler Switches, Modulators and Filters Using Alternating $\Delta\beta$ Techniques", IEEE Trans. Circuits Syst. CAS-26:1099 (1979).

30. R. V. Schmidt and P. S. Cross, "Efficient Optical Waveguide Switch/Amplitude Modulator", Opt. Lett. 2:45 (1978).

31. I. P. Kaminow and E. H. Turner, "Linear Electrooptical Materials", in: "Handbook of Lasers", Chemical Rubber Co., Cleveland, OH (1971).

32. E. Garmire, "Semiconductor Components for Monolithic Applications in Integrated Optics", Springer, Berlin, Germany (1975).

33. F. K. Reinhart, "Monolithic Optical Integration", in: Proc. 8th Conf. Solid State Devices (Tokyo, Japan), p. 357 (1980); R. A. Logan and F. K. Reinhart, "Integrated GaAs-Al$_x$Ga$_{1-x}$As Double Heterostructure Laser", IEEE J. Quantum Electron. QE-11:461 (1975).

OPTICAL BISTABILITY AND THE OPTICAL TRANSISTOR

USING SEMICONDUCTORS

S.D. Smith and D.A.B. Miller

Physics Department
Heriot-Watt-University
Edinburgh, Scotland, U.K.

Optical bistability (OB) is, as its name implies, the existence of two stable states for one set of optical input conditions. The concept has been in existence for some twelve years since the suggestion of Szöke et al.[1], but was first demonstrated by Gibbs et al.[2] using Na vapour as the active medium inside a Fabry-Perot cavity. The first type of OB to be proposed was the so-called "absorptive" OB in which a saturable absorber is placed inside a Fabry-Perot cavity tuned on-resonance, the idea being that as the field inside the cavity increased, the absorber would "bleach", thus further increasing the field inside the cavity by increasing the finesse (or "Q-factor"). Ideally this process would be self-sustaining leading to a "switching-on" action at one intensity with a "switching-off" only occurring at lower intensities, thus creating a bistable region between the switch-on and switch-off intensities. However, this turns out to be difficult to achieve because the requirement on the degree of bleaching is rather severe (ie. a ratio of > 8 between low and high intensity absorption limits). In practice, it is much easier to observe OB through intensity dependence of the refractive index as was first observed by Gibbs et al.[2], usually called "dispersive" (or "refractive") OB; the most common method is to use the non-linear medium inside a Fabry-Perot, this time tuned somewhat off-resonance. Then as the incident intensity is increased the cavity is pulled towards resonance by the intensity-dependent refractive index changing the optical length. However, as the cavity is pulled towards resonance a larger fraction of the incident light gets into the cavity, thus enhancing the change in optical length. This process also can be self-sustaining, leading to a switching action with "switch-on" and "switch-off" intensities defining the boundaries of the bistable region. A model, with a simple graphical interpretation of the bistable region, was

proposed by Felber and Marburger[3] almost simultaneously with the
observation of dispersive OB by Gibbs et al.[2] (who also presented
a simple model).

A large amount of theoretical activity was stimulated by the
work of Bonifacio and Lugiato[4] who developed an analytical technique
for handling the coupled Maxwell-Bloch equations. This work is
generally concerned with the rather specific conditions obtainable
with two-level atomic systems and will not concern us further here.

The subject of OB has grown very rapidly since 1976, stimulated
both by its fundamental theoretical aspects and the opportunity it
offers for a new class of all-optical devices capable of logical
operations in a fashion loosely analogous to the electronic transistor. Despite this interest, there have been comparatively few
observations of "intrinsic" OB, i.e. truly all-optical devices with
no electronic component. Of those which have been demonstrated,
those based on near-bandgap nonlinear refractive effects in semiconductors[5-9], arguably currently offer the best prospects for
practical devices especially for low-power logic operations because
they offer small size and potentially fast operation at low switching energies combined with some prospects that they might be useable
in integrated optical circuits although presently in both cases the
materials have to be cooled to liquid cryogen temperatures. Optical
bistability has also been reported at room temperature in a system
using two tellurium crystals[10,11] mentioned the observation of some
nonlinear Fabry-Perot effects in $Hg_{1-x}Cd_xTe$ in their room temperature experiments on degenerate four-wave mixing. Gibbs et al.[12]
prepared a GaAlAs-GaAs-GaAlAs "sandwich" structure, of thickness
$0.21\mu m - 4.1\mu m - 0.21\mu m$, by molecular beam epitaxy on a GaAs substrate and etched away a 1-2mm diameter hole leaving the thin GaAs
layer. This layer was coated for 90% reflectivity on each side, and
the resulting cavity could be tuned by moving the focussed spot
across the crystal which was slightly non-parallel after the etching.
The laser spot was ~10μm diameter and OB was observed at intensities
equal to ~50-100KW/cm^2 (ie. ~100mW laser power) at 15°K. The device
could switch in ~40ns, corresponding approximately to the excitonic
lifetime but the turn-on time could be greatly reduced by using a
short pulse to switch on. Working at a "holding" intensity within
the bistable region the device turned on in ~1nS using an additional
200pS 5900Å pulse with an effective absorbed switching energy of
~24pJ in the active region. They predicted a limiting switch-on
time of ~1pS, based on the measurements of Shank et al.[12] (discussed
above in Section 3), and limiting switching energies of ~1fJ (10^{-15}J)
on the assumption that a $(0.2\mu m)^3$ device could be made to switch (ie.
one cubic wavelength inside the material). The proposed mechanism
for the nonlinear refraction in this dispersive bistability is associated with the saturation of exitonic absorption.* At the

---

*See general references[23,24].

intensities used for bistability the excitonic resonance is almost
totally saturated.

In confirmation of this interpretation, this relatively fast OB
could be observed only up to ~120°K (where kT is ~2 times the exciton
binding energy). Thereafter a slower thermal effect took over as the
samples were not heatsunk. The existence of bistability due to
excitonic saturation has also been considered theoretically by Goll
and Haken[13].

Simultaneously with the GaAs work (the Bell and Heriot-Watt
papers were both received on June 15, 1979!), Miller, Smith and
Johnston[7] demonstrated non-linear Fabry-Perot action in InSb at 5°K
using a 500μm thick InSb sample, polished plane-parallel and using
only the natural reflectivity of the crystal faces (36%) to form the
cavity mirrors. With this simple system, they obtained five successive nonlinear orders of the Fabry-Perot cavity (ie. 5λ/2 change in
effective optical length) with increasing intensity, with clear
bistability in the fifth nonlinear order, using spot sizes ~200μm
diameter and laser powers up to 500mW (Figure 1). Subsequently they
were able to demonstrate bistability in two successive nonlinear
orders (3rd and 4th) in both transmission and reflection at powers
~100-200mW in similar samples (Miller, Smith and Seaton (1981 (a)
and (b))and in a 130μm crystal at 77K coated to 70% reflectivity,
bistability was seen in the first nonlinear order at ~8mW (~80W/cm^2)
(Figure 2)[8,14]. The measured switching times were detector-system
limited but at ~500nS consistent with interband recombination as the
longest time constant in these devices. The proposed mechanism for
the nonlinear refraction causing this dispersive OB is the bandgap
resonant effect is discussed by Miller, Seaton, Prise and Smith[9].
Miller and Smith (1979) extended the observations of non-linear
Fabry-Perot action in InSb to demonstrate true two-beam "optical-
transistor" action. They focused two beams, a main beam and a weak
beam, at slightly different angles, coincidently on the sample. They
made small changes in the incident weak beam power and observed the
change in the transmitted main beam power, ratioing the latter to the
former to define a "gain". The results of this experiment taken
under otherwise indentical conditions to the results in Figure 1 are
shown in Figure 3. At the powers corresponding to the sharp rises
in the results in Figure 1, the gain reaches a peak, ultimately >1,
showing real optical signal gain of up to 10. This measurement of
two-beam gain is physically different from the one-beam differential
gain which would result from the derivative of the results in Figure
1, as in the presence of two beams and a cavity degenerate four-wave
mixing is possible and has indeed been observed in InSb at 5°K at
similar powers[16,22]. Because the device operates mainly by "transfer
of phase thickness" it was termed the "transphasor" by analogy with
the transistor.

Gibbs et al.[5] and Miller, Smith and Seaton[8] have considered
various ways in which device switching times and energies may be

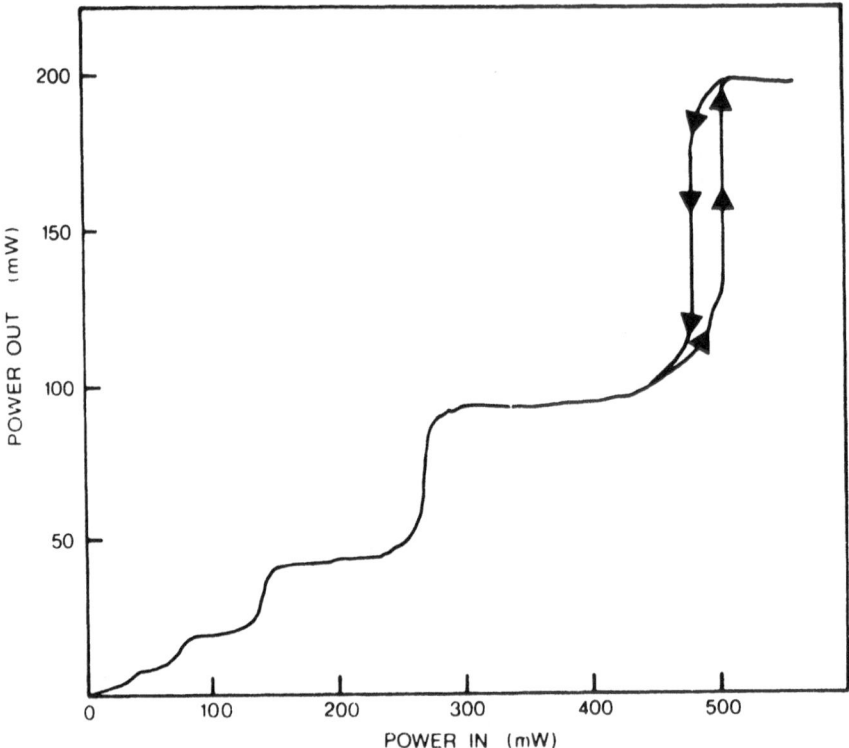

Fig. 1.   Transmitted power plotted against incident power for a CW
           CO laser beam (wavenumber 1895cm^{-1}, spot size ~180μm)
           passing through a polished plane-parallel InSb crystal
           (5 x 5mm x 560μm thick, N_D-N_A~3 x 10^{14}cm^{-3} (n-type)) at~5$\overset{\circ}{K}$.

altered, including alteration of switch-off times by doping or
diffusion. Miller[17] has analysed the design of nonlinear Fabry-Perot
cavities in the presence of linear absorption (as is present in
practice in the InSb devices) and deduced the effective material
figure of merit for minimum switching intensity as $n_2/\alpha\lambda$ where α is
the linear absorption coefficient. The ratio of $n_2/\alpha\lambda$ is due to the
fact that for a more highly absorbing material, the cavity has to be
made shorter to preserve the finesse and consequently n_2 has to be
larger. This parameter $n_2/\alpha\lambda$ is physically identical (except for
fundamental constants) with δ, the refractive index change for one
excited system per unit volume (see Miller, Smith and Seaton[8]. On
the basis of this analysis[8] they deduced limiting switching energies
also of ~10^{-15}J with 1pJ being more readily feasible. It is important to emphasise that the switching energy is not dependent (at
least in these simple models) on the switch-on speed desired; faster
switch-on requires higher power, but this may be achieved by using
a shorter pulse of the same energy. The limits on switch-on time
are liable to depend on the speed of intraband processes.

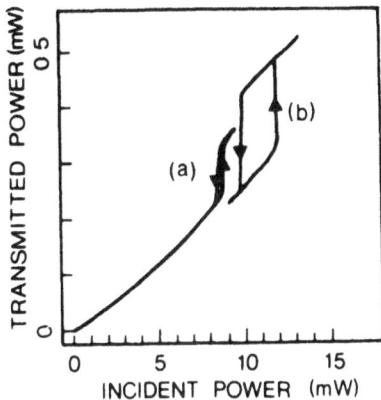

Fig. 2.   Transmitted power plotted against incident power for CW CO
           laser beam (wavenumber 1827cm⁻¹, spot size ~150μm) passing
           through a polished polycrystalline InSb slice (5 x 5mm x
           130μm thick) coated to ~70 percent reflectivity on both
           faces, held at ~77K. Onset of bistability is seen at ~8mW
           (trace (a)) with clear bistability at slightly higher
           powers with different cavity detuning (trace (b)).

     Whether it is possible to scale these devices to faster switch-
off and/or higher temperature operation depends to some extent on
the "robustness" of the nonlinearity. Excitons favour pure, low-
temperature materials, and it may not be possible to dope heavily
or increase the temperature. The nonlinearity in InSb has however
survived to 77°K (already ~20 times the theoretical exciton binding
energy) and on the model of Miller, Seaton, Prise and Smith[9] should
exist at room temperature, weakening only in proportion to temperature. The observations of OB at 77°K by Miller, Smith and Seaton[8,14]
were also made on relatively impure polycrystalline material. However, it seems likely that the cavity field build-up times can be
kept very short in semiconductors with ~1.2ps calculated for GaAs
and ~20ps for InSb in the devices already demonstrated, so cavity
times should not be a problem.

     Neither the GaAs or InSb devices have been optimised and much
work remains to be done to develop a practical and useful bistable
device. Nevertheless, the observations in these two materials offer
considerable promise, especially as they result from apparently
different physical mechanisms thus offering a variety of ways of
achieving the necessary nonlinearity and suggesting the existence of
related nonlinearities in other materials. Miller, Smith and Seaton[8]
and Miller, Seaton, Prise and Smith[9] have considered the scaling of
the interband saturation contribution to the nonlinearity and conclude that on this contribution alone similar effects should be
observable for similar energies in other materials, because although
the interband saturation nonlinearity decreases as $1/E_G^2$ (E_G is the

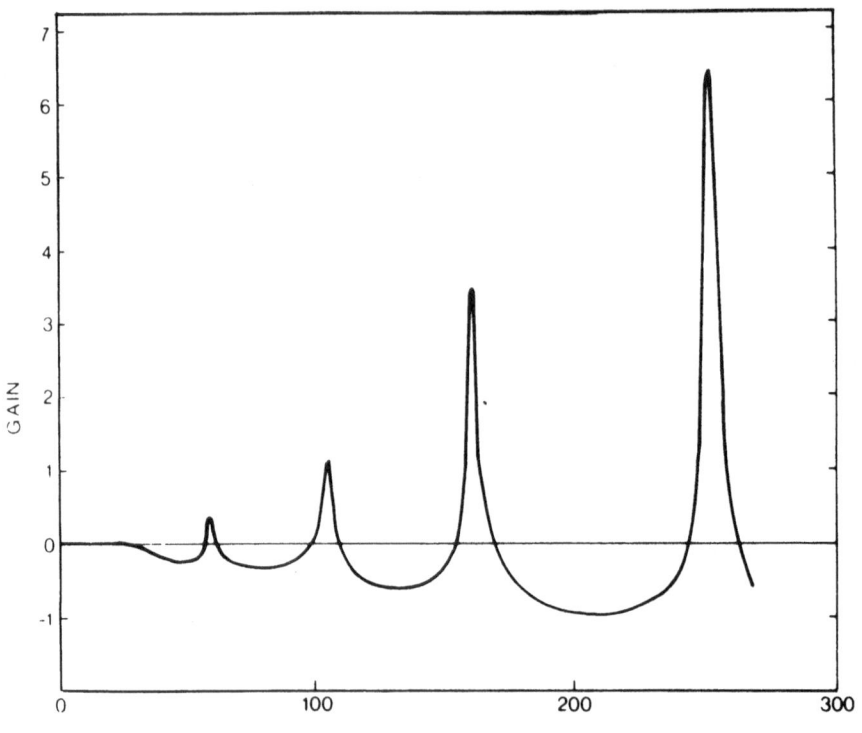

Fig. 3.   Transphasor gain (see text) plotted against main beam
power. Sample, beams and temperature are similar to
those for Figure 1.

bandgap energy) in going to wider gap materials, this is exactly
compensated by the possibility of tighter focusing (by a factor $1/\lambda^2$
in intensity for a given power) due to weaker diffraction at shorter
wavelengths thus making the effective material figure of merit
$n_2/\alpha\lambda^3$ or δ/λ^2 . The scaling of other effects (eg. excitonic and
bandgap-renormalisation) is not yet clear, although it is likely
they will be of a different form and hence it is highly unlikely
that the various effects will always conspire to cancel one another.
Therefore despite the current limited understanding of the bandgap-
resonant nonlinearities, the future is very promising for device
applications. Recently, the possibility of OB utilising yet another
nonlinearity near the bandgap of wider band direct-gap semiconductors,
namely that due to the creation of excitonic molecules in for example CdS .or CuCl, has also been suggested by Koch and Haug[18].

A different approach to optical bistability has been demonstrated
using the semiconductor Tellurium[10]. Two crystals of Tellurium were
used, both pumped by the same Q-switched CO_2 laser pulse; one crystal

was arranged to generate second harmonic radiation through the conventional passive $\chi^{(2)}$ of Tellurium whereas the other crystal was used to absorb by two or three photon absorption various combinations of 10.6μm photons (directly from the CO_2 laser) and 5.3μm photons (produced by the second harmonic generation in the first crystal). The bandgap of Tellurium (~0.35eV) is such that three 10.6μm photons or one 5.3μm + one 10.6μm photon or two 5.3μm photons can bridge the gap. The second crystal was arranged as a Fabry-Perot cavity, by polishing the crystal plane-parallel and utilising the natural reflectivity of the crystal faces. The creation of free carriers in the crystal by multiphoton absorption was presumed to alter the refractive index through the usual Drude-model free carrier refraction and hence tune the cavity. With both the 5.3μm and 10.6μm beams coincident on the crystal they were able to demonstrate a variety of modulation effects at intensities 1-10mW/cm^2 and also showed clear hysteresis in the output which they ascribed to optical bistability. Although this system is a compound one it still represents an all-optical method of optical modulation, utilising at least three different nonlinear optical processes simultaneously in one material.

DISCUSSION

The Fabry-Perot resonators discussed in this paper rely on the internal field enhancement at resonance for their non-linear characteristics. Several schemes using guided waves have already been proposed[19-21] to give similar effects and semiconducting materials have yet to be exploited. Effects in which a refractive index change of the order of unity can be readily achieved heralds the beginning of "Milliwatt power nonlinear optics". From a device point of view switching, amplifying and logic elements are emerging with a similar range of operating parameters. If it proves possible to combine milliwatt power levels with near-picosecond operating time constants the possibilities for all-optical electronics will be considerable.

ACKNOWLEDGEMENTS

We thank Monika Pfannkuchen and MÜTEK GmbH, Diessen for preparation of this manuscript.

REFERENCES

1. A.Szöke, V. Daneu, J. Goldher and N.A. Kumit. App.Phys.Lett. 15:376 (1969).
2. H.M. Gibbs, S.L. McCall and T.N.C. Venkatesan. Phys.Rev.Lett. 36:1135 (1976).
3. F.S. Felber and J.H. Marburger. App.Phys.Lett. 28:731 (1976).
4. R. Bonifacio and L.A. Lugiato. Opt.Comm. 19:172 (1976).

5. H.M. Gibbs, T.N.C. Venkatesan, S.L. McCall, A.C. Gossard,
    A. Passner and W. Wiegman, App.Phys.Lett. 35:451 (1979 (a)).
6. H.M. Gibbs, S.L. McCall and T.N.C. Venkatesan, in: "Optical
    Bistability", C.M. Borsden, M. Ciftan and H.R. Robl, eds.,
    Plenum Press, New York p. 109 (1981).
7. D.A.B. Miller, S.D. Smith and A.M. Johnston, App.Phys.Lett.
    35:658 (1979).
8. D.A.B. Miller, S.D. Smith and C.T. Seaton, IEEE J.Quant.Elect.
    QE17:312 (1981 (b)).
9. D.A.B. Miller, C.T. Seaton, M. Prise and S.D. Smith, Phys.Rev.
    Lett. 47:197 (1981).
10. G. Staupendahl and K. Schindler, Proc.2nd Int.Symp. on Ultra-
    fast Phenomena in Spectroscopy, Jena, p.437 (1980).
11. R.K. Jain and D.G. Steel, Appl.Phys.Lett. 37:1 (1980).
12. H.M. Gibbs, T.N.C. Venkatesan, A. Passner and W. Wiegman, App.
    Phys.Lett. 34:511 (1979 (b)).
13. G. Goll and H. Haken, Phys.State Sol. (b) 101:489 (1980).
14. D.A.B. Miller, S.D. Smith and C.T. Seaton, in: "Optical Bis-
    tability", C.M. Borsden, M. Cifter and H.R. Robl, eds., Plenum
    Press, New York, p.115 (1981 (a)).
15. D.A.B. Miller and S.D. Smith, Opt.Comm. 31:101 (1979).
16. D.A.B. Miller, R.G. Harrison, A.M. Johnston, C.T. Seaton and
    S.D. Smith, Opt.Comm. 32:478 (1980).
17. D.A.B. Miller, IEEE J.Quant.Elect. QE17:306 (1981).
18. S.W. Kock and H. Haug, Phys.Rev.Lett. 46:450 (1981).
19. P.W. Smith, Opt.Eng. 19:546 (1980).
20. A. Kaplan, Optics.Lett. 6:360 (1981).
21. D. Sarid and G. Stegeman: to be published.
22. S.D. Smith and D.A.B. Miller, J.Phys.Soc.Japan. 49:Suppl.1 A,
    597 (1980).
23. A. Miller, D.A.B. Miller and S.D. Smith, in: "Dynamic Nonlinear
    Optics in Semiconductors", Advances in Physics (in press),
    Taylor and Francis, sections 3,5,6 (1981).
24. E. Abraham and S.D. Smith, in: "Optical Bistability and Related
    Devices", Progress in Phys. 1981, (to be published).

INTEGRATED OPTICAL SIGNAL PROCESSORS: DESIGN AND

IMPLEMENTATION CRITERIA

G. C. Righini

Istituto di Ricerca sulle
  Onde Elettromagnetiche, C.N.R.
50127 Firenze, Italy

INTRODUCTION

A very promising application field for integrated optical (IO) devices is represented by signal processing. In the IO format it is limited to one-dimensional signals: however, this characteristic is of no consequence when signals as a function of time are involved. Utilization of these devices in radar, communication and computer processing systems is envisioned.

The potential advantages of IO processors, that include the capability of parallel handling of large amounts of information, the compact, rugged optical design and the low electrical drive power requirement, have been recognized a long time[1]. Their implementation in a miniaturized form has been made possible only recently[2,3] due to the availability of high quality AlGaAs laser diodes, of near-diffraction-limited waveguide lenses, of high-quality waveguides, and of suitable photodetector arrays. Moreover, significant advances have been made in the surface-acoustic-wave (SAW) technology, that permits the fabrication of acousto-optic Bragg modulators and deflectors with Gigahertz bandwidth.

Current research interests are focused on two classes of devices: the first one includes RF spectral analyzers and correlation and convolution filtering devices. A typical processor of this kind consists of three main parts: 1) an input transducer, converting a time window of an electrical signal into a spatial optical distribution, 2) a guided-wave optical system, fed by a laser diode, performing the Fourier transform of the converted signal or its correlation (convolution) with a reference signal; and 3) an output transducer, which reconverts the processed optical signal into an electrical signal;

to be read out in a parallel or a serial format. The second class
of processing devices concerns analog to digital converters and
optical logic circuits, that generally consist of a number of electro-
optic modulators and switches interconnected by single-mode waveguides. In a typical logic device light from one or more lasers is
coupled into channel waveguides on a electrooptic substrate, and
propagates through logic gates (switches and modulators) activated
by electrical signals corresponding to data inputs. Intensity-
modulated light coupled out of the waveguides is detected and amplified to provide the electrical data outputs.

     This paper intends to describe the current status of analog
processors, pertaining to the former class of devices. A simple
analog processor, that utilizes most of the elements which would be
found in an integrated optical circuit, is the Integrated Optic
Spectrum Analyzer (IOSA); the experimental operation of this device
has been already demonstrated[2,3]. Therefore we may discuss the implementation criteria and the performance limits of this processor
without losing generality.

THE INTEGRATED OPTIC SPECTRUM ANALYZER: DESIGN CONSIDERATIONS

     The IO device considered herein is illustrated in Figure 1. It
requires a laser source (single transverse and longitudinal mode)
and a waveguide lens to expand and collimate the beam. A transducer
that consists of an interdigital finger electrode array, and is driven
by the incoming radar signal (mixed with a local oscillator and amplified), excites surface acoustic waves (SAW) traversing the optical
waveguide. Propagation of the SAWs creates both a moving index
grating in the waveguide and moving corrugation gratings at both the
air-waveguide and the waveguide-substrate interfaces. In most practical cases the index grating is the dominant mechanism for the diffraction of the incident optical beam. By fulfilling the conditions
for the Bragg-type diffraction, a portion of the optical beam will be
deflected at an angle proportional to the acoustic frequency, with an
intensity proportional to the applied signal power. This diffracted
beam is focused by a second lens onto a photodetector array, placed
at the lens focal plane. Each detector cell (pixel) acts as a frequency channel of the analyzer. Serial readout of the signal-power-
density spectrum contained in the detector array can be accomplished
by coupling it to a charge-coupled-device (CCD).

     A more detailed analysis of the theory of operation of the IOSA
is beyond the scope of this paper. We report in the following the
few equations expressing the physical constraints on the laser source
and on the optical and acoustical circuit design. The reader inter-
ested in the complete mathematical formulation is referred to the
literature.[4-6]

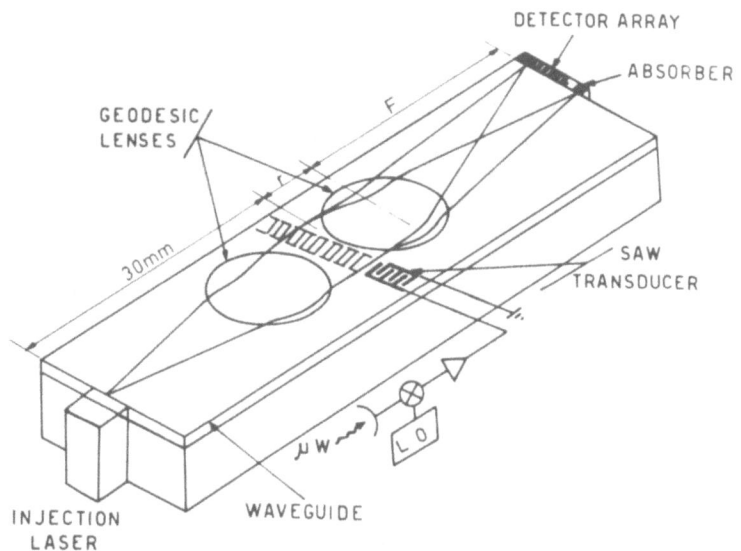

Fig. 1.   Schematic of an Integrated-Optic Spectrum Analyzer.

Optical Resolution Requirement

Given a Fourier-transforming lens of aperture D (ie. the second lens in Figure 1), the smallest resolvable interval in the FT line can be determined as the half-width between nulls of the diffraction pattern corresponding to a Gaussian plane wave impinging on an aperture D:

$$\Delta f = g \frac{\lambda F}{nD} \tag{1}$$

where g is a weight factor depending on the amount of truncation of the Gaussian beam and on the definition of the optical spot size. A typical value of g is 1.41.[5]   The other symbols are defined in Table 1.   Equation (1) sets the lower limit of the center-to-center spacing S of the detector array:

$$S \geq g \frac{\lambda F}{nD} \tag{2}$$

RF Resolution Requirement

Let us consider an RF-signal frequency departure δf from the signal center frequency f_o.   The corresponding diffracted wave undergoes an angular displacement $\delta\theta$

Table 1. Symbols List

S	detector cell size, or pixel pitch
F	focal length of the FT lens
D	optical beam width (lens aperture)
λ	optical wavelength in free space
n	effective mode refractive index
Λ	wavelength of the SAW
v	propagation velocity of the SAW
$f = \frac{\Omega}{2\pi}$	frequency of the SAW in Hertz
f_o	center frequency of the SAW transducer

$$\delta\theta = \frac{\lambda}{nv}\,\delta f \tag{3}$$

and consequently a spatial displacement δy in the focal plane

$$\delta y = F\,\delta\theta = \frac{F\lambda}{nv}\,\delta f \tag{4}$$

This expression sets the upper limit of the detector cell size S:

$$S \leq \frac{F\lambda}{nv}\,\delta f \tag{5}$$

Conditions for AO Bragg Interaction

In order to achieve an efficient Bragg diffraction, the acousto-optic (AO) parameter $Q = 2\pi\lambda\, L/n\,\Lambda^2$ must be larger than or equal to 4π.[7] This condition sets the lower limit for the interaction length L, ie. the aperture of the acoustic wave:

$$L \geq \frac{2n\Lambda^2}{\lambda} \tag{6}$$

When the undiffracted and diffracted light propagate in the same waveguide mode, the Bragg condition may be written:

$$\sin\theta_i = \sin\theta_d = \frac{\lambda}{2n\Lambda} = \frac{\lambda f}{2nv} \tag{7}$$

Thus the diffraction angle is equal to the incidence angle, and both angles increase linearly with the acoustic frequency. Within the small-angle approximation the deflection angle δ is therefore given by

$$\delta = \theta_i + \theta_d = \frac{\lambda f}{nv} \tag{8}$$

Since the acoustic frequency here is at most 2 GHz, ie. very low in comparison to the optical frequency, no significant frequency shift of the optical beam occurs.

Effects of Laser Characteristics

The previous results refer to an idealized Gaussian plane-wave incident onto the SAW grating. In practice, however, the optical wave would have both a finite beam spread and a finite frequency linewidth. Le us consider the effects of these two factors, related to the beam expansion and collimation system and to the laser emission respectively.

Due to a beam spread α, the various input-wave angular components will be diffracted at different angles. Their effect on the analyzer's resolution will be negligible only if $\alpha < \delta \theta_{min}$, $\delta\theta_{min}$ being the angular deflection corresponding to the frequency resolution δf. By using eq.s (2) and (3) it turns out that

$$\alpha < \frac{\lambda}{nv} \delta f = \frac{\lambda}{nD} \tag{9}$$

that is the input optical wave must be near diffraction-limited. This condition is very severe, as it requires a collimated beam having an angular spread typically lower than 0.05mrad. However, at present it is more likely that the ultimate resolution of the IOSA will be limited by the value of the detector cell size S. In this case the condition (9) may be reduced to the point where $\alpha < S/F$; thus, for S = 12μm and F = 27mm, a value α = 0.44mrad will not compromise the resolution of the device.

As to the laser emission linewidth $\delta\lambda$, it will cause again a change $\delta\theta$ in the output diffraction angle, which can be expressed as

$$\delta\theta = \frac{f_o}{nv} \delta\lambda \tag{10}$$

Frequency resolution will not be adversely affected as long as that change is less than $\delta\theta_{min}$, ie.

$$\frac{\delta\lambda}{\lambda} \leq \frac{\delta f}{f_o} \tag{11}$$

A graphical display of the dependence of S/F on δf and D, and of $\delta\lambda$ on δf is presented in Figure 2. The line A represents the law S/F = $\lambda\delta f/nv$ (Eq.(5)), the hyperbola B represents S/F = $g\lambda/nD$ (Eq.(2)), and the line C corresponds to $\delta\lambda = \lambda\delta f/f_o$ (Eq.(11)). In the calculations we have assumed g = 1.41, λ = 0.83μm, n = 2.17, v = 3500m/sec, and f_o = 600MHz.

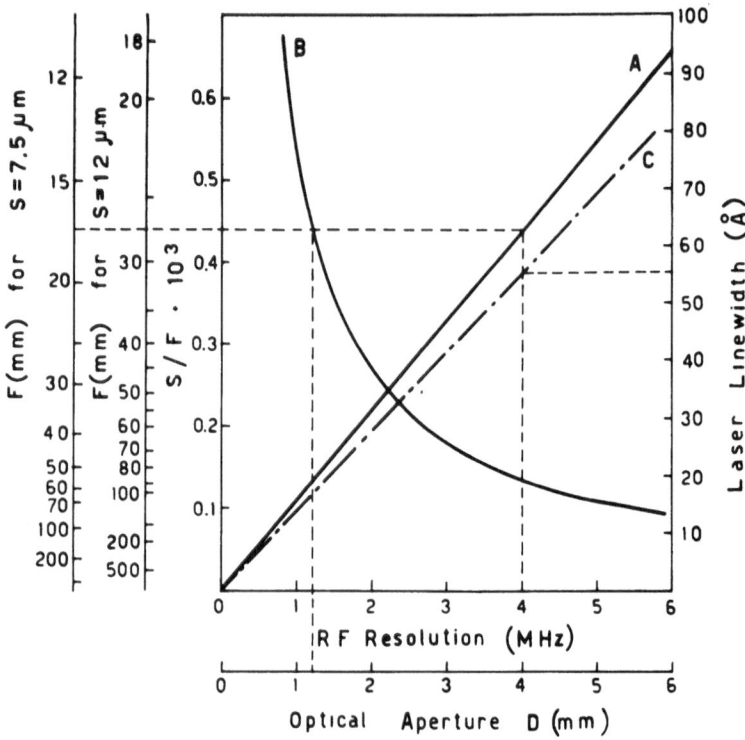

Fig. 2.   Graphical display of the relationships existing between
          optical aperture, focal length, detector cell size, laser
          linewidth and RF resolution.

        This design chart can be utilized as follows. If an RF frequency
resolution of 4MHz is desired, as in the devices implemented so far,
the corresponding value of S/F turns out to be equal to $0.44 \ 10^{-3}$,
which is related by the curve B to a minimum optical diameter D =
1.2mm.   The minimum focal length would be 27.27mm for a pixel pitch
S = 12µm, or 17.05mm for S = 7.5µm.   At the same time, the 4MHz
resolution would require a laser source having an emission linewidth
less than 55.4Å.

        It appears clearly that an increase in resolution to 1MHz would
set severe problems: with S/F equal to $0.11 \ 10^{-3}$, the focal length
should be 109.09mm or 68.18mm depending on the value of S, while the
minimum optical aperture would be ~5mm and the maximum laser linewidth 13.6Å.   Since detector arrays currently available have a pixel
pitch of the order of 12µm,[8] the 1MHz resolution would require a FT
lens with focal length of 109.09mm and therefore a substrate longer
than 145mm.   The realization of such a substrate in $LiNbO_3$ would be
difficult and costly.

        It is likely that in the future photodetector arrays with smaller
pixed pitch will be available.   However, another possible approach
to increase the resolution for a fixed value of S consists in util-

izing a third lens, that gives a magnified output distribution.[9] As
an example, with a magnification X = 5 and S = 12µm, an analyzer
having 1MHz resolution can be fabricated in a LiNbO$_3$ substrate of
approximately the same size (80mm x 15mm) as the current 4MHz devices.

COMPONENTS CONSIDERATIONS

The hybrid integration is a viable approach to the implementation
of a waveguide signal processor. On the contrary, the fabrication
of all the necessary active and passive components on a single-
material substrate (monolithic integration) is not currently feasible:
laser fabrication requires an AlGaAs/GaAs material structure, and the
present technology does not allow us to construct all the other components on it with the required quality.

Two material systems are well suited for the optical circuit
implementation: silicon and lithium niobate. The Si substrate offers
the advantage of a natural integration of the waveguide with the
detector-CCD system, requiring only the external laser chip. However, silicon is not piezoelectric, and the deposition of a single-
crystal ZnO film on it is required. LiNbO$_3$, on the contrary, is one
of the best acoustooptic materials, exhibiting low acoustic propagation loss and high electromechanical coupling constant. High-
quality waveguides are easily fabricated in it. The only disadvantage' is that both the laser and the detector are to be fabricated on
different chips and then butt-coupled to the LiNbO$_3$ waveguide; the
resulting losses are not negligible.

In the IOSA implementation the Si approach has been suggested
at Rockwell,[5] while the LiNbO$_3$, structure has been employed both at
Hughes[3] and Westinghouse[2]. Since the latter approach has given good
experimental results, in the following we will limit the discussion
to the components fabricated in the LiNbO$_3$ substrate, examining the
effect of their limitations on the device performance.

Laser Source

Significant advances have been made in the development of high
performance diode laser sources both for 0.8 to 0.9µm[10,11] and for
1.5 to 2.0µm[12] wavelength ranges. Transverse-mode stabilization and
longitudinal-mode control are among the most important goals in the
semiconductor laser fabrication, together with reliability improvement. Various laser cavity structures have been successfully introduced to stabilize the laser transverse mode, as in buried-hetero-
structure (BH) lasers and transverse junction stripe (TJS) lasers.
Longitudinal mode control is possible by grating feedback, but the
relevant technology is still immature.

GaAlAs lasers operating near 0.83μm are commercially available from several manufacturers. One of them, suitable for signal processing applications, is the Hitachi HLP-1400: it adopts a double heterojunction (DH) structure, and has excellent features such as a maximum output power of 15mW, and a beam divergence of typically 10° x 30°. This beam divergence is defined as the full beam width at half maximum intensity points parallel and perpendicular to the junction plane.

Typical emission spectra of an HLP-1400 diode at various output power levels under CW operation are shown in Figure 3.[13] At an output power of 0.5mW, the diode is just above threshold, and a number of longitudinal modes are observed. As the output power is increased, the envelope of the spectra narrows until most of the emitted power is concentrated in a single longitudinal mode. This happens for powers above 2 - 3mW. The lasing spectrum at an output power of 10mW is shown in Figure 4 at a different intensity scale, so as to display the fine mode structure still existing. Each peak corresponds to a longitudinal mode, and the wavelength separation between two of them is typically of the order of 3Å. As to the transverse mode of this Hitachi laser, the far-field patterns show that it emits in the single, fundamental, transverse mode up to reasonable output levels.

Experimental measurements of the broadband spontaneous emission in the HLP-1400 and other lasers (General Optronics model GOLS-1, Laser Diode Labs SCW-21, and Mitsubishi ML-4307) have been reported recently.[14] Figure 5 shows the spectral width measured at 50 dB below the peak intensity as a function of single-mode (longitudinal and transverse) output power. The minimum spontaneous emission bandwidth is 200Å at approximately 4mW output power. The effect of this bandwidth on the performance of the spectrum analayzer has been evaluated, and as a result Figure 6 reports the dynamic range of the IOSA expressed as the ratio of the signal reaching the n-th nearest neighbor pixel to the signal reaching the 0-th order pixel. This ratio has been calculated for an IOSA with 4MHz resolution, operating with a 400MHz acoustooptic bandwidth centered at 600MHz. At the fourth nearest neighbor the signal due to the spontaneous emission is down 37 dB at 800MHz and 42dB at 400MHz; at the second nearest neighbor the noise signal is down 31dB at 800MHz and 35dB at 400MHz. Thus a true signal having power level below these values will not be detected; efforts are therefore necessary to further decrease the spontaneous emission bandwidth in diode lasers.

Optical Waveguide

An optical waveguide layer in $LiNbO_3$ is commonly fabricated by the deposition of a thin Ti film on the polished surface of the substrate and the in-diffusion of the Ti.[15] The accompanying out-diffusion of Li_2O can be suppressed by suitable means, eg. by annealing

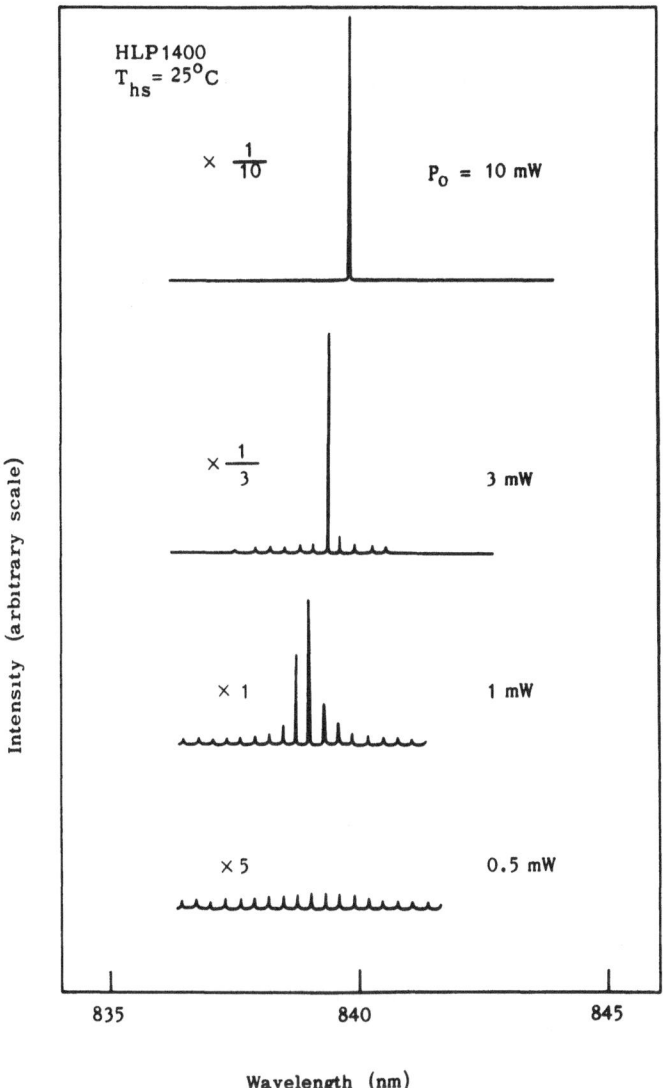

Fig. 3. Typical emission spectra of an Hitachi HLP-1400 Laser
         diode at various output power levels[13].

the crystal substrate in $LiNbO_3$ powder at 900°C in a flowing oxygen
environment.[16] Typically, single mode waveguides with absorption
loss less than 1dB/cm are formed by diffusing 200Å thick Ti at 950°C
for 4 hours. The corresponding diffusion depth is about 2μm.

     Optical scattering in the waveguide is expected to be one of
the key factors limiting the IOSA dynamic range. Experimental studies
of the in-plane scattered energy distribution have been performed in
order to determine and subsequently eliminate sources of scattering

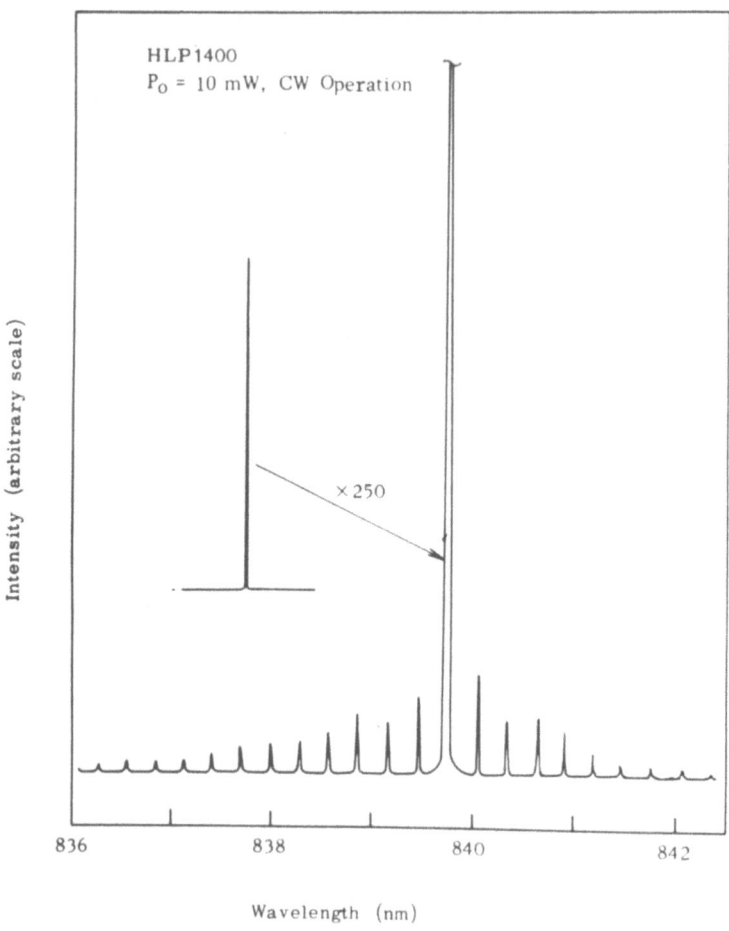

Fig. 4. Typical lasing spectrum of HLP-1400 at a 10mW output
         power[13].

in the LiNbO$_3$ waveguides. The observations made so far have allowed
Vahey[17] to conclude, among other things, that bulk scattering is far
more important than surface scattering, and that it increases in
proportion to the amount of Ti used to make the waveguide but is
rather insensitive to the time and duration of the diffusion treatment. For the best waveguides the ratio of the scattering intensity
at 1° from the forward direction to the unscattered light is of the
order of 45-50dB.

    The waveguide scattering problem has been faced theoretically
by some authors.[18-20] In particular, Boyd and Anderson[19] have evalu-

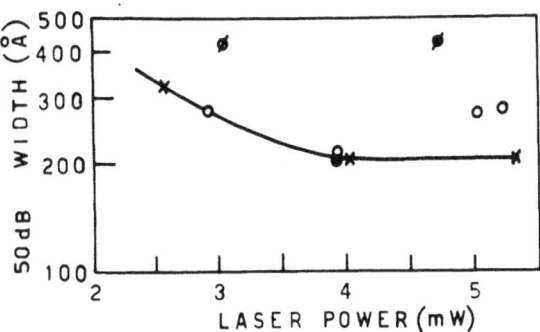

Fig. 5. Spectral full width at 50dB below the peak intensity versus
         the output power for different lasers (x GOLS-1/TB46;
         o GOLS-1/7034; o SCW-21; ∅ HLP-1400; ∅ ML-4307)[14].

ated the effect of scattering on the IOSA dynamic range. Their results indicate that the dynamic range decreases with increasing the
waveguide attenuation, while it is nearly independent of the surface
roughness correlation length for a fixed attenuation. For a waveguide attenuation decrease from 1dB/cm to 0.1dB/cm the dynamic range
may increase from 25dB to 38dB. A dynamic range of 40dB or higher
is required in most of the applications.

Waveguide Lenses

     Optical waveguide lenses can be fabricated by following different
design approaches.[21-23] Geodesic lenses are the best currently feasible solution with the high-index LiNbO$_3$ substrate. They are based
on the possibility of distorting the two-dimensional guide into the
third dimension by a suitable shaping of the surface of the substrate.
Various methods have been suggested to design perfect or corrected
lenses; they will be discussed in another paper of this issue.

     The geometrical profile of perfect geodesic lenses is strongly
aspheric and it may be difficult to fabricate them with the accuracy
(better than 0.5µm) required to achieve the theoretical diffraction-
limited performance. Conventional grinding techniques,[24] as well as
the ultrasonic impact grinding,[25] are probably inadequate because the
final accuracy in these methods depends to a great extent on the
amount of material removed during the final polishing. The most
promising technique is therefore the single point diamond turning,
that can ensure a form accuracy better than 0.5µm and a surface
roughness less than 0.01µm,[26] and has been successfully used in
LiNbO$_3$ substrates.[27]

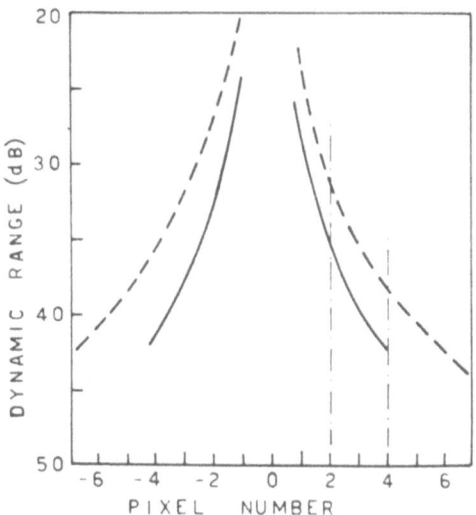

Fig. 6.   IOSA dynamic range as a function of the n-th pixel, as
          limited by the 200Å spectral width. Continuous and
          dashed lines correspond to an acoustic bandwidth of 800MHz
          and 400MHz respectively[14].

SAW Transducer

    So far, for sake of simplicity, we have considered a single SAW
transducer. However, some form of frequency-selective beam steering
of SAWs is required in order to match the Bragg-angle condition and
to keep the efficiency reasonably constant across the acoustic bandwidth. This can be accomplished by using different device configurations that are characterized by large bandwidth and high diffraction
efficiency.[7] One of these configurations consists of an array of
parallel fed ID transducers with staggered center frequencies, tilted
with respect to each other in acoustic propagation direction. The
tilt angle between each pair of adjacent transducers is set equal
to the difference of the two Bragg angles at the two corresponding
center frequencies. A further possibility is that of varying continuously the interelectrode spacings across the transducer.[7,28]
The spacing law in the specific chirp transducer depends upon the
shape of the passband desired. SAW transducers are made by optical
or electron-beam lithography, and therefore complex structures are
not much more difficult to fabricate than a simple transducer.

    The IOSAs implemented so far exhibit a 400MHz bandwidth with a
center frequency of 600MHz, but 1GHz bandwidth with an 1mW electrical
drive power per megahertz bandwidth and 50% diffraction efficiency
should be realizable in the near future.

Detector Array

The photodetector array to be used in the IOSA is required to exhibit minimal pixel pitch, low crosstalk, large signal dynamic range, and adaptability to a hybrid IO structure. Detector arrays have been designed for specific application in waveguide processors. One of the first devices[29] contained 8 arrays of 19 elements each. The center-to-center spacing was 32μm, due to the single photodiode width of 25μm (length 115μm) plus 7μm of isolation between adjacent elements. Recently an array has been tested,[30] consisting of 7 groups of 20 photodiodes each, with a 12μm pixel pitch and no dead spaces: pixel size is 12 x 360μm. Both these arrays have been coupled to CCD shift registers to perform the spatial to temporal conversion so that the RF spectrum is read out serially in time. The latter array (Westinghouse 5050) has an access time (integration time) of 2μsec, and an electronic dynamic range of the order of 50dB.

Curved detector arrays might be necessary for an IOSA with resolution better than 1MHz and bandwidth larger than 600MHz, due to the field curvature of rotationally symmetric geodesic or Luneburg lenses.

A different readout scheme has been also suggested, employing a very high scan rate analog deflector to convert the deflected beams into fast-scanning focused light spots.[7] A single high-speed photodetector, incorporating a narrow slit and located in the focal plane, is then used to detect and convert the scanning focused light spots into an electrical spectra which is already serial in time. In some preliminary experiments using laser light at 0.6μm and a center frequency of 300MHz, a frequency resolution of 3MHz and a bandwidth of 180MHz have been demonstrated.

INTEGRATION PROBLEMS

From the previous analysis of the components of a typical IO processor it appears that technological problems still exist, limiting the quality and availability of the devices. For example, perfect geodesic lenses are not yet routinely fabricated, and on the other hand the available detectors have a pixel pitch larger than the focal spot size which can be achieved with corrected lenses. The scattering in waveguides, as well as the light coupled in substrate modes, the laser spontaneous emission, and other factors contribute to degrade the performance of an IOSA as far as the dynamic range is concerned. Further problems arise from the integration of the source and the detectors with the $LiNbO_3$ substrate. The butt-coupling of the three chips deserves careful attention; its efficiency affects the power requirement of the source and the sensitivity requirement of the detector.

Light coupling to a planar waveguide by end excitation has been investigated by a number of researchers.[31-33] In a recent analysis concerning specifically the butt-coupling of a GaAlAs laser diode with a Ti:LiNbO3 waveguide, the maximum achievable efficiency has been evaluated to be less than 40% because of the mismatch between the waveguide mode and the laser radiation distribution.[33] Detailed calculations and experimental measurements of the coupling efficiency as a function of the transverse displacement Δ, of the angular mis-alignment θ, and of the longitudinal separation z (Figure 7) have been reported. Within an efficiency range from 25 to 30 percent, as occurs in most practical cases, the coupling efficiency is not very sensitive to such errors. In particular it turns that a slightly higher efficiency is achieved at a certain angle θ, of the order of 10°, rather than at $\theta = 0$. The exact value of θ at which the ef-ficiency is a maximum varies, depending on the separation z and the mode profiles.

Methods for improving the coupling efficiency may require: (i) availability of laser diodes with less divergence, (ii) fabrication of LiNbO3 waveguides which confine the light more closely to the waveguide surface, and (iii) use of a suitable cylindrical lens to achieve a better match between the laser field and the waveguide mode. In the present situation, however, approximately half of the laser output power will be coupled into substrate modes. A not neg-ligible fraction of this power will undergo total internal reflections in the LiNbO3 substrate and will likely reach the detector array, so contributing to lower the dynamic range. In order to partially sup-press the substrate modes as well as the radiation scattered from the waveguide into the volume of the substrate, absorbing layers formed near the back side of the substrate by H implantation have been successfully tested.[34] Also the use of absorbing multilayer inter-ference coatings (dark mirror coatings) on the back side of the LiNbO3 substrate have proven effective in reducing reflected sub-strate modes at $\lambda = 0.83\mu m$ over a braod range of incident angles.[35]

The problem of coupling the light from the waveguide to the detectors has received less attention so far. Anti-reflection coat-ing of the cross-sectional area of the Si array and an angular mounting at a 45° angle to the LiNbO3 waveguide edge have been suggested to increase the coupling efficiency.[30] However, at present no quantitative data are available to us.

OTHER WAVEGUIDE PROCESSORS

Other approaches to the IOSA implementation have been proposed, concerning different fabrication methods of either the optical system or the input-signal transducer. In the first case the use of grating components, to be fabricated by holographic and electron-beam litho-graphy, is envisioned.[36] The second approach represents an electro-

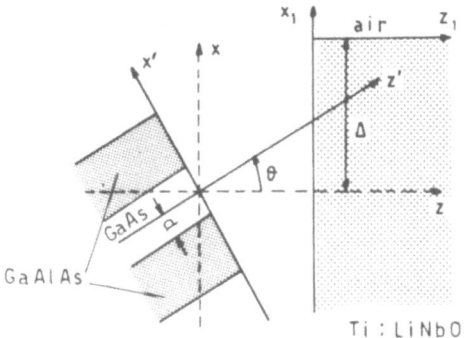

Fig. 7. Coupling geometry for an injection laser and a
planar waveguide.

-optic alternative to the AO analyzer and is based on the use of an
electrode array giving a weighted Discrete Fourier Transform of the
sampled input voltages.[37]   In addition, processing devices performing
different functions are under development in some laboratories.[7,9,
38-40]   The implementation and performance considerations that have
been made in the previous sections apply to most of them.

     As an example of the freedom in design offered by the guided
wave technology, let us consider a potentially attractive device as
the AO time-integrating correlator (AOTIC).[41]   Different designs of
the geodesic optical system can be used to match the device to the
particular application one has in mind. Figure 8 shows an hybrid-
structure AOTIC.[7] Here the signal to be correlated $s_1(t)$ is used
to modulate the intensity of the LD laser source. The modulated
light is then collimated by the lens L_1 and diffracted by an acousto-
optic modulator, driven by the reference signal $s_2(t)$. The diffracted
light is collected by the lens L_2, filtered and imaged by L_3 onto a
detector array. It can be shown that, under proper conditions, the
output of the photodetectors contains the correlation of the signals,
which can be read out in time using a CCD array.

     A more compact optical design, that had been suggested previously,[9]
is sketched in Figure 9. With this configuration the field curvature of the lenses has a negligible effect, at least when a pulse
response is expected. The Fourier transform of the signals is correct,
apart from a quadratic phase factor. The output line, where the detectors are to be placed, corresponds to a circle whose radius depends
on the focal length of the second lens and on the distance between
the AO transducer and the focal line of the first lens. In this case
particular attention has to be devoted to the design of the AO transducer, due to the curvature of the incident optical beam.

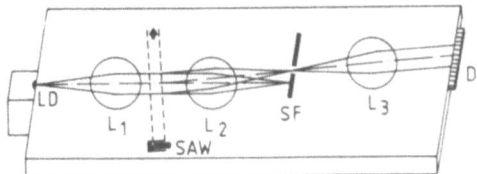

Fig. 8.   A waveguide AO time-integrating correlator (AOTIC)[7].

A further approach to the AOTIC implimentation is based on the
focusing and imaging properties of spheric geodesic optical systems.
It is known that a quarter of spheric surface acts like a perfect
focusing element. Therefore a three-lens AOTIC having the same
structure as the device of Figure 8 can be fabricated in a very
rugged configuration, that consists of three quarters of a sphere.[39]
Such a substrate is very easy to build, while it may be more difficult
to realize the other components on the curved surface.

CONCLUSIONS

The theoretical and experimental results so far obtained in many
research laboratories confirm that waveguide signal processors are
viable applications of IO technology. In particular, IO circuits
offer a way to achieve coherent optical signal processing in a rugged
package, almost not affected by environmental factors such as vibration and temperature variation. Anticipated uses of these devices
are in radar, computer, and communication systems.

Limitations to the performance of the single components and of
the complete device still exist, but most of them are mainly technological. It is therefore reasonable to expect that, due to the development work, significant improvements of the current performance
level of IO processor will not be far behind.

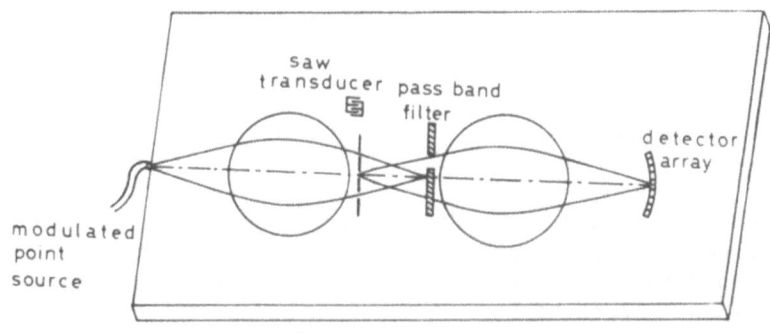

Fig. 9.   Schematic of a two-lens AOTIC[9].

REFERENCES

1. R. Shubert and J.H. Harris, Optical surface waves on thin films and their application to integrated data processors, IEEE Trans. Microw.Theory Techniques, MTT-16:1048 (1968).
2. D. Mergerian, E.C. Malarkey, R.P. Pautienus, J.C. Bradley, G.E. Marx, L.D. Hutcheson and A.L. Kellner, Operational integrated optical R.F. spectrum analyzer, Appl.Opt. 19:3033 (1980).
3. T.R. Ranganath, T.R. Joseph and J.Y. Lee, Integrated-optic-spectrum analyzer: a first demonstration, in: "IOOC '81 Techn. Digest," San Francisco, p.114 (1981).
4. M.K. Hamilton, D.A. Wille and W.J. Miceli, An integrated optical R.F. spectrum analyzer, Opt.Engin. 16:475 (1977).
5. D.B. Anderson, J.T. Boyd, M.C. Hamilton and R.R. August, An integrated Optical approach to the Fourier Transform, IEEE J. Quantum Electron. QE-13:268 (1977).
6. M.K. Barnoski, B.U. Chen, T.R. Joseyh, J.Y.M. Lee and O.G. Ramer, Integrated-optic spectrum analyzer, IEEE Trans.Circuits and Systems, CAS-26:1113 (1979).
7. C.S. Tsai, Guided-wave acoustooptic Bragg modulators for wide-band integrated optic communications and signal processing, IEEE Trans. on Circuits and Systems, CAS-26:1072 (1979).
8. G.M. Borsuk, A. Turley, G. Marx and E. Malarkey, Photosensor array for integrated optical spectrum analyzer systems, Proc. SPIE, 176:109 (1978).
9. G.C. Righini, V. Russo and S. Sottini, Signal processing in integrated optics employing geodesic lenses, Proc.SPIE, 164: 20 (1979).
10. I. Melngailis, Laser sources and detectors for guided wave optical signal processing, Opt.Engin. 19:941 (1980).
11. M. Nakamura and S. Tsuji, Single-mode semiconductor injection lasers for optical fiber communications, IEEE J.Quantum Electron. QE-17:994 (1981).
12. G.H. Olsen, Laser diodes for the 1.5µm-2.0µm wavelength range, J.Opt.Commun. 2:11 (1981).
13. "Hitachi Laser Diode Application Manual", June 1979.
14. W.K. Burns and R.P. Moeller, Effect of laser diode spontaneous emission on IOSA operation, App.Opt. 20:913 (1981).
15. R.V. Schmidt and I.P. Kaminow, Metal-diffused optical waveguides in LiNbO$_3$, Appl.Phys.Lett. 25:458 (1974).
16. B. Chen and A.C. Pastor, Elimination of Li$_2$0 and LiTaO$_3$, Appl. Phys.Lett. 30:570 (1977).
17. D.W. Vahey, In-plane scattering in LiNbO$_3$ waveguides, Proc.SPIE 176:62 (1979).
18. D. Marcuse, Mode-conversion caused by surface imperfections of a dielectric slab waveguide, Bell Syst.Tech.J. 48:3187 (1969).
19. J.T. Boyd and D.B. Anderson, Effect of waveguide optical scattering on the integrated optical spectrum analyzer dynamic range IEEE J.Quantum Electron. QE-14:437 (1978).

20. D.G. Hall, Comparison of two approaches to the waveguide scattering problem, Appl.Opt. 19:1732 (1980).

21. S.K. Yao and D.B. Andersen, Shadow sputtered diffraction-limited waveguide Luneburg lenses, Appl.Phys.Lett. 33:307 (1978).

22. S. Sottini, V. Russo and G.C. Righini, General solution of the problem of perfect geodesic lenses for integrated optics, J.Opt. Soc.Am. 69:1248 (1979).

23. W.S.C. Chang and P.R. Ashley, Fresnel lenses in optical waveguides, IEEE J.Quantum Electron. QE-16:744 (1980).

24. G.C. Righini, V. Russo and S. Sottini, Aspherics in integrated optics, Proc.SPIE. 235 (1981).

25. B. Chen, E. Marom and R.J. Morrison, Diffraction-limited geodesic lens for integrated optics circuits, Appl.Phys.Lett. 33:511 (1978)

26. See, for instance, the papers in "Proc. of the Conference on Aspheric Optics" (London, 1980).

27. D. Mergerian, E.C. Malarkey, R.P. Pautienus and J.J. Bradley, Diamond-machined geodesic lenses in LiNbO3, Proc.SPIE. 176:85 (1979).

28. T.R. Joseph and B.U. Chen, Broadband modified chirp transducers, in: "Tech.Digest of Topical Meeting on Integrated and Guided-Wave Optics," Incline Village, Paper ME-2 (1980).

29. J.T. Boyd, Integrated optoelectronic silicon devices for optical signal processing and communications, Proc.SPIE. 139:167 (1978).

30. G.E. Marx and G.M. Borsuk, Evaluation of a photosensor for integrated optics spectrum analyzers, in: "Tech. Digest of Topical Meeting on Integrated and Guided-Wave Optics," Incline Village, paper ME6 (1980).

31. H.P. Hsu and A.F. Milton, Single-mode coupling between fibers and indiffused waveguides, IEEE J.Quantum Electron. QE-13:224 (1977).

32. R.G. Hunsperger, A. Yariv and A. Lee, Parallel end-butt coupling for optical integrated circuits, Appl.Opt. 16:1026 (1977).

33. C.T. Mueller, C.T. Sullivan, W.S.C. Chang, D.G. Hall, J.D. Zino and R.R. Rice, An analysis of the coupling of an injection laser diode to a planar LiNbO3 waveguide, IEEE J.Quantum Electron. QE-16:363 (1980).

34. W.K. Burns, J. Comas and R.P. Moeller, Applications of ion implantation to integrated optical spectrum analyzers, Opt.Lett. 5:45 (1980).

35. A.H. Singer and R.T. Holm, Absorption of reflected substrate modes in Ti:LiNbO3 waveguide systems, Opt.Lett. 6:53 (1981).

36. V. Neuman, C.W. Pitt and L.M. Walpita, Guided wave holographic grating beam expander - fabrication and performance, Electron. Lett. 17:165 (1981).

37. L. Thylen and L. Stensland, Electrooptic approach to an integrated optics and spectrum analyzer, Appl.Opt. 20:1825 (1981).

38. D.W. Vahey, C.M. Verber and R.P. Kenan, Development of an Integrated-optics multichannel data processor, Proc.SPIE. 139:151 (1978).

39. G.C. Righini, V. Russo and S. Sottini, in: "Optical thin film
     processor for unidimensional signals," U.S. Patent, 4:222,628
     (1980).
40. C.M. Verber, R.P. Kenan and J.R. Busch, Correlator based on an
     integrated optical spatial light modulator, Appl.Opt. 20:1626
     (1981).
41. R.A. Sprague and K.L. Koliopoulos, Time integrating acoustooptic
     correlator, Appl.Opt. 15:89 (1976).

GEODESIC OPTICS: OVERVIEW AND PERSPECTIVES

V. Russo

Istituto di Ricerca sulle
  Onde Elettromagnetiche, C.N.R.
Firenze, Italy

INTRODUCTION

Geodesic optics can be defined as the particular field of guided optics where the light propagation occurs on a non planar waveguide of uniform thickness. According to the Fermat principle, the rays follow the geodesics of the curved guiding surface. This property can be utilized to form high quality passive components in integrated optics.

Generally, a geodesic component consists of a depression (having rotation symmetry) created in the substrate, covered by a homogeneous film of constant thickness (Figure 1). Because its operational characteristics are determined purely geometrically, a geodesic component has the advantage of being compatible with any type of film and substrate. It works the same for any wavelength as well as for any mode supported by the waveguide. Geodesic lenses, for instance, are the best solution whenever crystal substrates of very high refractive index must be used. In this case, building a Luneburg or a Fresnel lens, based on a substantial change in the refractive index of a planar guide is rather difficult for the limited range of suitable materials.

As is well known, the problem of geodesics has been widely studied in microwaves optics, because it was of interest in the field of rapid scanning antennas. There microwaves antennas, either constituted by conflection elements or by continuous curved parallel plates were investigated. Much has been done in guided optics to design geodesic components, following the directions given by Kunz[1] and Toraldo[2] in their fundamental works on geodesic microwave optics.

267

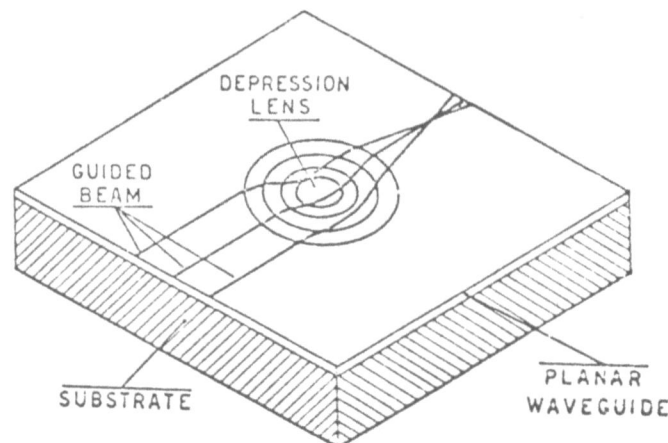

Fig. 1. Waveguide geodesic lens realized as a depression in
         a substrate covered by a homogeneous guiding film of
         constant thickness.

     Geodesic optics was introduced in integrated optics in 1972
with two very simple thin film elements: the 1/4 of spherical
surface which operates as a perfect lens, and the hemispherical
surface which is a perfect imaging device.[3] Then attention was
drawn to the possibility of obtaining both perfect aspheric lenses
without discontinuity to the planar circuit and conflection components constituted by more than one developable surface.[4]

     Some years later when the possibility of constructing thin
film integrated optics spectrum analyzers became more realistic,
geodesic optics attracted renewed attention and the associated
fabrication problems were approached. Besides lenses, that have
been widely investigated due to their employment as Fourier transform elements, other geodesic components performing different operations from focusing or imaging have been designed.

     I will first review a class of elements which usually employ
sections of intersecting cones, that is surfaces developable onto
a plane. Their optical performances are determined by the conflection angle made by the rays at the sharp bends between the intersecting surfaces. Then I will consider the other geodesic components, which present in general rotation symmetry, and where, contrary to the above class, sharp bends that cause scattering and
radiation losses and mode conversions are avoided. Two design
procedure are mainly used to obtain such components. The first is
based on the principle of equivalence between inhomogeneous refractive index distributions and the corresponding geodesic components.[1] The second one is based on a general theorem on the

geodesics of a rotation surface.[2] It directly approaches and solves
the problem of designing both perfect aspheric lenses without discontinuity and a set of new components.

Without going into detail about the design procedures, which
are well established in the literature, an overview of the geodesic
components so far produced in guided optics will be given, trying
to point out the differences from the operational point of view and
keeping in mind practical applications. Much attention will be given
to the practical problems which arise and the difficulty in finding
a fabrication technique that ensures the required precision. Some
predictions of the effects of fabrication errors can be envisaged
utilizing the analytic procedure used for designing perfect aspheric
lenses.

Experimental data on the optical characteristics of the geodesic
lenses so far produced will be reported. Finally aberration-free,
rounded-edge components able to deflect, split or mix guided beams
which have been recently proposed will be discussed.

CONFLECTION ELEMENTS

This class of components is characterized by the fact that they
are formed by surfaces which are developable onto a plane. Let us
consider one such element as shown in Figure 2.[5] Its operational
characteristics are determined by the "law of conflection" of the
optical rays, which is well known in microwave optics. It states
that a geodesic crossing a sharp bend must have equal angle with
the tangent to the bend in both sides (Figure 3). As a consequence
a parallel beam propagating over the two folds suffers a net deflection with respect to the original direction. Very recently an
elastic rubber waveguide optical deflector with a geometry similar
to that of Figure 2 has been constructed and tested.[6] A transparent
rubber film waveguide rests flat and unsupported over the cuneiform

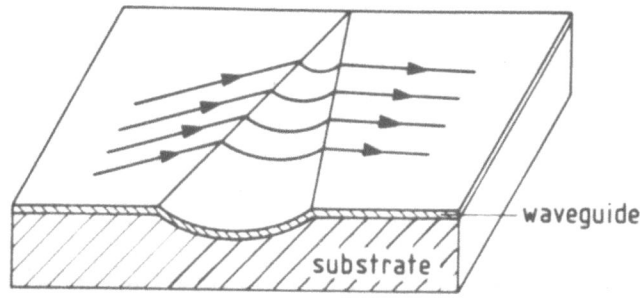

Fig. 2.   Sketch of a parallel beam deflector, constructed by
a conic surface intersecting a plane.

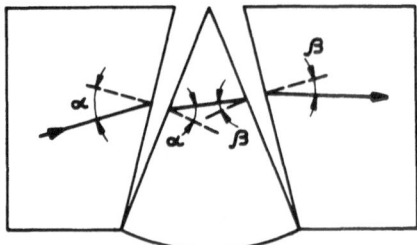

Fig 3. "Conflection law" for the optical rays crossing
the developable surfaces of Figure 2.

depression. When forced in intimate contact with it, for example
by pneumatic pressure, a continuous large angle of deflection is
obtained. Analogously a cylinder or two crossing cylinders can
shift or split a parallel beam,[5] (Figure 4).

Conflection components can be proposed only for multimode
propagation because the sharp transition between one surface and
the other causes high losses and mode conversion. On the other hand
any edge rounding yields to a deterioration of its behavior.

A number of more complex elements can be proposed in guided
optics by suitably combining cones and cylinders, giving rise also
to correct focusing devices.[4,7] Such devices have non coincident
input and output planes and are more difficult to construct, however
they would deserve to be investigated more thoroughly.

GEODESIC COMPONENTS

The first geodesic component to be studied was a focusing lens
and it is so far the most widely investigated because of its ability
to perform Fourier transforming operations.

For application in IO signal processors, a geodesic lens is
required to exhibit diffraction-limited performances and single mode
operation. This means that it must be aberration-free and matched
without discontinuity to the planar circuit. A smooth transition
from the planar to the curved guide is necessary to prevent significant radiation, scattering and mode conversion.

The simplest geodesic lens consists of a quarter of spherical
surface, but it cannot be inserted in a planar circuit. On the other
hand every portion of a spherical surface focuses with strong spherical aberration. Some efforts have been made to compensate for this
aberration with an overlay region of slightly higher effective
refractive index,[8] following a well known method in bulk optics and

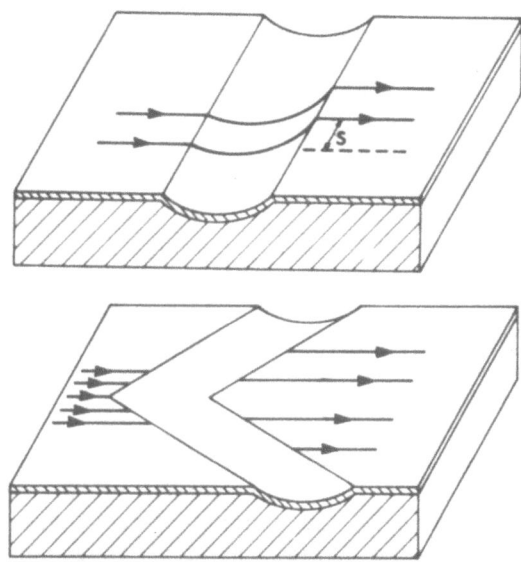

Fig. 4. Parallel beam conflection shifter and splitter.

microwave conflection antennas. Diffraction limited performances
can be achieved by using a non circular overlay of uniform thickness;[9]
however the image spot size deteriorates rapidly with increasing off
axis angle. Geodesic lenses corrected with overlays can be proposed
only for multimode operations.

A direct approach to solve the problem of the aberrations is to
give the depression an aspherical shape. Significant work has been
done on aspherical lenses presenting sharp transition with the plane
[10,11,12] and also on the effects of an a posteriori edge rounding.
[13,11] Since an arbitrary edge rounding affects the focal properties
of a perfect lens, the only way to obtain aberration-free lenses
without discontinuity is to include the edge rounding problem in the
design procedure. A first family of geodesic lenses which meets the
above requirement was proposed in guided optics.[4] They perfectly
focused a plane beam at the smooth transition with the plane. However in view of practical application, generalized lenses, focusing
outside the transition, are more convenient.

Two different design procedures have been followed to generate
aberration-free rounded-edge generalized geodesic lenses. The first
is based on the principle of equivalence between geodesic lenses and
flat refractive index variation[1] lenses. This principle states that
two lenses are equivalent if their Fermat matrices are equivalent.
In other words two lenses are equivalent if their indexes and surface
profiles satisfy the condition of equal optical path length (where
the optical path is determined by the Fermat principle). The design

procedure by means of a numerical approach, entails the conversion
of a rounding function to a refractive profile of an outer ring of
a Luneburg lens. The refractive index distribution of the inner
portion of the lens is calculated to achieve aberration free focusing.
Then the refractive index profile is transformed into an aspherical
geodesic profile.[14,15]

The second design procedure makes use of a general theorem of
geodesic of rotation surfaces.[2] An analytic procedure[16] allows one
to derive the general expression for the profile of aspherical
geodesic lenses able to form perfect geometrical images of the points
of two given concentric circles on each other. The depression is
divided in two parts. The first part consists of an outer shell
that joins together the external plane surface and the true lens.
It is specified arbitrarily but subject to well determined conditions
in order to avoid discontinuities. Then the profile of the inner
part is uniquely derived, which constitutes the true lens because
only the rays that enters the central part of the depression can be
perfectly focused in the image. Equations have been given that
characterized a family of lenses having two conjugate foci external
to the lens depression. A typical lens profile of an imaging lens
is shown in Figure 5. As another example, Figure 6 shows three
meridional curves of lenses that perfectly focus a collimated beam
on the plane at the same focal distance f = 9mm. They have the same
radius (5mm) of the lens depression while the rounding widths as well
as the F/numbers are different. The advantage of this analytical
method with respect to the numerical one is represented by the clear
theoretical approach that does allow a large flexibility without requiring long and expensive calculations.

Once the lens profile is found, one can visualize the behaviour
of the lens by means of the ray tracing technique or by means of the
method of the beam propagation, recently suggested.[17] It is based
on the angular spectrum of plane waves through an inhomogeneous
medium and it allows one to calculate directly the propagation of an

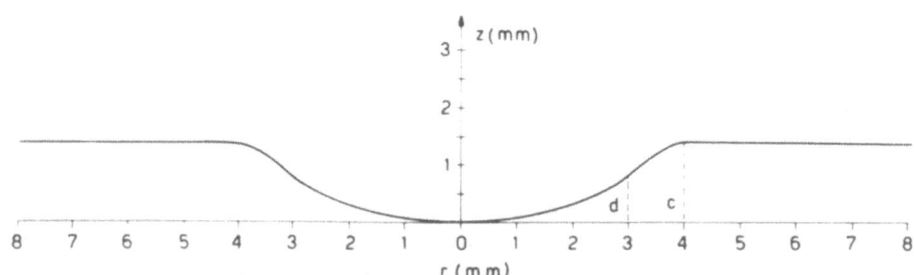

Fig. 5.   Profile of a generalized geodesic lens with focal length
          f = 9mm and F/number = 1.5 It perfectly images two
          circles, of radius 11.25 and 45mm respectively on each
          other.

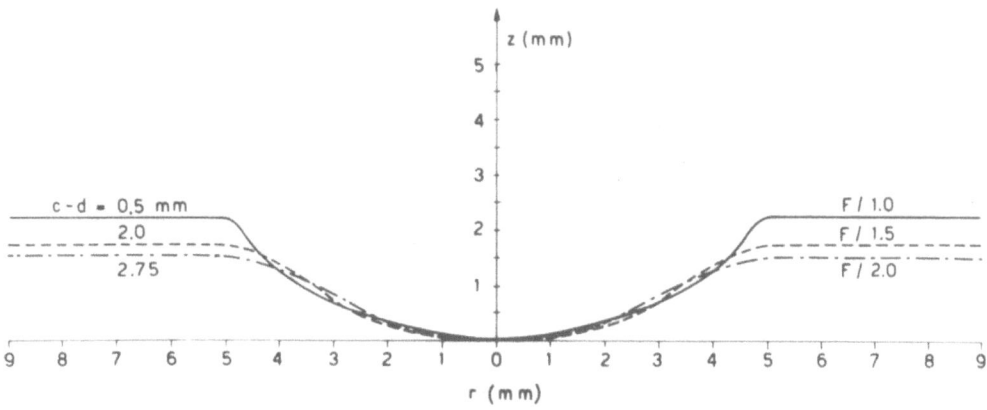

Fig. 6.   Three profiles of generalized geodesic lenses perfectly
focusing a collimated beam at the same focal distance
f = 9mm.   The lenses have the same depression diameter
too, while the rounding widths, as well as the F/numbers
are different.

arbitrary light beam through the lens and to get the complete field
at the output plane. The calculation is based on the Fast Fourier
Transform. It has been applied to two of our lenses (Figure 7 and
Figure 8).

FABRICATION TOLERANCES AND TECHNIQUES

        Aspherical geodesic lenses, designed either with the equivalence
principle or with the analytic procedure, have been constructed in
view of applications to integrated optical processors. For the time
being they are proposed for use in R.F. spectrum analyzers (see Ref.
18,19,20,21). However translating the design in a working component
particularly when it is required to have diffraction limited performances is not an easy task in practice.

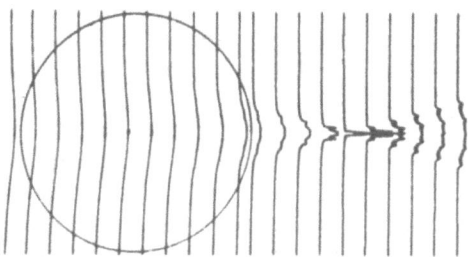

Fig. 7.   Propagation of Gaussian beam through a geodesic lens
with depth = 480μm; radius = 1mm; focal length = 1.8mm.

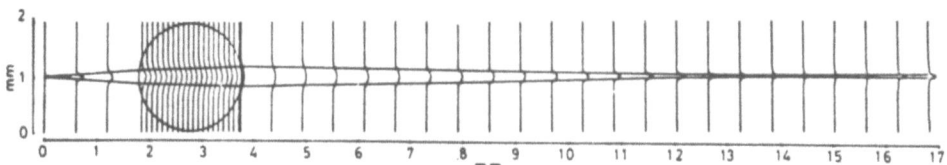

Fig. 8. Propagation of Gaussian beam through geodesic lens with
         two external foci. Depth = 575μm; radius = 1mm; focal
         lengths = 281mm, 11.25mm.

At present three different techniques are under testing. The
first, which we have used in workshop, is the conventional one of
grinding and polishing a glass substrate. A reference template is
first obtained either in stainless steel plate 0.3mm thick by photo-
litographic and chemical cutting or in an alumina plate 0.6mm thick
by means of a computer controlled CO_2 laser cutter. The accuracy
of the template is 10μm. An iron tool having approximately the same
profile as the template is then fabricated and used to grind the
depression by spinning the glass substrate. The final polishing
with Syton powder is done with the aim of fitting as well as possible
the depression profile with the reference template. With this
technique we have realized a lens (D = 9mm, f = 4.5mm) that turned
out to be able to resolve collinear beams, 2.5mm width separated by
3mrad, corresponding by a linear separation of 14mm. This corresponds to an arc separation of 14μm, this means that the focal spot
size is less than this figure. Another rough measurement indicated
that no focal shift greater than 50μm was obtained. Other lenses
have been constructed and a precision control of the achievable
accuracy is under investigation.

The second technique exploited by Chen et al.[18] is the ultrasonic impact grinding. Ultrasonic energy transmitted through the
tool, having the intended shape of the depression, agitates the
abrasive slurry, which removes material from the surface of the
sample. This technique would be capable of replicating the shape
of the tool with 1/4μm tolerance. However the operation require
more than one tool in order to reduce the effect of tool wear that
changes substantially the final depression shape. Aspherical lenses
both in glass and $LiNbO_3$ substrates have been machined and the profile was measured.

Single point diamond turning is a precision machining technique
used for production of high quality optical bulk components in crystalline and policrystalline materials. The surface roughness is of
the order of 40Å. With an experimental machine of the Moore Special
Tool Company, Bridgeport, Connecticut, equipped with a numerical
control system for its movements, geodesic lenses into $LiNbO_3$ crystal
substrate have been fabricated.[20] A single point diamond-tool is

mounted in place the usual grinding wheel. It rotates in a planetary orbit so that the cutting edge is held at all times normal to the contour being machined into the workpiece. It is worthwhile to note the fact that the machine is designed so that the contour accuracy achieved can be measured at any time while the workpiece is mounted on the machine. The achieved accuracies far exceed those achievable by other techniques. An overall figure accuracy of better than 1.5μm total indicator and a final depth error of only 0.25μm have been obtained in the best case. This technique is the most attractive for crystal materials but cannot be used for glass substrates where the problem has not at present a satisfactory solution. It should be pointed out that testing these lenses during or after the manufacture presents problems not encountered in bulk optics. The lens cannot focus unless used with its subsequently applied waveguide. Thus the profile must be assessed before forming the waveguide.

An analysis of the effects of the fabrication errors on the optical properties of a lens is of great help to the designer in order to evaluate the performance of the entire optical circuit.[22,23] Generally, two possible fabrication errors are considered the edge and the depth changes with respect to the ideal shape. Although the cases reported in literature are few, it seems clear that the depth error is more difficult to avoid. Its effect is usually small on the image spot size while it is relevant for the shift of the focal line. Therefore, taking advantage of our analytic design method we have performed an extended numerical analysis on lenses with the same aperture and rounding but with different depth at the lens center z_0, $z_0 + \Delta z$ (Figure 9). The aim was that of relating the depth change z_0, which can be easily measured, to the consequent focal length change Δf.[23] We have found the simple relationship $\Delta f/f = Q. \Delta z_0/z_0$, where the parameter Q keeps almost constant for all the lenses we have considered. The most probable value of Q is ~1.9. For Δf less than 3μm, this rule cannot be used, however this case corresponds to fabrication errors usually tolerated.

Such rule of thumb has been also used to evaluate the increase in the focal spot size at a prespicified detector line. This problem

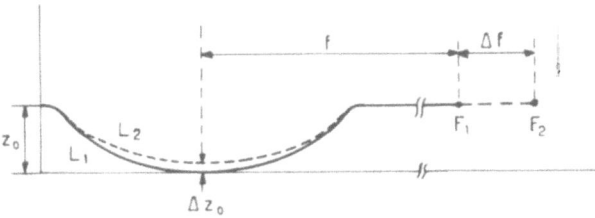

Fig. 9. Two perfect lenses, with the same aperture and rounding but with different depth at the center z_0, $z_0 + \Delta z_0$. Lens L_1 has focal length f, lens L_2 has a longer focal length $f + \Delta f$.

is of great importance for a spectrum analyzer. We have found that
in order to build geodesic lenses with resolution compatible with
a spacing of 12μm of the detector array we have to get a precision
control of the lens profile of the order of 0.5μm during the fabrication. This goal seems to be achievable with the single point
diamond turning technique.

It has been also estimated that in order to have a spot size
near diffraction limit values, the ideal shape must be maintained
within 10^{-4} x diameter of the lens.[22]

Measurements of the focal properties of the geodesic lenses so
far constructed (Table I) confirm that while the spot size does not
change very much from the expected one, the major drawback is the
shift of the focal line. This problem can be overcome by means of
a control of the profile, during the fabrication. An alternative
solution could be that of measuring the actual error Δz_o made during
the fabrication. Then evaluate the focal shift by means of the rule
of thumb and consequently to vary the position of the other elements
of the circuit, before forming the waveguide.

A first attempt have been also reported to construct geodesic
lenses on a silicon substrate by means of anisotropic etching.[24]

Other phenomena which can degrade the performance of a $LiNbO_3$
Fourier transform lens are leaky mode due to propagation in non
axial propagation direction and radiation losses associated with the
curvature of the depression. Numerical calculations have shown that
the curvature losses are dominant and require fabrication of the
waveguide having well confined fundamental modes.[25]

As a conclusion we can say that at present, no problem seems to
exist for designing perfect, rounded edges, focusing and imaging
lenses. On the contrary, the realization of the depression is a
difficult task when diffraction limited performances are required,
mainly because the position of the remaining components must be
decided before the waveguide is fabricated. Fabrication techniques
and fabrication tolerances are still today subject of great interest
and field of future work.

In the meantime, taking advantage of the wide theoretical investigation, research is progressing towards the design of new components performing operations different from focusing or imaging.

NEW COMPONENTS

A variety of new components with diffraction limited performance
and rounded edges have been designed[26] following the analytic procedure used for the generalized geodesic lenses. As an example,

Table I.

D(mm)	f(mm)	F/N	z_o (mm)	Δz_o(μm)	Δf(μm)	Q	Δz_o(μm) (Q=1.9)
2.5	7.5	3	0.54957	0.50	13.29	1.947	0.51
3.5	17.5	5	0.51244	0.45	30.15	1.962	0.46
5	30	6	0.49746	0.34	41.25	2.012	0.36
3	4	1.3	0.93259	0.86	6.79	1.841	0.83
4	12	3	0.72397	0.59	19.33	1.977	0.61
6	9	1.5	1.37647	0.64	7.83	1.870	0.63
2.5	3.5	1.4	0.82099	1.30	10.26	1.851	1.24
4.5	6.5	1.4	0.95881	0.83	10.42	1.851	0.80
7	22	3.1	1.04512	0.67	27.25	1.932	0.67
2.5	2.5	1	0.75653	1.86	10.87	1.768	1.73
4.5	9.5	2.1	0.78281	1.06	24.70	1.919	1.07
6.5	26.5	4	0.76176	0.79	54.20	1.963	0.82

Figure 10 shows the profile of a geodesic corner reflector that has been constructed and tested. Figure 11 shows the guided beam, taper coupled into an epoxy resin waveguide formed on a glass substrate, which is retroreflected. This component is the analogous one of the inhomogeneous refractive index Eaton lens.

Another interesting component is the 270° deflector (Figure 12). It has been found that all the beam deflectors have the same equations for their profiles. They differ only for a parameter which takes into account the wanted steering angle. It is to be noted that this component can also work as a power beam splitter (Figure 13) when the input beam is symmetric to the meridian of the depression. If it is not this component works as a directional coupler.

Other geodesic components could be created by means of the analytic method provided that a suitable function which mathematically represents their optical characteristics is found.

Very recently, similar components (splitters, resonators, deflectors) have been designed following a different theory that does not foresee any rounding of the sharp transition to the plane of the circuit. As an example, Figure 14 shows the profile of a beam divider from a point source (Figure 15) which can also operate as a mixer.[27]

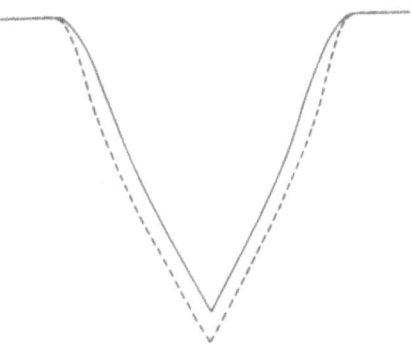

Fig. 10. Profiles of two GCR. They have the same radius c = 5mm
          of the depression and different toroidal connections
          limited by parallels of radius d = 3mm (continuous line)
          and d = 4mm (dashed line). The depth of the vertices
          turned out to be z = 9.665 and 10.643mm, respectively.

Fig. 11. Top view of the GCR realized in the laboratory. The
          collimated beam, 2.2mm wide, enters the guide of the
          guide edge and then U-turns in the depression.

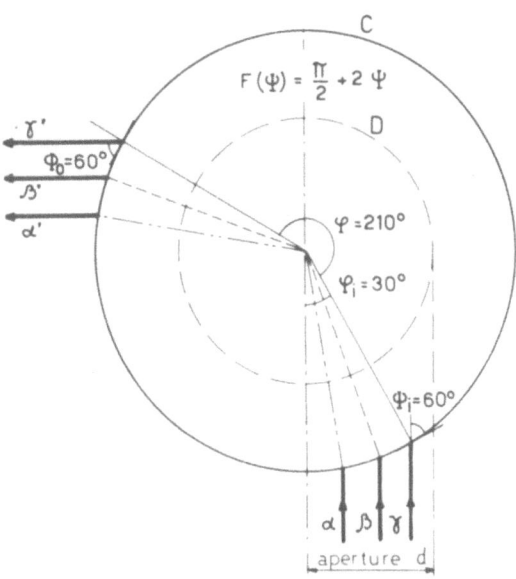

Fig. 12. Top view of a 270° deflector where the input and
          output beams are sketched.

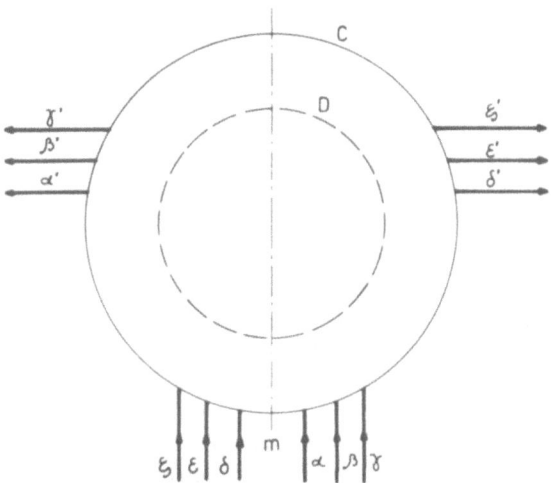

Fig. 13. Sketch of a 270° deflector working as a power beam
          splitter (or Y branch). The input beam, symmetric
          with respect to the meridian m of the depression,
          is split into two output beams propagating in
          opposite directions.

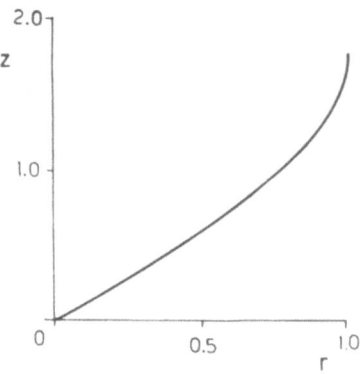

Fig. 14.   Profile of a geodesic mixer.[27]

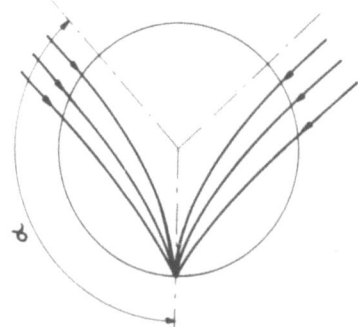

Fig. 15.   Beam divider or mixer with point focus.[27]

CONCLUSIONS

Geodesic optics in thin film optical waveguides has offered the possibility of constructing passive components with near diffraction-limited performances. The spectrum analyzers under investigation in several laboratories foresee the use of the geodesic lenses as they represent the best solution with crystal substrates of high refractive indices. The theoretical work, carried out on different bases, allows one today to design high quality lenses matched without discontinuity to the plane of the IO circuit. Either focusing or imaging lenses can be designed, as well as splitting, deflecting or mixing geodesic components. The problem to solve are connected with the fabrication of the depression having as much as possible the ideal shape. No batch production of geodesic components seems to be foreseen in the near future. Thus a component must be fabricated one at a time. Among the fabrication techniques, the single point diamond turning seems to be at the moment that which assures

the higher precision, the machine including a profile control during
the fabrication. An estimate of the effects of the fabrication
errors can be of great help for knowing where by the other components
must be positioned on the optical circuit before the waveguide is
formed. The effects of the fabrication errors make difficult also
the realization of more complex optical systems, which could be obtained combining more geodesic lenses. Nevertheless, correlators
employing more complex geodesic optical systems are suggested.[28,29,30]

REFERENCES

1. K.S. Kunz, Propagation of microwaves between a parallel pair
    of doubly curved conducting surfaces, J.Appl.Phys. 25:642 (1954).
2. G. Toraldo di Francia, Un problema sulle geodetiche del le
    superfici di rotazione che si presenta nella tecnica delle
    microonde, Atti.Fondaz.Ronchi. 12:151 (1957).
3. G.C. Righini, V. Russo, S. Sottini and G. Toraldo di Francia,
    Thin film geodesic lens, Appl.Opt. 11:1442 (1972).
4. G.C. Righini, V. Russo, S. Sottini and G. Toraldo di Francia,
    Geodesic lenses for guided optical-waves, Appl.Opt. 12:1477
    (1973).
5. W.L. Chang and E. Voges, Geodesic components for guided wave
    optics, Arch.Elek.Ubertragung. Vol.34:385-393 (1980).
6. M. Johnson, Elastic rubber-waveguide geodesic optical deflector,
    Appl.Phys.Lett. 37:123 (1980).
7. D.B. Anderson, Optical waveguide lenses, Rokwell International
    Corporation, Anaheim, California AFAL-TR - 76 - 54.
8. E. Spiller and J.S. Harper, High resolution lenses for optical
    waveguides, Appl.Opt. 13:2105 (1974).
9. G.E. Betts and G.E. Merx, Spherical aberration correction and
    fabrication tolerances in geodesic lenses, Appl.Opt. 17:3969
    (1978).
10. D.W. Vahey and V.E. Wood, Focal characteristics of spheroidal
    geodesic lenses for integrated optical processing, IEEE J.
    Quantum Electron. QE-13:129 (1977).
11. W.H. Southwell, Geodesic optical waveguide lens analysis, J.Opt.
    Soc.Am. 67:1293 (1977).
12. G.E. Betts, J.C. Bradley, G.E. Marx, D.C. Schubert and H.A.
    Trenchard, Axially symmetric geodesic lenses, Appl.Opt. 17:
    2346 (1978).
13. V.E. Wood, Effects of edge-rounding on geodesic lenses, Appl.
    Opt. 15:2817 (1976).
14. D. Kassai and E. Marom, Aberration-corrected rounded-edge
    geodesic lenses, J.Opt.Soc.Am. 69:1242 (1979).
15. B. Chen and O.G. Ramer, Diffraction limited geodesic lens for
    integrated optic circuit, IEEE J.of Quantum Electronics. QE-
    15, 9:853 (1979).
16. S. Sottini, V. Russo and G.C. Righini, General solution of the
    problem of perfect geodesic lenses for integrated optics,

J.Opt.Soc.Am. 69:1248 (1979).

17. J. Van der Donc and P.E. Lagasse, Analysis of geodesic lenses
     by beam propagation method, Electronics Letters. 16:292 (1980).

18. B. Chen, E. Marom and R.J. Morrison, Diffraction-limited
     geodesic lens for integrated optics circuits, Appl.Phys.Lett.
     33:511 (1978).

19. G.C. Righini, V. Russo and S. Sottini, A family of perfect
     aspherical geodesic lenses for integrated optical circuits,
     IEEE J.Quantum Electron. QE-15, 1, (1979).

20. D. Mergerian, E.C. Malarkey, R.P. Pautienus and J.C. Bradley,
     Diamond-machined geodesic lenses in LiNbO$_3$, Proc.SPIE. 176:85
     (1979).

21. G.F. Doughty, R.B. Wilson, J. Singh, R.M. De la Rue and S.Wright,
     Aspheric geodesic lenses in an integrated optics spectrum
     analyser, Proc.SPIE. 235 (1981). (in press).

22. J.C. Bradley, E.C. Malarkey, D. Mergerian and N.A. Trenchard,
     Theory of geodesic lenses Guided wave optical by systems and
     devices, SPIE. 176:75 (1979).

23. S. Sottini, V. Russo and G.C. Righini, Fabrication tolerances
     in geodesic lenses: a rule of the thumb, IEEE Trans.Circuits
     & Systems. CAS-26, 1036, (1979).

24. A. Naumaan and J.T. Boyd, A Geodesic optical waveguide lens
     fabricated by anisotropic etching, Appl.Physics Lett. 35(3):234
     (1979).

25. D.W. Vahey, R.P. Kenan and W.K. Burns, Effects of anistropic
     and curvature losses on the operation of geodesic lenses in
     Ti: LiNbO$_3$ waveguides, Appl.Opt. 19:271 (1980).

26. S. Sottini, V. Russo and G.C. Righini, Geodesic optics: new
     components, J.Opt.Soc.Am. 70:1230 (1980).

27. S. Cornbleet and P.J. Rinous, Generalized formulas for equivalent
     geodesic and nonuniform refractive lens, IEEE Proc. 128:95 (1981).

28. G.C. Righini, V. Russo and S. Sottini, Signal processing in
     integrated optics employing geodesic lenses, Proc.SPIE. 164:
     20 (1979).

29. C.S. Tsai, Guided-wave acoustooptic Bragg modulators for wide-
     band integrated optic communications and signal processing,
     IEEE Trans.Circuits & Systems. CAS-26:1072 (1979).

30. G.C. Righini, V. Russo and S. Sottini, Optical thin film pro-
     cessor for unidimensional signals, U.S. Patent. n.4, 222, 628.

INTEGRATED OPTICAL SPECTRUM ANALYSERS USING GEODESIC LENSES ON

LITHIUM NIOBATE

R.M. De La Rue

Department of Electronics and Electrical Engineering
University of Glasgow
Glasgow G12 8QQ

I. INTRODUCTION

This lecture will be entirely devoted to integrated optical spectrum analysers (IOSA) using waveguides of titanium diffused into substrates of single crystal lithium niobate (LiNbO$_3$). The only integrated lens type considered will be geodesic lenses created by precision machining of depressions into the LiNbO$_3$ substrates. This concentration of a very limited situation is certainly not intended to imply that other material systems and lens types are irrelevant or uninteresting, but merely that the lithium niobate/geodesic lens combination is the most well-developed technology for guided wave acousto-optical spectrum analysis and that it has the most immediate promise for practical systems application. It is certainly the lithium niobate/geodesic lens approach with which this lecturer is personally best aquainted.

II. ELEMENTS OF ACOUSTO-OPTIC SPECTRUM ANALYSIS

The Figure 1 shows schematically the essential features of an integrated acousto-optic spectrum analyser. A surface acoustic wave (SAW) launched by an interdigital transducer (IDT) acts as a weak but very thick diffraction grating and Bragg-diffracts light into, ideally, a single first diffraction order. The feature used in spectrum analysis is that the angle between the diffracted beam and the undeflected, zeroth-order, beam is given by[1]:

$$2\beta = \sin^{-1} \frac{\lambda f}{2V}$$

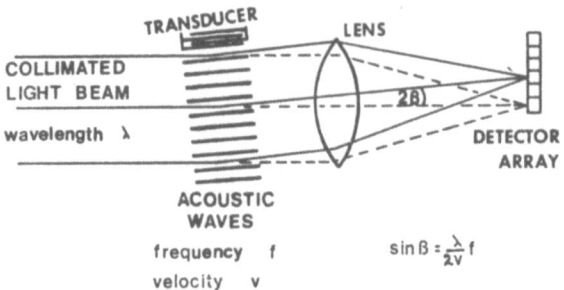

Fig. 1.   Basic spectrum analysis arrangement by acousto-optic
          Bragg deflection.

where λ is the optical wavelength in the waveguide medium, f is the
acoustic wave frequency and V is the acoustic wave velocity. In
most, but not all practical situations, $\frac{\lambda f}{2V}$ is small and therefore
the angle 2β is linearly related to the acoustic frequency and hence
the frequency of the electric signal applied to the transducer.

For coherent light, a perfect thin lens produces a focal-plane
distribution which is the Fourier transform of the input light distribution. This means that the lens focusses the first diffracted order
beam into a 'spot' separated by a distance d from the spot produced
by the focussed zeroth order, where:

$$d = \frac{\lambda f}{V} \cdot L_f$$

and L_f is the focal length. The light distribution within the spot
is determined by the aperture function for the input light beam.
For example, a Gaussian input beam distribution produces a Gaussian
spot distribution, while a uniform finite-width input beam produces
a sinx/x spot distribution (ie. one with distinct side-lobes). The
size of the 'spot' s depends quite simply on the local length of
the lens, L_f, and the width of the optical beam, D, ie:

$$s = \frac{\lambda L_f}{D} \cdot \alpha$$

where the factor α has a value of about one – the exact value depending on the actual form of the light distribution and on the resolution criterion to be used.

But the optical beam width is related to the acoustic velocity,
V, and the transit time of the acoustic wave across the optical beam,
τ, by the relation:

$$D = V\tau$$

while the range of angles through which the acoustic-optics first diffraction order varies, $\Delta\beta$, is related to the bandwidth of the device, Δf, by:

$$\Delta\beta = \frac{\lambda\Delta f}{V}$$

The number of resolved spots, N, is obtained by dividing the scanned distance $\Delta\beta.L_f$ by the spot size, s. Taking α equal to unity gives:

$$N = \frac{\Delta\beta.L_f}{s} = \Delta f.\frac{D}{V} = \Delta f.\tau$$

The number of resolved spots is therefore given by a time-band-width product and the frequency resolution, δf, is the inverse of the acoustic transit time τ. As an example, suppose that a value of Δf = 1GHz is available, then achievement of N = 1000 requires τ = 10^{-6}sec and an optical beam width of D = 3.4mm (assuming an acoustic velocity V = $3.4.10^3$m/sec). This is already quite a wide optical beam. A more stringent resolution criterion, say α = 2, would imply the need for a beam as wide as 7mm - approaching the limit of what seems practical at present. The spot size, s, for a focal length L_f = 2.10^{-2}m is only 2.3μm. Such a small size represents a major challenge to detector array technology.

III. THE FUNDAMENTAL OPTICAL POWER-ACOUSTIC POWER RELATIONSHIP

Before describing in more detail the construction and operation of an IOSA and, in particular, the fabrication of geodesic lenses for the IOSA, we shall consider, briefly, the relationship for the dependence of the diffracted light intensity on the acoustic power level and beamwidth, in the Bragg regime. This relationship has an important bearing on the overall capabilities of a spectrum analyser in several ways.

Consider a plane optical wave of wavelength λ encountering an acoustic beam of width L and wavelength Λ. Provided that the value of the parameter Q, difined by the relation $Q = \frac{2\pi\lambda L}{\Lambda^2}$, is large enough, acousto-optic interaction takes place in the Bragg-regime, where the incident light beam produces a single first diffraction-order beam as well as the undiffracted, zero-order, beam.

Analysis[1] shows that the ratio of the power in the diffracted first-order beam to the power in the incident light beam is given by:

$$I_1/I_0 = \sin^2\left[\frac{\pi}{\sqrt{2}.\lambda} L \sqrt{M_2 P_s}\right]$$

where M_2 is the material-determined Figure of Merit ($M_2 = \frac{n^6 p^2}{\rho V^3}$) and P_s is the acoustic power density. For spectrum analysis it is important that the relationship between the diffraction

efficiency, I_1/I_0, and the acoustic power density, P_s, be as linear
as practicable (a consideration which does not necessarily apply
in other applications of the acousto-optic interaction). Provided
that P_s is sufficiently small, the relationship does become linear -
but with the penalty that only a small fraction of the input light
is made available in the first diffraction-order. The linearised,
small power, relationship is:

$$I_1/I_0 = \frac{\pi^2 L^2}{2\lambda^2} M_2 P_s$$

Clearly the acoustic power required can be reduced by increasing
the acoustic beamwidth L and by increasing the Figure of Merit M_2,
ie. in the latter case by choosing a different material. In the
IOSA the acoustic power density available with surface acoustic waves
is large and, by appropriate transducer design, the acoustic beam
width L can also be readily increased while maintaining broadband
interaction and reasonably small electrical mismatch. In consequence
the value of the Figure of Merit becomes less important - and other
material-dependent considerations such as broadband transducer efficiency, acoustic and optical propagation losses are much more
important.

The essentially non-linear nature of the I_1/I_0 vs P_s relationship
becomes particularly important when consideration is given to possible
spurious signals which may occur when two or more acoustic signals
are present simultaneously. The reason why the bandwidth specification for an IOSA is restricted to octave bandwidth is simply that
any second order signals generated at the sum or difference frequencies of two in-band signals will lie outside the band and can
therefore readily be eliminated. Hecht[2] has considered in detail
the problem of 'third-order intermodulation products', ie. spurious
signals generated in-band by difference terms at frequencies such
as $2f_1 - f_2$. Such intermodulation products may well be a major
limiting factor on the dynamic range obtainable in the IOSA. For
an example with two signals present at equal strength Hecht's analysis
indicates a 50dB spurious-free dynamic range if the low intensity
threshold is at $I_1/I_0 = 10^{-7}$, ie. two signals at $I_1/I_0 = 10^{-2}$ will
produce a spurious third-order signal at the 10^{-7} level. Clearly
the situation where many signals are present simultaneously becomes
very complicated and the possibility of significant spurious signals
increases rapidly with the number of genuine signals.

IV. INTERACTION OF GUIDED LIGHT AND SURFACE ACOUSTIC WAVES

The results of the previous section can be carried over, in
broad terms, to the integrated optics situation where the light beam
and acoustic wave are non-uniform and are confined to a region close
to the surface of the substrate. However, variation in the optical
and acoustic field strengths with depth must be taken into account

if reasonably accurate estimates of diffraction efficiency coefficients in LiNbO$_3$ the 'indirect' acousto-optic effect can be as
large as the direct (strain-optical) effect. The anisotropic nature
of the crystal medium must be taken into account explicitly. Calculation of the diffracted light intensity requires calculation of a
number of 'overlap-integrals'[3] which - for the appropriate photo-
elastic, piezoelectric and electro-optic coefficients - give estimates
of how effectively the different strain and electric field components
of the acoustic wave modify the dielectric constants seen by the
guided optical wave. For the IOSA, such overlap integral calculations
become important at higher frequencies, where the penetration depth
of the acoustic wave becomes comparable with that of the optical
wave and leads to significant acoustic frequency dependence in the
diffraction efficiency. At 1GHz and 2GHz the acoustic wavelengths
are 3.4μm and 1.7μm respectively - while the optical guided wave
depth could well be significantly more than 1μm, depending on the
diffusion parameters of the waveguide.

V. LAYOUT OF THE IOSA

    Figure 2 shows more realistically the arrangement envisaged
for a hybrid-integrated spectrum analyser. A precisely positioned
semiconductor laser is edge-coupled into the diffused waveguide on
the LiNbO$_3$ substrate. The diverging optical beam is collimated by
a geodesic lens to a wide parallel beam which interacts with the
surface acoustic wave. The surface acoustic wave is launched by a
special IDT structure which has a large acousto-electric bandwidth,
maintains the Bragg angle for the acousto-optic interaction over
this frequency range, and for which there are a number of alternative
designs. As already stated earlier, the second geodesic lens Fourier
transforms the light after the acousto-optic interaction and the
focal 'plane' light distribution is imaged at the edge-coupled detector array. The literature on broadband, beam-steering, SAW IDT con-

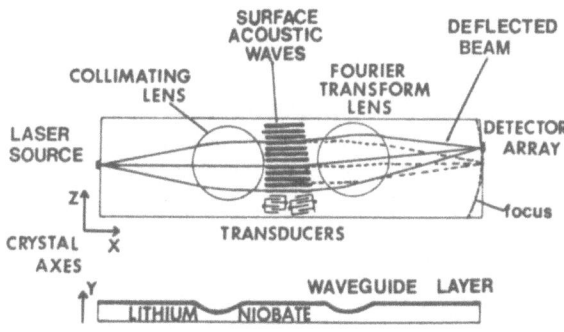

Fig. 2.   Geodesic lens integrated optics hybrid spectrum analyser
          schematic.

figurations is reasonably accessible[4,5] and will not be dealt with
explicitly in the published version of this lecture. Laser source,
detector array and input and output coupling are dealt with briefly
in Sections VII and VIII. The bulk of the lecture is concerned
with geodesic lens fabrication by single-point diamond-turning —
dealt with in Section XI.

VI. IMPORTANT PROPERTIES OF LiNbO$_3$

    In the following we list some of the advantages and disadvantages
of lithium niobate. Advantages are: (a) that it is strongly piezoelectric, (b) that reasonably low-loss waveguides are obtainable
for optical wavelengths in the visible and near infra-red, (c) that
LiNbO$_3$ has an adequate acousto-optic Figure of Merit, (d) that LiNbO$_3$
has very low acoustic propagation losses, and (e) it is strongly
electro-optic. Of this list, only (e) is not immediately essential
for the spectrum analyser — but nevertheless it is relevant because
the indirect acousto-optic effect via piezoelectricity and the
electro-optic effect may well be stronger than the direct, strain-
optical, effect.

    Disadvantages of LiNbO$_3$ are: (a) its high refractive index,
(b) that (with possible qualification) it cannot form sources or
detectors, (c) that it is relatively easy to produce reversible
optical damage through the photo refractive effect.

    A useful reference work giving data on lithium niobate is the
book by Milek[6]. Detailed surface acoustic wave properties are documented in the report by Slobodnik and Conway[7] now widely circulated
in the Western World (and elsewhere?) in Samizdat format. The most
obvious choice of substrate crystal orientation for the IOSA is a
Y-cut substrate with SAW launched along the Z-direction and optical
waves propagating along the X-axis. However, Mergerian et al[8] have
preferred to use X-cut substrates.

VII. THE LASER SOURCE AND SOURCE-TO-WAVEGUIDE COUPLING

    The presently favoured wavelength for the IOSA is 0.85μm because
good quality stripe-geometry lasers are available at this wavelength.
The arguments which favour use of longer wavelength (up to ~1.6μm)
in fibre optical systems are much less important for the IOSA — and
there are distinct advantages in using the shorter wavelength because
of reduced acoustic power requirements and the much better developed
detector array technology.

    The small size of the stripe-geometry laser emission region
(~1μm x 6μm) implies the need for very precise alignment to achieve
efficient end-fire coupling into the specially prepared input face

of the Ti:LiNbO$_3$ waveguide. Hall et al.[9] have investigated this
coupling problem and shown that use of a SiO$_2$ quarter-wave layer
greatly decreases fluctuations in the power coupled into the waveguide due to the external Fabry-Perot cavity effect in the gap between
laser and waveguide. An estimated input coupling efficiency of 50%
has been achieved and predictions of 80% coupling efficiency have
been made. The fact that the output beam from a stripe-geometry laser
is strongly divergent, particularly normal to the junction plane,
means that the laser-waveguide spacing must be small (ideally <5µm)
for efficient coupling. For the IOSA, the strong in-plane divergence
is, however, beneficial, since a collimated beam serveral millimetres
wide is required for the acousto-optic interaction region and this
can be obtained with a single collimating lens which has its front
focal position at the input waveguide edge.

VIII. THE PHOTO-DETECTOR ARRAY AND WAVEGUIDE-DETECTOR COUPLING

Linear CCD detector arrays are readily available from companies
such as Reticon and have up to 1000 dectector elements with 15 to 25µm
spacing. The size of the detector elements represents a major practical constraint on attainable IOSA performance - and it does not
seem likely that a 1000 spot IOSA would use such a detector array
because of the excessively long focal lengths implied. The possibility of path-folding to obtain long focal lengths within practical
substrate sizes (maximum length 75mm) should, however, not be ignored.

Even for the immediate, present-day, targets of 100 spots, the
leading groups[8,10] working on the IOSA have felt the need to develop
their own dedicated detector arrays - in particular arrays with
smaller detector element spacing and more rapid response. For the
hybrid integration approach the detector array is placed very close
to the polished output edge of the substrate - which is assumed to
coincide with the focal surface of the Fourier Transform Lens of
the IOSA. In this situation the relevant Bragg-angle is that within
the lithium niobate substrate. Mergerian and Malarkey[11] have inclined
the plane of the detector array at 45° to the substrate plane to reduce the effects of background scattered light and multiple reflections. It should be reasonably obvious that problems of alignment
to the detector at the waveguide output are less critical than
alignment problems from a semiconductor laser source to the waveguide
input.

IX. WAVEGUIDE FABRICATION AND WAVEGUIDE CHARACTERIZATION

The production of waveguides for integrated optics by diffusing
titanium into lithium niobate at high temperatures is now a well-
established technique. Considerable effort has gone into understanding and controlling the physical and chemical processes involved.

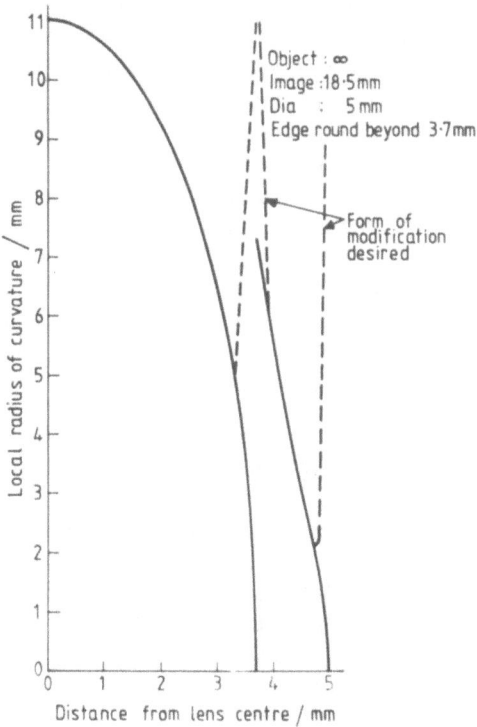

Fig 3.   Local curvature of Sottini (SRR) lens profile.

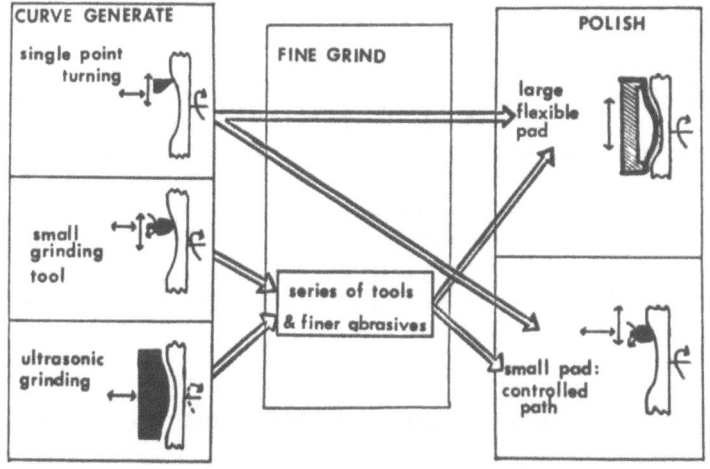

Fig. 4.   LiNbO$_3$ aspheric lens fabrication techniques.

For the published version of this lecture, only a short list of matters to be considered under this heading will be given as follows:

(a) thickness of initial titanium film, diffusion temperature and time,
(b) atmospheric conditions in the diffusion furnace and control of out-diffusion,
(c) processes occurring diffusion,
(d) waveguide refractive index distribution, modal field distribution and effective depth,
(e) substrate preparation, polishing and cleanliness,
(f) waveguide inhomogeneity, surface roughness and light-scattering.

Results of work at Glasgow University are described in reference[12].

X. GEODESIC LENS FABRICATION TECHNIQUES

No attempt will be made to cover the theoretical aspects of rotationally symmetrical geodesic lenses for integrated optics as these will be dealt with in other lectures in the school. It is sufficient for the moment to note that all theories indicate the need for a diffraction-limited geodesic lens to be strongly aspherical, with a smooth transition between the lens surface and the surrounding planar regions.

Work at Glasgow University has concentrated on the fabrication of lenses having the profile described by Sottini, Righini and Russo[13], henceforward referred to as the SRR profile. This profile can be calculated and described in a reasonably simple way but nevertheless gives a significant challenge to precision fabrication because it has two points where, although the describing function is continuous, the local radius of curvature goes to zero. This is illustrated in Figure 3. The cure for the problem is to bridge the theoretical zero-radius region with a practical, short region of infinite radius. But it is then essential to evaluate the resulting effect on the ray-tracing performance of the lens and find, if possible, an acceptable compromise between lens aberration and excessively small radii of curvature.

Figure 4 indicates some of the possible techniques involved in lens fabrication. Three possible primary processes are:

(i) single-point diamond-turning,
(ii) grinding with a small rotating (diamond-impregnated) tool,
(iii) ultrasonic impact grinding. Two approaches to final surface are:
(iv) use of a flexible tool covering the whole surface, and
(v) controlled-path polishing with a small polishing pad.

All of these approaches are to be covered in work at Glasgow
University. A machine for controlled path polishing is currently
under development. However, we now turn to the fabrication process
on which work has been concentrated - single-point diamond-turning.

XI. SINGLE-POINT DIAMOND TURNING OF GEODESIC LENSES IN LITHIUM
NIOBATE

Wilks[14] has given an excellent review on diamond turning, con-
centrating on the problems of the single-crystal diamond tool,
including the strongly anisotropic wear, chipping and cracking which
can occur during machining.

An important conclusion is that diamond tools vary greatly
depending on their preparation and origins so that, particularly
with a relatively untried material such as lithium niobate, it is.
worth experimenting with a number of different tools. The diamond
tools used have two distinct faces meeting at about 90° and a sharp
edge which has a circular profile typically of 0.75mm radius.

For the IOSA programme at Glasgow University, diamond-turning
work has been carried out at several different organisations including
the Moore Company in the USA and Bryant Symons Ltd., in the UK. At
Moore, work was performed on an experimental numerically controlled
(CNC) machine capable of the required submicron precision. The
specially prepared lithium niobate substrates, approximately 3mm
thick, were mounted on a special vacuum chuck, mounted in turn on a
horizontal axis air-bearing spindle and rotated at 1000rpm. The
diamond for machining was mounted onto a reasonably massive tool-post
on a stepper motor driven, high-precision, X-Y-θ movement. The
X-Y-θ movement allowed the tool to be moved across and into the substrate being machined - with θ variation being used to keep the tool
axis normal to the surface at all times.

The geodesic lens profiles to be machined had a depth of 1.3144mm
and an overall diameter of 1cm. Substrates were prepared by Syton-
polishing to a high degree of flatness (less than 2μm spherical sag
over a length of 55mm) and parallelism (front and back faces within
10 seconds of arc). This approach was considered preferable to the
alternative possibility of diamond turning both the planar waveguide
regions and the lens regions because less damage to the diamond
should occur and Syton-polishing surfaces are significantly smoother.
The lens profiles were specified radial distances from the lens axis.
With the rotational speed of 1000rpm the maximum relative speed
between diamond tool tip and work piece (at 0.5cm radius) is about
0.5m/sec. Considerably higher tip speeds appear to be permissible.

A number of coarse cuts were made, removing 50μm depth of material at a time. The surface finish produced was definitely rougher

than the final finish obtained after subsequent fine cuts of 3µm depth and a final cut made at exactly the 2.5µm depth required to achieve the overall design depth. Several checks of depth machined were made during the whole operation, with an electronic dial gauge.

The tool motion, during machining, was practically continuous across and into the substrate and each traverse took a total of 30 seconds - implying a tool translation speed of about 10µm per revolution.

Two basic alternative approaches exist to approximating a profile using an X-Y-θ motion to control tool position. One possibility is to specify the profile using a set of circular arc sections of specified radii and centre coordinates. For our work, however, the alternative of linear interpolation was used. Specification of 96 pairs of data points was adequate to keep the machined profile identical to the exact theoretical profile, within machine accuracy (0.2µm error).

An optical micrograph (100 times magnification) of the central region of a machined geodesic lens depression in a substrate of Y-cut lithium niobate is shown in Figure 5. The surface certainly does not appear completely smooth! Of particular interest are two obvious features (i) the circular machining marks - with, in places, quite regular radial spacing of about 15µm and (ii) quite distinct sectoral variations in apparent roughness. Measurements performed with a Talystep machine in the roughest appearing area indicate that the roughness is in the region of 30-40nm (300-400Å), peak to valley.

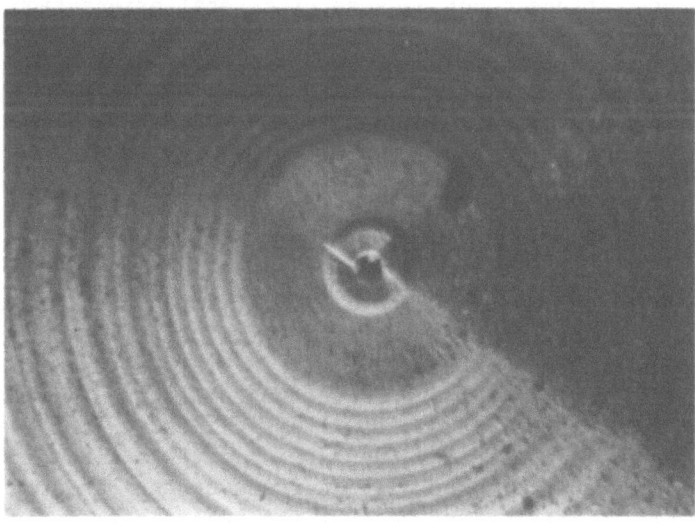

Fig. 5. Diamond-machined geodesic lens surface on LiNbO$_3$.

Measurements in the apparently less rough area indicate roughness
in the region of 20-30nm (200-300Å) peak-to-valley. In the latter
case the roughness shows the relatively long horizontal period
characteristic of the machining marks, while in the former, rougher,
area the horizontal 'period' is much shorter.

As machined diamond surfaces with roughness on the scale shown
in the micrograph are not acceptable for immediate diffusion to form
the waveguide. Superior machined surfaces in $LiNbO_3$ may well be
possible. Workers at Westinghouse have reported the need for a short,
gentle, polishing operation after diamond-machining. The objective
of such a polishing operation is presumably to remove about 50-100nm
of rough surface material (ie. 0.05-0.1μm) and produce a surface
with considerably less than the 2nm roughness which our experience
indicates to be typical of a $LiNbO_3$ surface when titanium has been
diffused into it. A short polishing operation is desirable because
it implies that the machined profile is modified to a very small
extent. However, it may be preferable to make a definite allowance
for material removal during final polishing and to deliberately control the final polishing operation by using a numerically controlled
machine with a small polishing tool. This possibility has been
allowed for in one of our machined lens surfaces by machining 1μm
too shallow. It is also worth remembering that sub-surface damage
occurring during machining and polishing could be an important source
of in-plane light-scattering.

The distinctive rough sector observed in the micrograph appears
to relate to crystallographic orientation - it is approximately symmetrical about the X-axis on this Y-cut substrate. Very recent work
performed for us at Bryant Symons appears to confirm that significant
machining variations occur because of crystal orientation in the
lithium niobate substrate. For this work, sample substrates with
X-cut and Y-cut surfaces showed clear sectoral variations in apparent
roughness, whereas the Z-cut surface gave a completley uniform but
apparently much rougher surface. Further work to evaluate these
samples will be carried out.

XII. FURTHER ANALYSIS OF THE IOSA SYSTEM

A lens which ray-traces to a perfect focus should give diffraction - limited performance. However it is certainly not true that
obtaining diffraction - limited focussing is the sole criterion of
performance - rather it provides the bottom line. Ignoring questions
of crystalline optical anisotropy it is evident that the circular
symmetry of the geodesic lens depression gives a circular focal surface which can only fit the plane surface of the detector array
imperfectly. If the output edge of the IOSA substrate is a straight

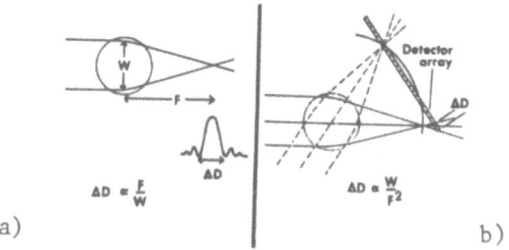

Fig. 6. Two factors determining resolution, (a) diffraction;
         (b) field curvature.

line, the best fit will be obtained by having the focal curve for
the diffracted beam intersect the straight edge at two points halfway
between the middle and ends of the detector array as illustrated in
Figure 6. Calculations[15] indicate that spot-defocussing at the middle
and ends of the array does not significantly degrade performance
in a practical 100 resolution spot IOSA system but does seriously
affect performance in a 1000 spot deflector.

The complete diffraction optical analysis of an IOSA represents
a major challenge. However, Van der Donk and Lagasse[16] have developed
a fast Fourier transform (FFT) technique which, by transforming the
three-dimensional system to an equivalent planar system, allows the
IOSA to be analysed without excessive computational demands. The
attraction of this technique is that a variety of different effects
can readily be investigated, eg. changes in size and distribution
in the laser source, defects at different points in the optical
system, effects of restrictive apertures and so on.

XIII. DISCUSSION AND FUTURE WORK

This lecture has concentrated in particular on the problem of
fabricating high-precision geodesic lenses for the IOSA. It appears
that considerable work remains to be done before IOSA sub-systems
with 1GHz bandwidth, 1MHz resolution and useful dynamic range becomes
available. No attempt has been made here to consider possible
alternative technologies but there are clearly grounds for believing
that the IOSA approach will be competitive[17]. Improvements on almost
all aspects of the IOSA will be required, however.

As the basic hybrid-integrated IOSA becomes established, other
related, more complex integrated optical devices for space - and
time - integrating correlation, memory correlation and multi-channel
convolution[17,18] are likely to emerge.

ACKNOWLEDGEMENTS

    IOSA work at Glasgow University is supported by a contract from
the Procurement Executive of the UK Ministry of Defence and sympath-
etically monitored by Dr B. E. Wheeler and Mr P. Williams. The work
has been guided and supported by Professor J. Lamb and ably carried
out by G. F. Doughty, J. Singh, J. F. Smith and Dr S. Wright.
Dr C. D. W. Wilkinson played a major role in initiating the work.
Valuable assistance has been received from Barr and Stroud Ltd.,
and Logitech Ltd., (Dr R. B. Wilson).

REFERENCES

1. R. Adler, IEEE Spectrum 4:42–45 (1967). For more detailed
       theory see "Elasto-optic light modulation and detection,"
       by E. K. Sittig in Progress in Optics, Vol. X, Wolf, ed.
2. D. Hecht, IEEE Trans. SU-24:7–18 (1977).
3. E. G. H. Lean, J. M. White and C. D. W. Wilkinson, Proc. IEEE
       64:779–788 (1976).
4. B. Kim and C. S. Tsai, Proc. IEEE 64:788–793 (1976).
5. M. K. Barnoski et al, IEEE Trans., CAS-26, 1113–1124 (1979) and
       C. S. Tsai, ibid, 1072–1098.
6. J. T. Milek and M. Neuberger, in: "Linear Electrooptic Modular
       Materials," I.F.I./Plenum, New York (1972).
7. A. J. Slobodnik Jr and E. D. Conway, in: "Microwave Acoustics
       Handbook," Vol. 1, Surface Acoustic Waves, USAF Cambridge
       Research Labs. (1970).
8. D. Mergerian et al, SPIE, Vol. 239, 121–126 (1980).
9. D. G. Hall et al, App.Optics, 19:1847–1852 (1980).
10. B. Chen et al, Topical Meeting on Guided Wave and Integrated
       Optics, Incline Village, Nevada, paper ME3, see also ibid,
       paper ME6, Jan(1980).
11. D. Mergerian and E. C. Malarkey, Microwave J. 23:37–44 (1980).
12. R. J. Esdaile et al, Topical Meeting on Guided Wave and
       Integrated Optics, paper WB3- A.D. McLachlan,Ph.D. Thesis,
       Glasgow, and J.Singh, University of Glasgow internal report,
       (1981).
13. S. Sottini, V. Russo and G. C. Righini, J.O.S.A. 69:1248–1254
       (1979).
14. J. Wilks, Precision Engineering 2:57–71 (1980).
15. G. F. Doughty et al, SPIE, Vol. 235, (1980).
16. J. van der Donk and P. E. Lagasse, Elect.Letters 16:292–294
       (1980).
17. M. C. Hamilton, SPIE, Vol. 139:144–150 (1978).
18. J. N. Lee, N. J. Berg and M. W. Casseday, IEEE Ultrasonics
       Symp.Proc. 34–39 (1979).

DEVELOPMENT OF INTEGRATED OPTICAL CIRCUITS IN THE DEPARTMENT OF ELECTRONICS AND ELECTRICAL ENGINEERING, THE UNIVERSITY OF GLASGOW

J. Lamb

J. Watt Professor of Electrical Engineering
The University of Glasgow
Glasgow G12 800, Scotland

INTRODUCTION

The purpose of this postgraduate tutorial seminar is to give an outline of research which has been in progress over the past 12 years in the author's laboratory, involving the combination of thin film technology and guided wave optics for optical device applications. Methods have been developed for the preparation of low-loss planar films and stripe light waveguides in a number of materials with control of thickness and stripe width to fractions of a μm. Earlier work was concerned with films of Corning 7059 glass sputtered by RF discharge on to a glass substrate of lower refractive index with subsequent etching through a mask to form e-dimensional stripe waveguide structures. An alternative to depositing the thin guiding-film on the surface of a substrate is to modify internally the substrate surface itself by raising the refractive index of the surface layer, thus creating an integral optical guide within the substrate. One such technique which has been employed is the exchange of silver ions from a molten silver nitrate bath with the sodium ions in a soda-lime glass substrate. Silver nitrate is molten at 212°C and may be used without decomposition at temperatures up to 350°C. The ion exchange and counter-diffusion of two different ion species establishes an electric field which, in turn, affects the diffusion rates and gives rise to a depth profile markedly different from the familiar error-function distribution. The substrate refractive index of $\sim$1.51 is raised to $\sim$1.60 at the surface and, below the surface, is proportional to the silver ion concentration. Linear stripe waveguides, Y-junctions, circular waveguides for ring-resonator filter structures or ion exchange grating patterns for beam deflectors and filters can be manufactured by restricting the silver-ion diffusion

to the uncovered regions of an aluminium mask deposited on the surface of the substrate by conventional photo-lithography, with subsequent removal of the aluminium lines by chemical dissolution.
Control of diffusion time with accompanying cost reduction of the
process is effected by using a 0.1% dilute melt of $AgNO_3$ in $NaNO_3$
with the accompanying requirement of a higher diffusion temperature
of 315°C than is necessarily required for pure $AgNO_3$ ($\sim$220°C).

Light guiding in glass structures is a convenient method for
the fabrication of passive devices for which there is a demand in
many optical applications, but materials other than glass must be
used for active devices, such as modulators and beam switches. The
requirement here is for a material in which the refractive index
can be suitably changed by an applied electric field produced by a
voltage difference across neighbouring metallic electrodes. This
can be achieved in lithium niobate which, because of its large
electro-optic coefficients combined with relatively low-loss, has
been universally employed. Optical waveguiding on the surface of
$LiNbO_3$ is produced by in-diffusion of titanium metal at a temperature
of $\sim$980°C: simultaneous out-diffusion of Li is prevented by satu-
rating the atmosphere inside the sealed furnace tube with lithium
vapour by the inclusion of a small amount of powdered $LiNbO_3$.

Photolithographic techniques are well-known from their extensive
application in the formation of masks and surface patterns for micro-
electronic devices and they likewise play a vital role in the fabrication of passive and active optical waveguide structures. However,
very smooth edges are required for optical waveguide applications
in order to reduce scattering losses and, whilst the width of optical
devices is less than ten optical wavelengths the corresponding length
is generally several thousand wavelenghts. There are two distinct
types of patterns required for optical devices:

(1) Periodic structures formed by holographic methods using inter-
    ference patterns generated by recombination of two light beams
    derived from the same laser source. Grating periods somewhat
    less than 200nm (0.2μm) have been fabricated by these means and
    ion-beam etched into the substrate to produce grating couplers,
    reflectors and filters.
(2) Non-periodic patterns, such as Y-junctions or ring structures
    formed by photographic reduction of large scale versions cut in
    coloured plastic film. Aspect ratio is maintained in this process
    but extremely good lenses are required for faithful replication
    on a reduced scale – a final waveguide 5μm wide and 1cm long
    requires an original pattern typically 0.2mm wide and hence 40cm
    long. Line width limitation using this technique is somewhat
    less than 1μm but adequate for the fabrication of monomode stripe
    waveguides, on the surface of which it is then possible to fabri-
    cate periodic frequency-selective grating reflectors, as described
    in (1) above.

Thin film optical devices have been mainly constructed as individual units formed in various materials and the essential problem is to devise ways of intercoupling such devices to build a complete system. Eventually, this should desirably be achieved by true integration on a common substrate, which may include the laser source. One limitation is the practical necessity to work with relatively long optical structures in which coupling and interaction occurs over several hundred optical wavelengths. In addition, the Y-junction is the basic means for beam splitting and/or switching in stripe waveguides and, owing to the very small permissible junction angle of a few degrees, it is physically necessary to extend the length of the arms of the "Y" in order to obtain access to the separate signals. Inevitably, these factors make it difficult to cascade many devices on the same substrate until problems of $180°$ path direction change have been solved but, as an interim stage in this development, the technique adopted in the author's laboratory has been to employ "hybrid" interconnection in which separate devices are linked by flexible passive optical bridges. Use has been made of the "sandwich ribbon", comprising a high index filament bonded to one side of a flat ribbon substrate of lower refractive index. Evanescent fields are available for directional coupling at the exposed surface of the ridge, whilst the reverse side of the ribbon can be handled with impunity. Optimum coupling can be adjusted in situ, followed by permanent adhesive bonding. Alternatively, bridging waveguides of higher refractive index can be formed as rigid overlay tapered connection between devices on the same substrate.

APPLICATIONS OF HOLOGRAPHICALLY FORMED DIFFRACTION GRATINGS

Grating Couplers

A light beam incident normally from air on to a regularly corrugated surface of a transparent material of refractive index, n_1, will be diffracted into m orders, for each of which the angle, θ_m, of the diffracted beam to the normal is given by the grating formula: $\sin \theta_m = m\lambda_1/D$, when λ_1 $(=\omega/v_1)$ is the wavelength in the unbounded medium and D is the grating spacing. If D is greater than λ_1 and the beam is incident at an angle θ_i then $\sin\theta_m = \sin\theta_i/n_1 + m\lambda_1/D$. If region n_1 constitutes a planar film guide supported on a substrate then it is possible to couple synchronously to a given mode of propagation in the planar guide by appropriate choice of D and of the angle of incidence, $\sin\theta_m = \beta/K_1 = \lambda_1/\lambda_{1g}$. Coupling efficiencies realised in this arrangement (Figure 1a) are generally less than 50% but efficiencies of 80% can be achieved by "reverse" coupling with $D<\lambda_1$ and the beam incident through the substrate (Figure 1b)[1,2].

If the grating is ion-beam etched into the surface the coupler is extremely rugged, being an integral part of the waveguide assembly.

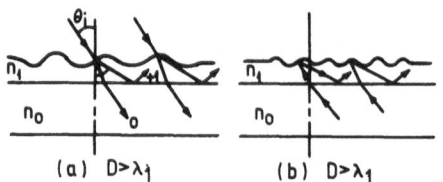

Fig. 1. Grating coupling to guided mode.

Grating Filters and Beam Deflectors

Consider a planar guide, the surface of which is regularly corrugated to form a diffraction grating. The angle to the normal of the incident beam, θ_i, for synchronous coupling is given from the above by:

$$\sin\theta_i = n_1 \beta/K_1 - m\lambda/D = \lambda/\lambda_{1g} - m\lambda/D$$

where λ_{1g} is the guide wavelength, $K_1 = 2\pi/\lambda_1$ and λ is the free space wavelength. Now, if $D = \lambda_{1g}/2$, $|\sin\theta_i| > 1$, since $\lambda > \lambda_{1g}$ is the condition for a propagating mode, and hence no coupling can occur from an incident beam via such a grating nor can outward coupling take place to a radiated beam. However, a wave already established in the planar guide will experience maximum reflection from a grating when incident at an angle ϕ to the grating normal given by $\lambda_{1g} = 2D\cos\phi$: the grating then acts as a Bragg rejection filter and wavelengths other than $\lambda_{1g} = 2D\cos\phi$ are transmitted in the direction of the incident beam. The response of two such filters, the gratings of which have been etched into the surface of the waveguide, are shown in Figure 2[3,4]. In the case of the ion-exchange waveguide (Figure 2b), the grating was etched into the surface of the substrate glass prior to carrying out the ion exchange, since excessive heating during ion beam etching causes the silver ions to migrate to the surface, forming colloidal silver. These results demonstrate clearly the possibility of achieving narrow bandwidth filtering for application in optical wavelength demultiplexing but, having mastered the technique for planar waveguides, the next stage is to fabricate the etched corrugations into the surface of a stripe waveguide.

The two main conditions which must be fulfilled in order to fabricate stripe waveguide filters successfully are:

(1) the grating must be othogonal to the longitudinal axis of the waveguide and
(2) the grating must be etched or otherwise implanted into the surface of the waveguide.

It is now well established that oblique incidence of the guided mode on to the grating of a planar waveguide filter causes TE to TM mode

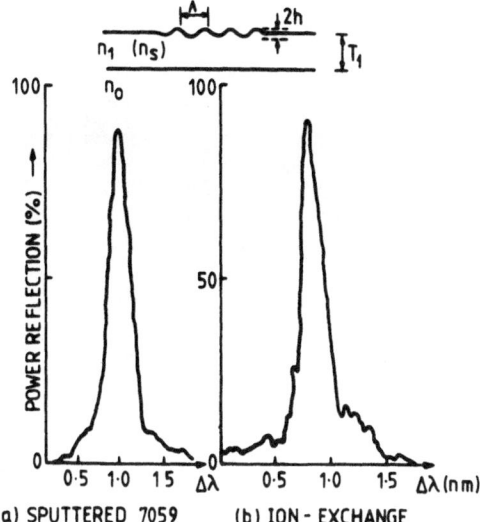

Fig. 2. Bragg rejection filter response in planar waveguide.

a)			b)	
n_0	1.51695			1.51615
n_1	1.56844		$n(s)$	1.5585
$T_1 (\mu m)$	0.9395		$d(m)$	2.1330
$N = \beta/k_0$	1.55045			1.54916
$L (\mu m)$	432.24			104.71
$h(nm)$	25.20			16.00
$\Lambda (nm)$	195.52			197.10
$\lambda_0 (nm)$	605.91			610.97

conversion in the reflected beam[5] (Figure 3): it also produces
interference across the beam due to multiple reflections within the
grating (Figure 4). The grating of Figure 4 is fabricated in a
photoresist overlay on a planar waveguide but similar effects have
been observed with implanted ion-exchange grating structures. By
inference, a grating filter on a stripe waveguide must be orthogonal
to the longitudinal axis: only small deviations from this condition
will reduce the overall efficiency of the filter and introduce a
broadening in the bandwidth due to mode conversion.

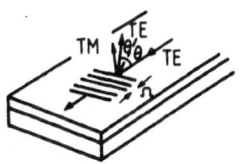

Fig. 3. Mode conversion at oblique incidence.

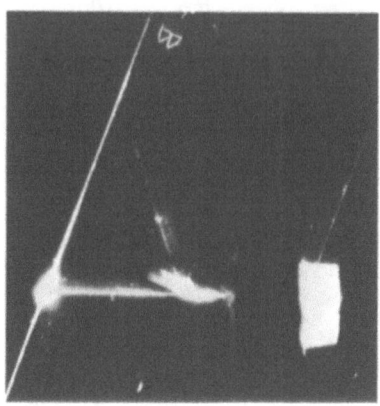

Fig. 4.   Interference effects due to multiple reflections within the
          grating.

IMPLANTED GRATING REFLECTORS

Gratings for use as thin-film Bragg filters have been formed
previously by periodically varying the thickness of the waveguiding
film, either by using an overlaid photoresist film with the grating
pattern recorded in the photoresist or by etching the pattern into
the waveguide surface. An alternative procedure is to produce a
periodic modulation of the refractive index of the waveguide structure itself, which can be accomplished by ion-exchange or by titanium
diffusion into lithium niobate. These techniques will now be discussed in turn.

Gratings in waveguides of constant depth but with periodically
varying refractive indices have two advantages over corrugated
devices:

(1) the surface finish should be unaffected, thereby significantly
    reducing the degree of scattering and
(2) the effect of the perturbation on the guided wave should be in-
    creased, since it occurs in the region of the field strength
    maximum in the guide.

Ion-Exchanged Gratings for Beam Deflectors and Filters

The use of a grating mask of aluminium strips on the surface
of the untreated soda-lime glass has the effect of partially inhibi-
ting the ion-exchange process in regions of the glass protected by
the aluminium strips. Since the presence of silver ions in the glass
has the effect of raising the refractive index, a waveguide and a
grating can be formed on the substrate in a single operation[5].

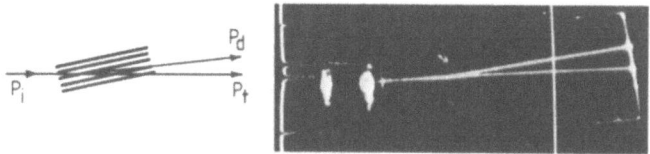

Fig. 5.   Ion-exchanged periodic grating deflector. $\eta = 80\%$.

Conventional photolithographic reduction of a computer-generated
line pattern has led to the successful production of grating patterns
in aluminium films which have been used as masks for ion-exchange.
The smallest line periodicity produced is 3μm. The glass slide
with alluminium pattern 0.75mm long was immersed for 30min in a dilute
melt of $AgNO_3$ in $NaNO_3$ (0.1%) at 330°C to form a single-mode waveguide
over the entire substrate surface. The aluminium strips were dis-
solved, leaving an area of waveguide of periodically varying refractive index. The input beam was deflected by 8° at the Bragg angle
and an efficiency of deflected power, P_d, to input power, P_i, of
80% was observed. (Figure 5).

Double diffusion techniques for improving the performance of
this structure are under investigation. The guiding-layer will first
be formed by conventional ion-exchange; a mask will then be produced
on the waveguide surface, followed by further diffusion from a more
concentrated melt or by selective out-diffusion into a sodium nitrate
bath.

Periodic Index Waveguide by Titanium Diffusion in Lithium Niobate

The refractive index distribution normal to the surface of a
waveguide formed by diffusion of titanium into lithium niobate follows
a Gaussian function: the maximum electric field is below the surface.
Hence, since the diffference between the value of the peak index
and that of the substrate is small, imposition of a surface grating
on such a structure leads to low diffraction efficientcy. Higher
efficiency can be obtained from periodic index waveguides in which
the perturbation extends across the guiding layer. Additionally,
there is the benefit of lower insertion loss, since radiation losses
into the surrounding media are sensibly eliminated.

Periodic index waveguides have been fabricated holographically
to produce passive beam splitters[6]. In the initial experiments, a
computer-generated mask was used to delineate the periodic pattern
in a resist film deposited on the surface of Y-cut $LiNbO_3$. After
development, samples were coated with a 200Å film of titanium by
vacuum evaporation, followed by lift-off of the resist lines using
acetone. Finally, the titanium films were diffused into the substrate

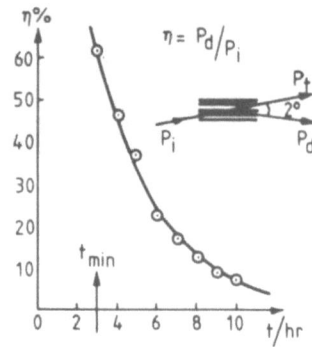

Fig. 6.   Efficiency of titanium diffused periodic grating in LiNbO$_3$.

at 1000°C. For a given thickness, T, of titanium there corresponds
a minimum diffusion time, t_{min}, which ensures that the titanium is
completely diffused into the substrate. Eight different devices
were fabricated for the same initial thickness of titanium (200Å),
using diffusion times ranging from t_{min} = 3hr to t = 10hr. Lateral
diffusion of the titanium ensured that the grating elements of the
original pattern of 3μm periodicity merged to form a smooth continuous
surface grating. Light from a HeNe laser was coupled in and out of
the guide by rutile prisms and the direction of propagation of the
guided TE$_0$ mode was adjusted to coincide with the Bragg angle. The
angle between the transmitted and reflected beams was 2° and the power
in the beams was monitored by a photodetector in conjunction with a
phase-locked amplifier and a mechanical chopper. A maximum efficiency of 60% was obtained for conditions such that t = t_{min}: for
longer diffusion times, t > t_{min}, the efficiency decreased to only
a few percent at 10hr (Figure 6). A refinement of this method is
to use the two stage diffusion technique of Figure 7, in which the
planar waveguide is first formed from a uniform 200Å thick Ti film
diffused for 8hr at 1000°C. The sample was then recoated with a
second film of Ti and spin coated with negative photoresist. A dark
field grating mask was then used to expose the resist film and

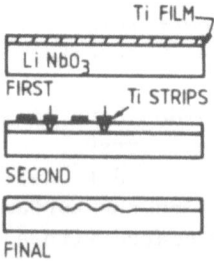

Fig. 7.   Two diffusion technique: Ti in LiNbO$_3$.

a buffer solution of HF was used to etch away the Ti film after
development of the resist. The residual grating mask in resist was
then removed and the Ti diffused for 5hr. Efficiencies comparable
with those obtained with the single diffusion were observed.

HIGH FREQUENCY Y-JUNCTION SWITCHES

A major interest in the Y-junction is in its utilisation as a
switch which implies fabrication in an electro-optical material such
as $LiNbO_3$. A titanium diffused waveguide structure in the form of
a Y-junction is fabricated in single crystal $LiNbO_3$. A metal electrode pattern is superimposed (Figure 8) and, when a voltage is
applied, one side of the input waveguide has an enhanced value of
refractive index at the expense of the other side. This causes the
beam to be deflected towards the region of higher refractive index,
so that, when it arrives at the Y-junction, it is preconditioned to
follow the corresponding output arm with continued assistance from
the biassing field. Reversal of the applied voltage causes the beam
to be deflected into the other arm. A thin isolating film of SiO_2
is sputtered on to the surface before the electrode structure is
deposited in order to reduce waveguide losses. The frequency response
is limited only by the capacitance and operation has been confirmed
up to 1.2GHz. The maximum isolation ratio achieved is 23dB and by
careful design the insertion loss can be less than 1dB.

In order to incorporate such Y-junction switches into fibre optic
communication systems it is necessary to achieve coupling between

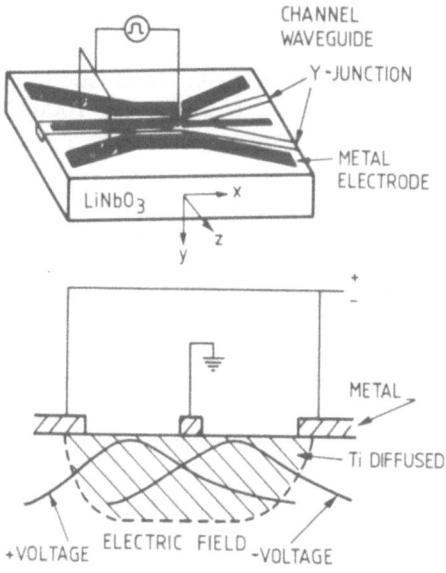

Fig. 8. Y-junction switch in lithium niobate.

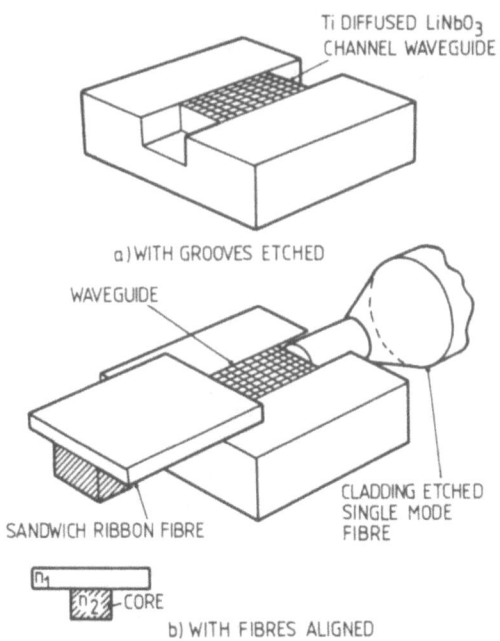

Ti DIFFUSED LiNbO₃
CHANNEL WAVEGUIDE

a) WITH GROOVES ETCHED

WAVEGUIDE

CLADDING ETCHED
SINGLE MODE
FIBRE

SANDWICH RIBBON FIBRE

CORE

b) WITH FIBRES ALIGNED

Fig. 9. Location of fibres with stripe guide by ion-milled slots.

glass fibres and the indiffused stripe guides. Butt coupling and
the "flip-chip" method have been employed with losses of 2 to 3 dB
per joint. A promising development has been to ion-mill grooves
directly in the substrate to locate the fibres in line with the
stripe waveguides (Figure 9). Relatively deep grooves with vertical
walls are required and hence a masking material with a low etch rate
is needed. Carbon masks have been used successfully with an ion-etch
rate less than 15% of that of LiNbO₃. A 1000Å layer of carbon is
first deposited and subsequently built up by carbon-arc deposition.

HYBRID INTEGRATION OF ACTIVE AND PASSIVE DEVICES

Whilst active optical waveguide devices may be constructed on
lithium niobate single crystal substrates, other important devices
such as gratings and filters can be readily fabricated on glass substrates. In order to form a hybrid integrated system, it is necessary
to devise a means of interconnection. Sandwich ribbon fibres have
been used to achieve transverse directional coupling between passive
waveguide circuits[7] and this technique has now been extended to couple
glass and lithium niobate waveguide systems. The main practical
difficulty arises from the fact that glass structures have relatively
low values of refractive index (circa 1.5) whilst corresponding values
for Ti-diffused LiNbO₃ waveguides are in the region of 2.2: in order

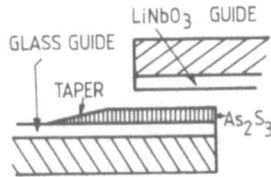

Fig. 10.   Coupling between glass and lithium niobate waveguides.

to achieve efficient coupling it is therefore necessary to overcome
the problem of obtaining an inherent phase match between the propagating modes of the fibre and the active device. We have used
arsenic trisulphide As_2S_3 as a coupling medium either as vitreous
film obtained by vacuum evaporation (MP = 308°C) to form a high index
overlay film (n = 2.45 at 1.15μm wavelength)[8] or as an extrudate
forming the guiding layer of a sandwich ribbon fibre. In the latter
case, the substrate employed is polyethersulphone (n = 1.65 at λ =
1.15μm): As_2S_3 and polyethersulphone pull together at 245°C to form
a stable but flexible optical fibre system[9].

     By using a high index overlay film deposited on a glass waveguide
(Figure 10) or on the core of a sandwich ribbon fibre (Figure 11),
the phase velocity of the propagating mode can be matched to that
of the guided mode in the lithium niobate system, thus allowing energy to be transferred efficiently. Figure 12 shows directional
coupling between planar Ti-diffused $LiNbO_3$ guides and phase-matched
As_2S_3 films on glass with a coupling efficiency of 70%. Figure 13
shows light of λ = 1.15μm launched into a planar silver-ion exchanged
waveguide which has been partly overlaid with a 0.25μm thick As_2S_3
film. Long tapered ends of 800μm on the arsenic trisulphide film
minimises radiation losses. The incident beam arrives at an oblique
angle to the boundary of the overlaid region and the effect of propagation in this layer of higher refractive index is clearly demonstrated by refraction of the beam at both boundaries of the tapered
overlay (Figure 13). The taper itself is fabricated by inserting a
knife edge in the path of the evaporated As_2S_3 beam with consequent
diffusion of the evaporated molecules into the shadow region of the
knife edge obstruction. For coupling of the fundamental mode at
λ = 1.15μm in planar As_2S_3 films on glass to a planar Ti-diffused

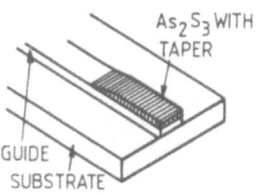

Fig. 11.   Sandwich ribbon fibre with As_2S_3 overlay.

Fig. 12.   Coupling from lithium niobate to glass waveguide via
           overlay film of Figure 10.

LiNbO$_3$ waveguide a typical coupling length is 50μm. The thickness
and refractive index of the film must be accurately controlled to
±0.01μm and ±0.01 respectively, in order to prevent loss of light
into substrate modes of the lithium niobate but these requirements
have been met by optical monitoring during deposition and post-
annealing. These methods are now being extended to achieve coupling
between stripe waveguides both by flexible sandwich ribbons and by
curved overlay guides on a rigid substrate.

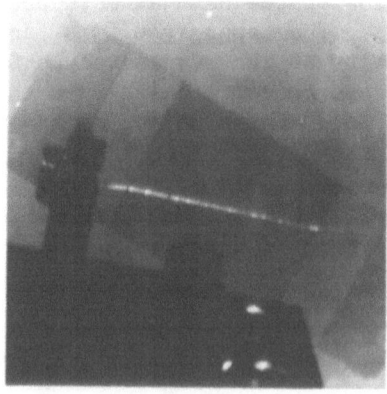

Fig. 13.   Ion exchange waveguide with overlay film of As$_2$S$_3$.

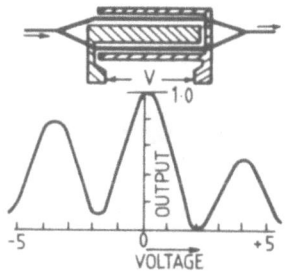

Fig. 14.   Mach-Zehnder interferometer.

MACH-ZEHNDER MODULATOR

Modification of the refractive index in an electro-optic light-guiding material by application of an electric field provides the opportunity to control the phase velocity of a guided wave. Combining this with equal power division at a Y-junction permits differential changes to be made in the relative phase between signals in the two areas. Subsequent recombination (Figure 14) gives rise to constructive and destructive interference in the output guide, converting the basic process of phase modulation into intensity modulation[10]. Extinction ratios in excess of 20dB have been obtained for applied voltages of a few volts using Ti-diffused lithium niobate waveguides.

HIGH RESOLUTION ELECTRON BEAM LITHOGRAPHY

Recent developments in very high resolution electron beam lithography[11,12,13] offer the prospect of fabricating optical wave guide junctions and couplers having dimensions comparable with the optical wavelength in contrast to existing devices several hundred wavelengths long. Moreover, owing to the short lengths involved, losses arising from metal boundary layers no longer preclude their use. Many problems have to be solved before optical equivalents of microwave devices can be realised, but it is instructive to review briefly recent developments which make this at least a possibility. Thus, platinum/palladium lines 17nm wide have been defined by exposure of a film of poly(methyl methacrylate) 60nm thick supported on a 30nm thick carbon film substrate followed by lift-off of overcoated metal. Line pitch may be reduced to 50nm before lift-off fails and pattern size accuracies of a few nano-meters can be achieved.

As illustrated in Figure 15 carbon films, 20-30nm thick, are deposited on soda lime glass by an evaporation technique and the surface is then spin coated with poly(methyl methacrylate) of molecular weight $M_W = 275,000$ ($M_W/M_N = 1.17$) to a thickness of 60nm and baked in nitrogen atmosphere at 150°C for 1hr. After scribing into

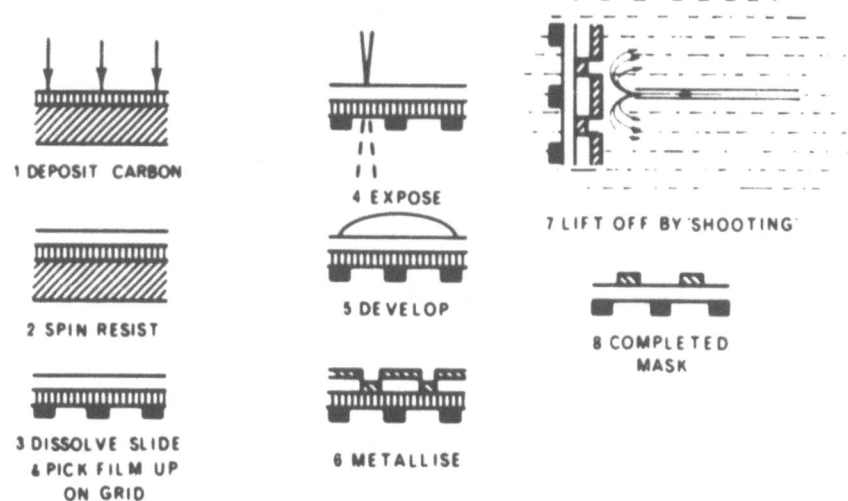

Fig. 15. Sequential processes for mask production by electron beam.

5mm squares, the cover slips are floated on the surface of a 13%
solution of HF in water at 70°C. This dissolves the glass, leaving
the carbon-supported PMMA squares floating on the liquid surface.
Each square is then picked up on a 3mm diameter copper grid of the
type used in transmission electron microscopy. Electron beam writing
takes place in a Philips PSEM500 Scanning Electron Microscope with
a microprocessor-driver digital scanning system. Exposed, degraded
lines in the resist are "developed" in a suitable solvent (1:3 solution of methyl isobutyl ketone in isopropyl alcohol) which does
not dissolve the unexposed high molecular weight PMMA. In this
process, drops of liquid from a hypodermic syringe are allowed to
fall on to the grid. After drying, a film of 80-20 Pt/Pd is evaporated over the resist surface and finally the unexposed regions of
the resist with metal covering are removed in a bath of methyl ethyl
ketone by turbulent agitation from a syringe needle immersed in the
solvent. This leaves the final mask pattern on the carbon substrate.
Enhanced lift-off and improvement in edge definition is achieved by
using two layers of PMMA resist, one spun on top of the other
(Figure 16). The top layer has a higher molecular weight than the

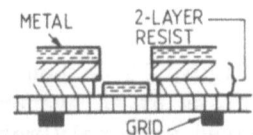

Fig. 16. Two layers resist fabrication of metal mask on carbon
          substrate.

 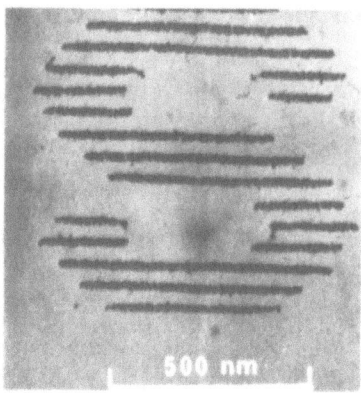

Fig. 17.   (a) 40nm thick Au/Pd lines, 20nm wide on 75nm pitch;
           (b) pattern of 20nm wide Au/Pd lines on 50nm pitch.

lower one which, because of its higher sensitivity, suffers extended
lateral degradation with consequent undercutting, as indicated on
the diagram. By this means, it is possible to form the Pt/Pd line
to a height of 50nm for a line width of 20nm (aspect ratio 2.5:1)
which is necessary for the second process of X-ray replication.
The extremely thin carbon film substrates are essential for fine
line writing by the electron beam, since they minimise back scattering of electrons. However, in order to replicate the pattern
on a dielectric or other substrate, the electron beam generated
pattern serves as a mask for subsequent X-ray replication on the
desired substrate, using a similar technique and the same PMMA resist.

    Typical patterns on a carbon substrate are shown in Figure 17,
whilst the present limitation on minimum line spacing to 50nm is
illustrated in Figure 18. Patterns have also been deposited on
silicon nitride membranes supported on a silicon grating and work
is continuing on the development of the X-ray exposure method for
replication. It is proposed to make short lengths (20 to 100µm)
of metal clad optical waveguides and to study their optical losses.
It is also intended to fabricate stripe waveguide demultiplexers
in the form of Y-junction s with superimposed metallic gratings,
which will form a mask for ion beam etching of the final grating into
the junction (Figure 19). In such an optical filter, light as wavelengths other than the Bragg condition will not be reflected into
the branch arm but will continue along the main arm. In principle,
a multiple wavelength demultiplexer should be feasible by adding
more branch arms, each having a grating of appropriate pitch for
the desired wavelength.

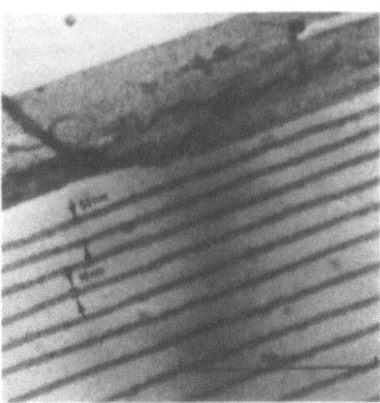

Fig. 18.   Resolution test pattern. Lines 16nm wide with decreasing
           pitch. Lift off falls below 50nm.

TITANIUM DIFFUSED LITHIUM NIOBATE STRIPE WAVEGUIDES

    In stripe waveguides the effect of the residual surface oxide
layer on waveguide propagation is small, except for short diffusion
times with relatively thick initial titanium films: the multiphase
surface oxide layer diffuses laterally away from the stripe waveguide.
Waveguide propagation has been analysed by the variational solution
method and computed results compared with experimental observation[14].
It has been assumed that modes are purely TE or TM and only TE modes
have been considered in the analysis with propagation in the X-
direction in Y-cut lithium niobate.

    Although prism-coupling yields information about both the number
of modes and their effective propagation constants whereas end-fire
coupling gives only the number of modes, end-fire coupling allows
the near field patterns to be observed. Moreover, since end-fire
coupling uses a laser beam focused to a small spot, the waveguides
of different widths can be placed very close to each other (50μm),
which has the advantage that the titanium thickness is sensibly
uniform over the set of experimental waveguides. Good agreement
has been found between the theoretical mode cut-off curves and the
experimental points, as illustrated in Figure 20. This diagram has

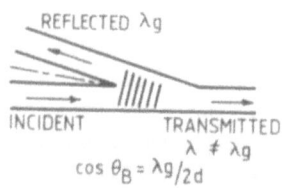

Fig. 19. Proposed stripe waveguide filter.

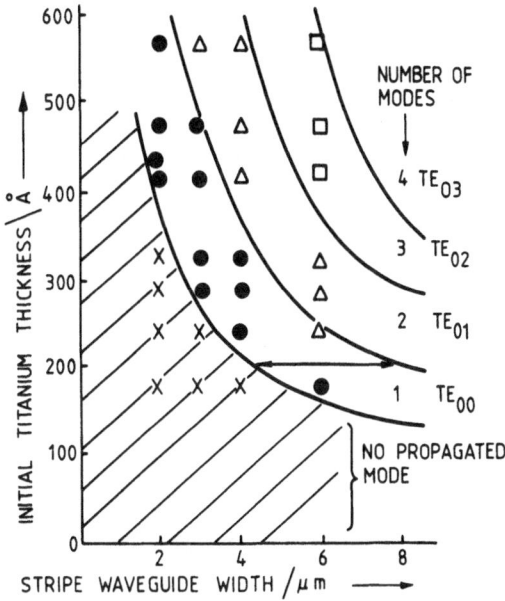

Fig. 20.   End-fire coupled waveguides.   Theoretical cut-off curves
           for 8hr diffusion at 1000°C.   Number of modes as a function
           of stripe width. ⟷ shows limits for design widths of
           Mach-Zehnder interferometer.   ✗ No propagated mode.
           ●, TE_{00};   △, TE_{01}; □ , TE_{02}.

important applications in the design of active $LiNbO_3$ devices.
Thus, for the Mach-Zehnder interferometer design (Figure 14) the
same titanium thickness is deposited throughout. However, the narrower side arms must not be below cut-off whilst the wider feed arm
must not be capable of supporting more than a single mode. Relative
dimensions must therefore lie on a horizontal line on Figure 20 between the cut-off curve for mode 1 and that for mode 2.

Cut-off Modulator

        The increase in refractive index of $LiNbO_3$ by indiffusion of
Ti is so small ($< 5 \times 10^{-3}$) that this can be completely counteracted
by application of an appropriate electric field, through the electro-
optic effect. Thus, if conditions are such that the lowest order
propagated mode is just above cut-off, the waveguide can be driven
below cut-off by an applied voltage between parallel strip electrodes
(Figure 21). Under these conditions, the wave enters the substrate
providing intensity-modulation of the light beam through the guiding
stripe. This principle was first demonstrated by Yariv in gallium

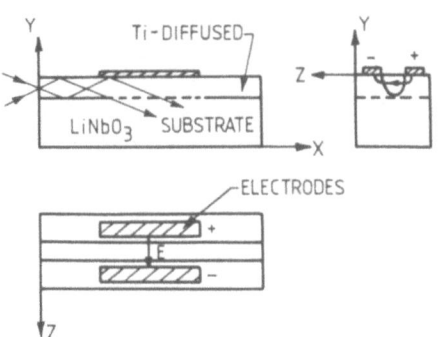

Fig. 21. Cut-off modulator.

arsenide and has recently been applied to lithium niobate. Amplitude
modulation with greater that 20dB extinction ratio has been produced
with less than 1V applied with operation up to several hundred MHz.

ACKNOWLEDGEMENTS

By the very nature of this description, the author has drawn
freely on the results of his colleagues. These are too numerous
to mention individually in the foregoing text and because research
at Glasgow is essentially team work, several individuals tend to be
associated with each project. It would be inappropriate not to list
those whose collective enthusiasms and abilities are without parallel
in the writer's experience. It is a privilege to work with such
able colleagues. Members of staff are Messrs P. G. Bower and
G. Boyle, Drs R. M. De La Rue, R. Dunsmuir, Mr R. Hutchins,
Drs P. J. R. Laybourn, C. D. W. Wilkinson and S. Wright: post-
doctoral Research Fellows: Drs S. P. Beaumont, M. Holbrook,
G. Stewart, R. G. Walker and A. Yi Yan: Research Students:
Dr A. D. MacLachlan, Messrs I. Andonovic, C. Binnie, B. Bjortorp,
T. Cullen, W. S. Mackie, D. R. MacLean, E. Pun, S. Rishton, R. Steele
and K. K. Wong. Collaborative research in electron beam and X-ray
lithography is carried out with Professor R. P. Ferrier and
Dr J. Chapman of the Department of Natural Philosophy, Glasgow
University, and related work in the Department of Electronics and
Electrical Engineering on molecular beam epitaxy is pursued by
Dr C. R. Stanley, Dr S. Yoshida, Dr M Akhter, Messrs R. Park and
T. Kerr.

To each and all of these the author is indebted.

REFERENCES

1. D. G. Dalgoutte, A high efficiency thin grating coupler for
    integrated optics, Opt. Commun. 8:124-127 (1973).
2. D. G. Dalgoutte and C. D. W. Wilkinson, Thin grating couplers
    for integrated optics: an experimental and theoretical study,
    Appl. Optics 141:2983-2997 (1975).
3. A. Yi-Yan, Frequency selective grating filters for integrated
    optics, Ph.D. Thesis, University of Glasgow (1978).
4. A. Yi-Yan, J. A. H. Wilkinson and C. D. W. Wilkinson, Optical
    waveguide filters for the visible spectrum, IEE Proc. 127H:
    335-341 (1980).
5. E. Y. B. Pun and A. Yi-Yan, Efficient non-collinear mode converter
    for thin-film optical waveguides, to be published.
6. A. Yi-Yan, I. Andonovic, B. Bjortorp and E. Y. B. Pun, Passive
    beam splitters for Ti:LiNbO$_3$ optical waveguides, to be pub-
    lished.
7. C. A. Millar and P. J. R. Laybourn, Coupling of integrated optical
    circuits using sandwich ribbon fibres, Opt. Commun. 18:80
    (1976).
8. G. Stewart, R. H. Hutchins and P. J. R. Laybourn, Hybrid inte-
    gration of active and passive devices, Integrated Optics and
    Optical Communications Conference Paper WE1, San Francisco
    (1981).
9. C. A. Millar and R. H. Hutchins, As$_2$S$_3$ - p.e.s. sandwich ribbon
    fibre for coupling high refractive index integrated optics,
    Electronics Letters 15:60 (1979).
10. H. Sasaki, Efficient intensity modulation in a Ti-diffused LiNbO$_3$
    branched optical waveguide device, Electronics Letters 13:693
    (1977).
11. S. P. Beaumont, P.G. Bower, T. Tamamura and C.D.W. Wilkinson,
    20nm wide metal lines by electron-beam exposure of thin poly
    (methyl methacrylate) films and lift-off, App. Phys. Lett. 38:
    436-439 (1981).
12. S. P. Beaumont, T. Tamamura and C. D. W. Wilkinson, A two layer
    resist technique for efficient lift-off in very high resolution
    electron beam lithography, Microelectronics Conference,
    Amsterdam (1980).
13. W. R. K. Clark, J. N. Chapman, R. P. Ferrier, S. P. Beaumont,
    T. Tamamura and C. D. W. Wilkinson, Profile determination of
    thin film resist masks for electron beam lithography, Electron
    Microscopy 1:322-323 (1980).
14. A. D. Maclachlan, Theroretical and experimental investigation
    of titanium diffused lithium niobate optical waveguides,
    Ph.D. Thesis, Glasgow University (1981).

INTEGRATED OPTICS IN CHINA

Hong-du Liu

Department of Physics
Peking University
Peking, China

INTRODUCTION

Integrated optics is in its beginning stage and still developing in China. Research activities in China, to my knowledge, involve the fabrication and measurement of optical planar waveguides, active and passive waveguide devices, and monolithically integrated optical circuits as well as some theoretical works. In this paper, recent progress in this field in China are reviewed and pertinent references are given. The research activities of semiconductor lasers, including ternary and quarternary I-V compounds, optical fibers and cables, and optical communication systems in China had been reviewed by C. M. Wang in an invited paper[1] at the 3rd International Conference on IOOC, April this year, so that these subjects will not be mentioned here.

Besides, there are some teaching activities of guided wave optics in some universities for graduate and undergraduate students.

OPTICAL PLANAR WAVEGUIDES: FABRICATIONS AND MEASUREMENTS

GaAs-GaAlAs multi-layer structures are widely used in semiconductor lasers and integrated optics. H. D. Liu et al. developed for the first time a new technique of masked and selective thermal oxidation of GaAs-GaAlAs structures[2,3]. Cr-Au film was successfully used as a mask for the thermal oxidation of GaAs. The effectiveness of the metal masking was verified by the specific contact resistance measurement of evaporated Cr-Au on heavily doped p-type GaAs samples after 50hr thermal annealing in an atmosphere of air. The oxidation rate of GaAlAs was found to be much smaller than that of GaAs; this

fact means that the thermal oxidation process is selective for the
GaAs-GaAlAs multi-layer structures. By means of this masked and
selective thermal oxidation (MSTO) technique, the stripe optical
waveguides of GaAs-GaAlAs and stripe-geometry DH lasers were fabricated. They also studied the optical-elastic effect, caused by the
difference in the thermal expansion coefficients of native oxide
and GaAs substrate, and checked the refractive index profile due to
the strain by the interference pattern and proposed a new method of
forming a stripe waveguide by the optical-elastic effect[4].

Ti-LiNbO$_3$ is one of the most popular waveguides in China.
Y. S. Li et al. have fabricated Ti-LiNbO$_3$ waveguide with optical
loss as low as 1dB/cm at 0.633μm wavelength[5]. In order to suppress
Li$_2$O out-diffusion, a closed-tube diffusion system was employed.
The Ti-sputtered LiNbO$_3$ wafer and fresh LiNbO$_3$ powder (0.5 to 1.0g)
were placed in a hermetically sealed rectangular (77 x 23 x 21mm^3)
white alundum boat. Since the Li$_2$O vapor pressure produced from
the LiNbO$_3$ powder was sufficient to maintain equilibrium with the
Ti-LiNbO$_3$ wafer, the diffusion process did not add new Li-vacancies
to the crystal.

Besides, for making the glass waveguides, RF sputtering[6,7] and
ion-exchange technique[8] were adopted. The native oxide film on
Silicon was also used as a substrate for planar waveguides[9].

As far as the measurement of waveguide parameters is concerned,
J. Li et al. proposed a method called "optimization of statistical
estimation" to determine the graded index profile of planar waveguide
by measuring the effective indices of several guided modes[10]. This
method has been applied to the multimode Ti-LiNbO$_3$ waveguide, and
its index profile is found to be described by the summation of two
Gaussian functions. F. Jin and Y. S. Li proposed a method called
"dual-polarization method" which can be used to determine the refractive index n and thickness w of single mode planar guides[11]. The
requirement is that the planar guide can support a TE$_0$ mode and a
TM$_0$ mode. After the effective indices of both TE$_0$ and TM$_0$ modes
are measured, the parameters of planar waveguide can be calculated.
The accuracy achieved with this dual-polarization method is $\Delta n \sim 1.3 \times 10^{-4}$ and $\Delta w/w \sim 1.3 \times 10^{-2}$. Besides these, C. H. Chen and G. L. Ding
have designed and constructed an instrument for measurement of multi-
mode planar guide parameters[12]. The accuracy they achieved is
$\Delta n \sim 1.2 \times 10^{-4}$ and $\Delta w/w \sim 1.6 \times 10^{-3}$.

Recently, in China, the digital controlled liquid-phase epitaxy
(LPE) system and molecular beam epitaxy(MBE) equipment have been
manufactured, some of them have been installed and have begun to
operate in some laboratories.

PLANAR OPTICAL DEVICES AND INTEGRATED OPTICAL CIRCUITS

Beam Coupler and Related Topics

There are several experimental works involving the periodic or grating structure. By using the holographic and preferential etching techniques, corrugation with periods of 0.33μm or 0.43μm have been formed on GaAs (100) plane by Z. W. Liu and G. T. Cao[13]. M. Xu and Y. Li also made a grating on AZ 1350 photoresist by holographic technique to form a grating coupler on a planar glass waveguide[14]. The coupling efficiency of laser beam to planar guide was measured to be about 20%. D. W. Xu constructed a grating coupler on the surface of glass waveguide by the duplicating technique with a period of 0.56μm form a master blazed grating[15]. In this case, it is easier to control the contour of grating grooves, so that the coupling efficiency which can be achieved is up to 66%.

Prism coupler is widely used in routine measurement of waveguide parameters and light guiding experiments. Y. S. Li and F. Jin studied experimentally the effect of the coupling gap between the prism and the dielectric film on the measurement error of waveguide parameters[16]. The coupling gap of prism-film coupler had been estimated by iso-thickness interference, and the mode propagation constants, refractive index and thickness of the optical waveguide had been measured at the same time at different coupling gaps. They found that the measurement errors of film parameters were minimized when the coupling gap was about $\lambda_0/3$, here λ_0 is the light wavelength in free-space.

Because TiO_2 prisms are rather difficult to get in China, Z. H. Cao et al. make use of a pizoelectric crystal $Bi_{12}GeO_{20}$ to fabricate a prism coupler[17]. This kind of crystal is optically homogeneous and highly transparent (T∼88% at 0.633μm) and its refractive index is about 2.536 at 0.633μm wavelength. But it is optically active which is estimated to be 20.7 degree/mm.

Waveguide Modulators

Both electro-optic and acousto-optic waveguide modulators are studied in some laboratories in China. Waveguide electro-optic prisms were made by F. Jin et al. for $Nb-LiTaO_3$ planar guide[18] and by Y. S. Li[19] for $Ti-LiNbO_3$. The electrode structure of the waveguide electro-optic prisms is similar to Kaminow's publication. The length of parallel electrodes is 1cm and their separation is 200μm. The width of the oblique electrode is 10μm. When the waveguide electro-optic prism is used as an intensity modulator, 100% modulation is achieved by applying ±19V bia-volts. The fundamental bandwidth is 1.5mW/MHz. When it is used as a bipolar switch, the switching voltage is ±19V and the switching time is estimated to be 0.1ns[19].

Preliminary experiments of acousto-optic diffraction in Ti-LiNbO$_3$ waveguides were carried out by B. R. Shi et al. for Raman-Nath diffraction[20] and by J. H. Wang for Bragg diffraction[21]. In the case of Raman-Nath diffraction, the frequency of SAW is about 60MHz and the modulation of approximately 100% for the zero-order beam is obtained 68mW acoustic power which is in good agreement with the calculated value of 63mW[20]. In the case of Bragg diffraction, the frequency of SAW is 120MHz and its wavelength is 28μm. The interaction length is about 7mm, and the modulation efficiency of the first order Bragg diffraction of TE$_0$ mode approaches 100% when the input electrical power is 320mW[21].

Monolithically Integrated Optical Circuits

Recently, GaAs monolithically integrated optical circuits were fabricated by H. Z. Pan et al.[22,23] containing two symmetrical mesa type active devices connected with each other through a piece of passive waveguide. Mesa A is forward biased and mesa B can serve as a detector. The absolute efficiency I_D/I_L is measured to be about 2%, here I_D is the output current of the dectector B and I_L is the input current applied to mesa A. When mesa B is also forward biased then a combination of laser and amplifier is obtained. Driving A and B with two synchronous pulse current generators, the light output at terminal B is enhanced appreciably and the maximum net gain is measured to be 17dB.

REFERENCES

1. C. M. Wang, Chinese Optical Communication, Proc. 3rd International Conf. on IOOC, San Fransisco, California, April 27-29, Paper MC3 (1981).
2. H. D. Liu, B. Zhang, D. H. Wang and W. X. Chen, "Masked and Selective Thermal Oxidation (MSTO) of GaAs-Ga$_{1-x}$Al$_x$As Multilayer Structures," Appl. Phys. Lett. 38, No.7:755 (1981).
3. H. D. Liu, B. Zhang, D. H. Wang and W. X. Chen, "Masked and Selective Thermal Oxidation (MSTO) of GaAs-Ga$_{1-x}$Al$_x$As Multilayer Structures - A New Technique for Integrated Optics and Stripe Geometry Double-Heterostructure Lasers," Proc. 3rd International Conf. on IOOC, San Francisco, Calif. April 27-29 1981. Paper MB4.
4. Z. C. Feng and H. D. Liu, "Optical -elastic Effect in Masked and Selective Thermal Oxidation GaAs-GaAlAs Multilayer Structure," to be published.
5. Y. S. Li, H. J. Liu, B. F. Ren and R. J. Yu, "Ti diffused LiNbO$_3$ Optical Waveguides with Low Loss," Kexue Tongbao (Scientific Review) 25, No.10:824 (1980).
6. "Planar Glass Optical Waveguide and Prism Coupler," Research Report of Changchun Institute of Physics, Academia Sinica (in Chinese) (1978).

7. Z. Y. Zhang and B. A. Zhong, Research Report on the Dielectric
    Optical Planar Waveguide with Step-index Profile; presented
    at 1st Conf. on Guided Wabe Optics, Guilin, China, June 2-5
    (in Chinese) (1981).
8. H. C. Dong, "Planar Glass Optical Waveguides by Ion-exchange
    Technique," presented at 1st Conf. on Guided Wave Optics,
    Guilin, China, June 2-5 (in Chinese) (1981).
9. X. X. Wang, S. F. Shi and D. W. Xu, "Optical Waveguide Experiment
    with Native Oside on Silicon as a Substrate," Research Report
    of Changchun Institute of Physics, Academia Sinica (in
    Chinese) (1981).
10. J. Li, Y. L. Chen and J. X. Fang, "A New Method for Determining
    Graded Index Profile of Optical Planar Waveguides," presented
    at 1st Conf. on Guided Wave Optics, Guilin, China, June 2-5
    (in Chinese) (1981).
11. F. Jin and Y. S. Li, "Dual-polarization Method for Measurement
    of Single Mode Planar Waveguide Parmeters," presented at 1st
    Conf. on Guided Wave Optics, Guilin, China, June 2-5 (in
    Chinese) (1981).
12. C. H. Chen and G. L. Ding, "Design and Construction of an
    Instrument of Measurement of Refractive Index and Thickness
    of Planar Waveguide," presented at 1st Conf. on Guided Wave
    Optics, Guilin, China, June 2-5 (1981).
13. Z. W. Liu and G. T. Cao, "The Formation of Grating Corugation
    on GaAs (100) Plane," Ji-Guang (Laser Journal), 8, No.1:27
    (in Chinese) (1981).
14. M. Xu and Y. Li, "Experimental Investigation on Grating Coupler,"
    presented at 1st Conf. on Guided Wave Optics, Guilin, China,
    June 2-5 (1981).
15. D. W. Xu, "A study of Thin Film Grating Coupler," Acta Physica
    Sinica (Chinese Journal of Physics), 29, No.9:1135 (1980).
16. Y. S. Li and F. Jin, "Effect of Coupling Gap for Prism-Film
    Couplers on the Errors of Film Parameters," Ji-Guang (Laser
    Journal), 7, No.1:42 (with English Abstract) (1980).
17. Z. H. Cao, G. L. Song and T. G. Li, "The $Bi_{12}GeO_{20}$ Prism Coupler,"
    presented at 1st Conf. on Guided Wave Optics, Guilin, China,
    June 2-5, (in Chinese) (1981).
18. F. Jin et al, "Nb-diffusion $LiTaO_3$ Waveguide and Planar Electro-
    optical Prism." Research Report of Changchun Institute of
    Physics, Academia Sinica, (in Chinese) (1978).
19. Y. S. Li, "Waveguide Electro-optical Prism," Acta Optica Sinica
    (Chinese Journal of Optics), 1, No.1:93 (with English Abstract)
    (1981).
20. B. R. Shi, C. J. Xu, Y. S. Li, H. J. Liu and P. Lu, "Acousto-
    optic Diffraction in Waveguides," Research Report of Changchun
    Institute of Physics, Academia Sinica, (with English Abstract)
    (1981).

21. J. H. Wang, Y. L. Zhan and Z. S. Zhang, "Acousto-optic Bragg
    Diffraction in Ti-LiNbO$_3$ Optical Planar Waveguides," presented
    at 1st Conf. on Guided Wave Optics, Guilin, China, June 2-5
    (in Chinese) (1981).
22. H. Z. Pan, Z. Y. Xiao, P. N. Shen and Z. Q. Chen, "Fabrication
    of a GaAs Monolithically Integrated Optical Circuit," Technical
    Digest of International Conf. on Lasers, Peking, China, May,
    P.108 (1980).
23. H. Z. Pan, P. N. Shen and Z. Y. Xiao, "Performance Study of a
    Twin-mesa Structure GaAs Monolithically Integrated Optical
    Circuit," (with English Abstract) to be published.

COUPLING TECHNIQUES: PRISM-, GRATING- AND ENDFIRE-COUPLING

E. Voges

FernUniversität, Nachrichtentechnik
P.O. Box 940
D - 5800 Hagen, Germany

1. INTRODUCTION

The investigation and application of integrated optic (IO)
devices with single mode film or strip waveguides require efficient
means for coupling laser beams into the planar waveguiding structure.
Four common coupling techniques are depicted in Figure 1. The prism-
coupler (Figure 1a) operates through a distributed coupling by frus-
trated total reflection at the base of a high-index prism when pressed
down onto a film or strip waveguide. The main advantages are a near
100% coupling efficiency which is achieved by proper construction,
a mode selective coupling, and an easy experimental realization-
at least for film waveguides and moderate coupling efficiencies.
The grating-coupler (Figure 1b) also operates by distributed coupling
due to diffraction at a periodic grating. It has emerged as a
potentially important coupler because of its planar nature. A 100%
coupling efficiency can be achieved. For this purpose, however,
gratings with proper shapes of the grooves are required which are
difficult to fabricate.

At present endfire- or butt-coupling techniques (Figure 1c, d)
are of particular interest. The endfire-coupling of a laser beam
has the advantage of in-line input and output beams, and plane output phase fronts. In addition, it is more easily to handle than
prism-coupling, especially when coupling into single mode strip
waveguides. Butt-coupling of single mode fibers to IO-devices is
very important in communication or sensor applications.

In the following, the above coupling techniques are described
together with experimental realizations. We restrict attention to
a more qualitative explanation of the coupling principles, since

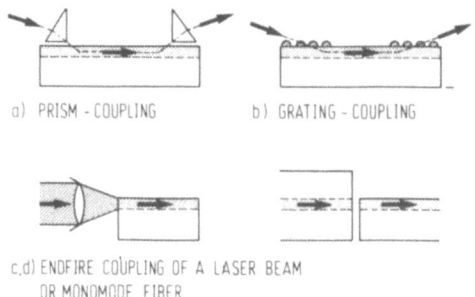

a) PRISM - COUPLING   b) GRATING - COUPLING

c,d) ENDFIRE COUPLING OF A LASER BEAM
OR MONOMODE FIBER

Fig. 1.   Schemes of common coupling techniques to integrated optic
          devices.

there exists an extensive theoretical work (see for example[1,2]),
and consider titanium indiffused waveguides in LiNbO$_3$ (Ti-LiNbO$_3$)
when presenting experimental results.

2. PRISM-COUPLING

The coupling of an incident laser beam by a rutile (TiO$_2$) prism
(indices at λ_o=0633μm : n_e=2.871, n_o=2.583) into a film waveguide
on LiNbO (indices at λ_o=0.633 m : n_e=2.2, n_o=2.29) and the output
coupling of an excited film mode are shown in Figure 2.

If the input beam is totally reflected at the base of the prism,
a standing wave distribution – with respect to the x-direction –
within the prism and an exponentially decaying field in the gap
region are generated (see Figure 3), which propagate with a phase
velocity

$$v_p = c_o/(n_p \sin \theta) \quad (c_o = \text{vacuum light velocity}) \quad (1)$$

parallel to the waveguide.

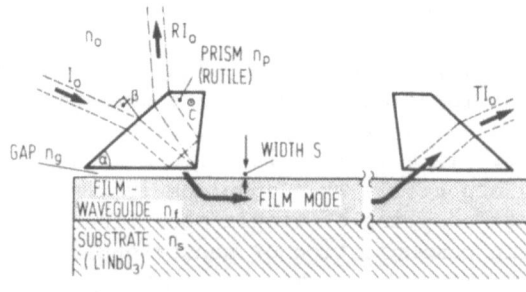

Fig. 2.   Prism-coupling to a Ti:LiNbO$_3$ film waveguide.

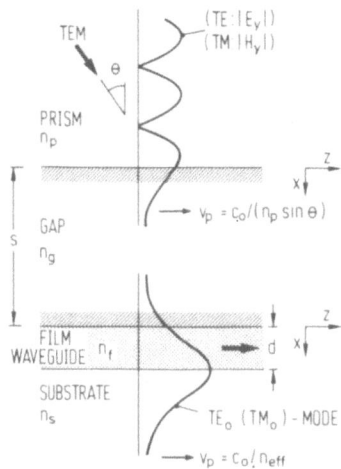

Fig. 3. Field distributions of a totally reflected laser beam at
the base of the coupling prism, and field distribution on
the fundamental film waveguide mode.

For a sufficiently narrow gap between prism and waveguide, the
field underneath the prism base reaches into the waveguide and excites
a waveguide mode, if this propagates synchronously, ie. with the
same phase velocity $v_p = c_0/n_{eff}$[3]. Here, $n_s < n_{eff} < n_f$ is the effective
mode index of a waveguide mode. We then have coupling angles θ to
waveguide modes within the range

$$n_s/n_p < \sin \theta < n_f/n_p \tag{2}$$

where θ is related to the external angle β (see Figure 2), yielding

$$n_{eff} = n_p \sin \{\alpha - \sin^{-1} (n_o \sin\beta/n_p)\} \tag{3}$$

One recognizes at once that the width s of the coupling gap must be
extremely small for efficient coupling, and that there exists an
optimum coupling length since the excited waveguide mode couples
back into the prism. Therefore, the input beam must be positioned
near the end of the prism.

Experimentally, it is advantageous to use prisms with a base
angle larger than 90°, since the incident and reflected beams are
separated in this case, and a measurement of the reflected intensity
yields an estimate of the coupled power. Figure 4 shows a typical
measurement of the reflected and transmitted intensities at HeNe-
laser wavelength for a Ti:LiNbO$_3$ film waveguide[4].

About 85% of the input power is coupled into the TE$_0$-mode, and
the transmittance is about 40%. The prism couplers are built up

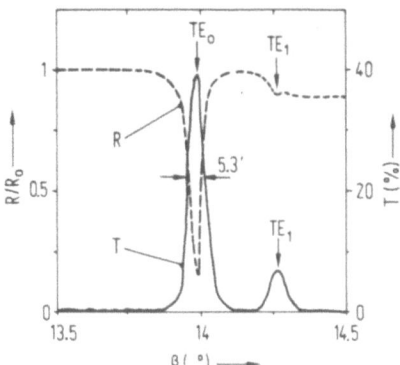

Fig. 4. Normalized reflected intensity R/R_o and transmittance T in
        dependence on the external angle β (prism angle $\alpha=45^o$) of
        a Ti:LiNbO$_3$ film waveguide, which is excited with TE-
        polarization and carries a TE$_0$- and a TE$_1$-mode.

according to the scheme of Figure 5, which allows rather high (91%)
coupling efficiencies[5]. For coupling efficiencies near 100% the
influence of the beam shape must be taken into account. Then, an
accurate analysis[6-8] is required, which is more easily performed
for output coupling. When a waveguide mode reaches the output prism
(Figure 6), it changes its character from a truly guided mode to a
leaky mode that continuously radiates power into an inhomogeneous
plane wave travelling at the synchronous angle θ into the prism.

This leakage of power into the prism leads to an attenuation
constant a_m of the waveguide mode, the radiated mode in the prism
will then simultaneously be attenuated, but transversely to its
direction of propagation. This one-sided decay may be closely approximated by an exponential function with a decay constant[2].

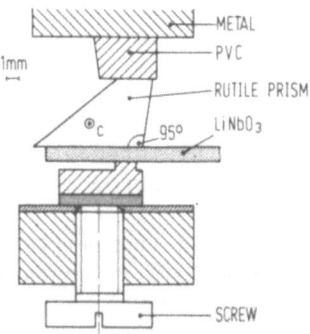

Fig. 5. Cross-section of efficient prism-film-couplers.

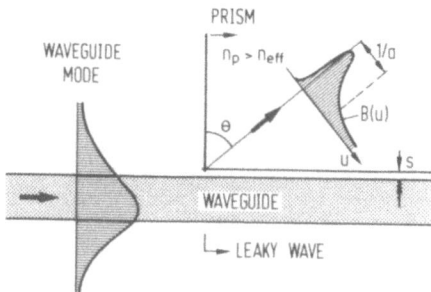

Fig. 6.   Waveguide mode and leaky mode with transverse attenuation
          within the output prism.

$$a = a_m/\cos\theta \sim \frac{4n_p n_f}{(n_p^2 - n_g^2)d} \exp\{-2k_0 s \sqrt{n_f^2 - n_g^2}\}, \quad k_0 = 2\pi/\lambda_0 \qquad (4)$$

Here, a step-index film waveguide of thickness d, weak guidance
$(n_f \approx n_s)$ and a large asymmetry $((n_s^2 - n_0^2)/(n_f^2 - n_s^2) \gg 1)$ of the waveguide
is assumed. The output coupling efficiency η_0 is then obtained as

$$\eta_0 = 1 - \exp\{-2 a_m L\} \qquad (5)$$

for a given length L of the prism coupler with $\eta_0 \to 1$ for $L \gg 1/a_m$.

Because of reciprocity an identical input coupling efficiency
η_i only obtains if the input beam is of the same transversely decaying
shape B(u). Otherwise, the input coupling efficiency is determined
by the normalized overlap integral[9]

$$\eta_i = |\int A(u) \, B^*(u) \, du|^2 / \{|\int A(u) \, du|^2 \cdot |\int B(u) \, du|^2\} \qquad (6)$$

where A(u) is the transverse distribution of the input beam. The
optimum widths and postitions of a Gaussian and a constant input
distribution are depicted in Figure 7. The optimum coupling efficiencies then are $\eta_i \approx 0.8$. For higher coupling efficiencies the
width s of the gap must be tapered to shape the output beam for a
closer match for example to a Gaussian beam. Figure 8 shows the
output coupling with a tapered gap.

Because of the decreasing width s of the gap, ie. increasing
radiated power, the transverse decay of the output beam can be compensated and near Gaussian beam shapes are obtained. In this way
prism-coupled Ti:LiNbO$_3$ devices with about 95% coupling efficiencies
can be achieved[10-12]. It should be noted that the high coupling
efficiency obtained by the arrangement of Figure 5 is also due to
a tapered gap, since the width s of the gap is never homogeneous,
and optimum beam positions for efficient coupling can be found by
a lateral shifting of the input beam along the edge of the prism.

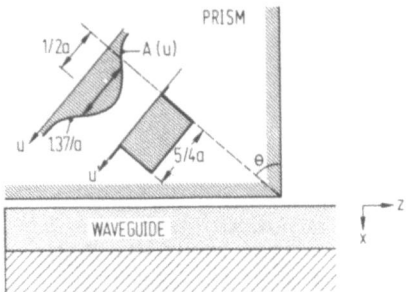

Fig. 7.   Input beams with Gaussian and constant transverse distri-
          butions for optimum coupling efficiencies.

For efficient prism-coupling into single mode strip waveguides
broad coupling horns are necessary. Figure 9 shows the top view of
a single mode Ti:LiNbO$_3$ waveguide[4] with parabolic coupling horns[13].
For this configuration 7dB transmission loss at 0.633μm including
the reflection losses of the prisms (2.5dB), waveguide losses, and
losses of the coupling horns have been achieved. Coupling horns
with an accurate geometry, however, are rather difficult to fabricate.

3. GRATING-COUPLERS

     The grating coupler as shown in Figure 1b seems to be at present
of less interest than prism-couplers or endfire-coupling techniques.
Therefore, its operation is only briefly described in a simple way.

     In a grating-coupler[14,15] the waveguide is covered by a dielec-
tric grating which leads to a distributed coupling between a waveguide
mode and a laser beam, if the scattering of the individual grating
lines adds up constructively. This is shown in Figure 10 for output
coupling.

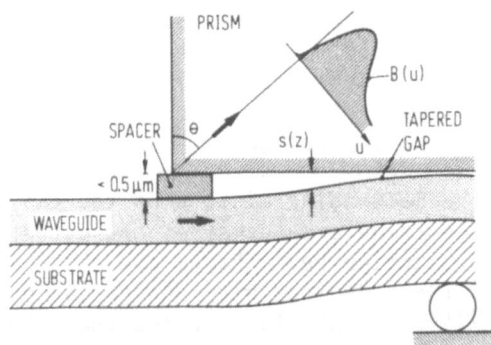

Fig. 8.   Output prism-coupling with a tapered gap and induced shaping
          of the output beam.

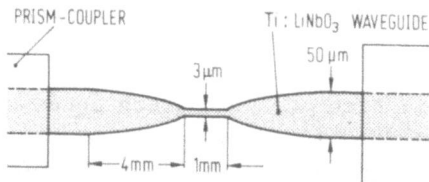

Fig. 9.   Prism-coupling into a single mode strip waveguide with
          additional coupling horns.

The incident waveguide mode excites a radiation field at each
grating line with a radiation characteristic as depicted by the
dashed curve in Figure 10. For weak coupling to the grating, the
contributions of two neighbouring lines have a phase difference
$w\,k_o\,n_{eff}$. All contributions sum up constructively for a direction
θ of the radiated field with

$$w\,k_o\,n_{eff} = w\,k_o\,\sin\theta + 2p\pi \; ; \; (p=1,\,2,\,3\,\ldots) \qquad (7)$$

Therefore, the grating radiates at directions

$$\theta_p = \sin^{-1}\{n_{eff}/n_o - p\,\lambda_o/w\} \qquad (8)$$

The main power is radiated in a direction where the individual radiation characteristic of a grating line has a maximum. For only one
main output beam the period w of the grating must be small, ie. λ_o/w
large, so that only one angle θ_p coincides with the maximum of the
individual radiation characteristic.

Detailed investigations (see for example[1,16,17]), which are
based on the coupled mode approach or the leaky wave concept reveal
a shape of the output beam similar to that of the prism-coupler.
Applying reciprocity again, the influence of the beam profile on
the input coupling efficiency can be determined. Output coupling
with 97% efficiency has been achieved[18].

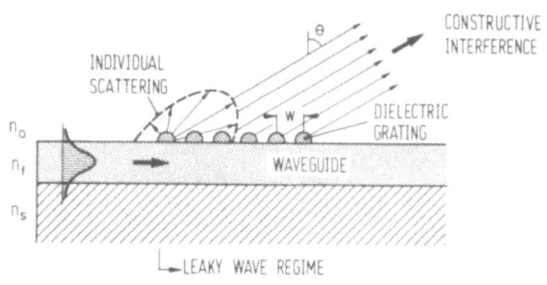

Fig. 10.   Schematic of a grating-coupler.

4. ENDFIRE-COUPLING OF LASER BEAMS AND FIBERS

The coupling of IO-devices to fibers is of particular practical importance, and efficient coupling techniques will decide on the utility of IO-devices. Today, the most practical approach for fiber to single mode strip waveguide coupling appears to be butt-joining (Figure 11). (See for example[19],[24]). Then, Fresnel reflexion losses and a mismatch between the mode profiles contribute to the coupling loss. The same loss mechanisms occur for endfire-coupling of a laser beam with the aid of a lens (Figure 12). Gaussian input beams can readily be assumed in both cases.

Figure 11 depicts the butt-coupling of strip waveguides to mono-mode fibers which are commonly aligned by Si-V-grooves. The lateral alignment can be very precise since the V-grooves can be delineated with the photomask for the strip waveguides. The vertical adjust-ment, however, is still a problem. Tapered fibers, aligned in orthogonal V-grooves, have been used for this purpose[19].

In order to assess the mode mismatch, a knowledge of the mode profile of a Ti:LiNbO$_3$ strip waveguide is required. Following[23],[24] the near field intensity of a Ti:LiNbO$_3$ strip waveguide is rather accurately described by a Gaussian distribution

$$I(x) = I_{max} \exp\{-2(x/\sigma_{//})^2\} \tag{9}$$

parallel to the surface, and a truncated Hermite-Gaussian function

$$I(y) = \begin{cases} I_{max} \dfrac{2y^2}{\sigma_{\perp}^2} \exp\{2(1-(y/\sigma_{\perp})^2\} & y > 0 \\[4mm] 0 & y \le 0 \end{cases} \tag{10}$$

perpendicular to the surface y = 0. (y > 0 points into the LiNbO$_3$ substrate). A comparison of eq. (9), (10) with experimental re-ults[23],[24] is shown in Figure 13.

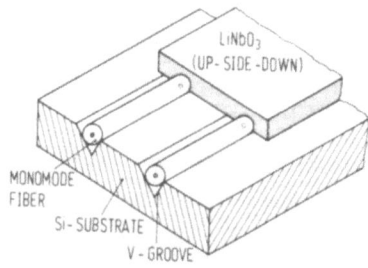

Fig. 11.   Butt-coupling of Ti:LiNbO$_3$ strip waveguides to monomode fibers which are aligned by Si-V-grooves.

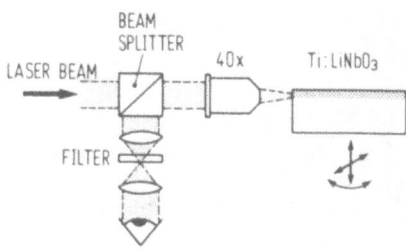

Fig. 12.   Endfire-coupling of a Gaussian laser beam into a Ti:LiNbO$_3$
            strip waveguide. The additional beam splitter and lens
            system is used to minimize the focal width.

The coupling loss due to mismatch of the mode profiles is then
determined from the overlap integral of the input intensity profile
and the intensity profile of the waveguide mode[19-24]. Theoretical
coupling factors up to 0.87, and experimental values up to 0.8 are
reported in[23]. The additional Fresnel reflection loss can be reduced
by antireflection coating.

A corresponding calculation can be used to determine and optimize the coupling efficiencies for endfire-coupled laser beams.
Experimentally, about 10dB transmission loss is observed at λ_0 =
0.633μm when endfire-coupling integrated Ti:LiNbO$_3$ Mach-Zehnder
interferometers to a laser beam[25], which reduces to about 7dB at
λ_0 = 1.06μm. Comparable transmission losses are observed when butt-
coupling single mode fibers[23].

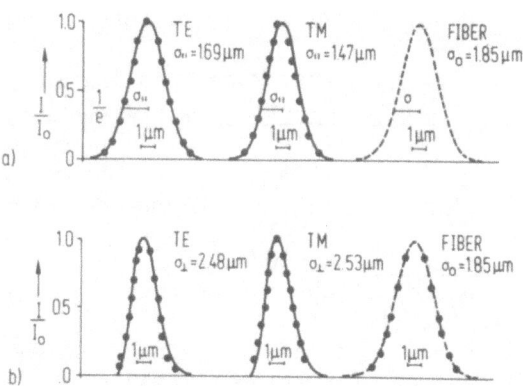

Fig. 13.   (a) Horizontal scan of the near-field profiles of single
            mode Ti:LiNbO$_3$ (c-cut) strip waveguides for TE- and TM-
            polarization, and near field profile of a single mode
            fiber. The solid curves are best Gaussian fits.
            (b) Same as a, perpendicular to the surface. The solid
            curves are best fits to eq. (10)[23,24].

REFERENCES

1. T. Tamir, Beam and waveguide couplers, in:"Integrated Optics,"
    T. Tamir, ed., Topics in Appl. Phys. Vol. 7, Springer, Berlin
    (1975).
2. H.-G. Unger, "Planar optical waveguides and fibres," Clarendon
    Press, Oxford (1977).
3. P. K. Tien, R. Ulrich and R. J. Martin, Modes of propagating
    light waves in thin deposited semiconductor films, Appl. Phys.
    Lett. Vol. 14:291-294 (1969).
4. A. Neyer, Diploma thesis, Institut für Physik, Universität
    Dortmund (1978).
5. W. Sohler and H. Suche, Efficient prism-coupling into titanium
    diffused lithium niobate optical waveguides, Wave Electronics
    3:269-275 (1979).
6. P. K. Tien and R. Ulrich, Theory of prism-film coupler and thin-
    film light guides, J. Opt. Soc. Am. 60:1325-1337 (1970).
7. R. Ulrich, Theory of the prism-film coupler by plane wave
    analysis, J. Opt. Soc. Am. 60:1337-1350 (1980).
8. J. E. Midwinter, Evanescent field coupling into a thin-film
    waveguide, IEEE J. Quant. Electron. QE-6:583-590 (1970).
9. R. Ulrich, Optimum excitation of optical surface waves, J. Opt.
    Soc. Am. 61:1467 (1971).
10. D. Sarid and D. Kermisch, Prism-waveguide coupling efficiency
    for waveguides with an arbitrary refractive index profile,
    Appl. Phys. Lett. 33:619-620 (1978).
11. D. Sarid, High efficiency input-output prism waveguide coupler:
    an analysis, Appl. Opt. 18:2921-2926 (1979).
12. F. Auracher and R. Keil, High deflection-efficiency low-insertion
    loss electrooptic Bragg reflector, 6th European Conf. on
    Opt. Comm. (ECOC) pp.272-275, IEE Conference Publication No.190
    (1980).
13. A. F. Milton and W. K. Burns, Mode coupling in iptical waveguide
    horns, IEEE J. Quant. Electron. QE-13:828-835 (1977).
14. M. L. Dakss, L. Kuhn, P. F. Heidrich and B. A. Scott, Grating
    coupler for efficient excitation of optical guided waves in
    thin films, Appl. Phys. Lett. 16:523-525 (1970).
15. H. Kogelnik and R. P. Sosnowsky, Holographic thin film couplers,
    Bell Syst. Techn. J. 49:1602-1608 (1970).
16. R. Ulrich, Efficiency of optical grating couplers, J. Opt. Soc.
    Am. 63:1419-1431 (1973).
17. C. C. Ghizoni, B.-U. Chen and C. L. Tang, Theory and experiments
    on grating couplers for thin-film waveguides, IEEE J. Quant.
    Electon. QE-12:69-73 (1976).
18. T. Aoyagi, Y. Aoyagi and S. Namba, High-efficiency blazed grating
    couplers, Appl. Phys. Lett. 29:303-304 (1976).
19. H. P. Hsu and A. F. Milton, Flip-chip approach to endfire coupling
    between single-mode optical fibres and channel waveguides,
    Electron. Lett. 12:404-405 (1976).

20. W. K. Burns and G. B. Hocker, Endfire coupling between optical fibers and diffused channel waveguides, $\underline{Appl.\ Opt.}$ 16:2048-2050 (1977).

21. J. C. Campbell, Coupling of fibers to Ti-diffused LiNbO$_3$ waveguides by butt-joining, $\underline{Appl.\ Opt.}$ 18:2037-2040 (1979).

22. M. Fukuma and J. Noda, Optical properties of titanium-diffused LiNbO$_3$ strip waveguides and their coupling-to-a-fiber characteristics, $\underline{Appl.\ Opt.}$ 19:591-597 (1980).

23. R. Keil and F. Auracher, Coupling of single-mode Ti-diffused LiNbO$_3$ waveguides to single-mode fibers, $\underline{Opt.\ Commun.}$ 30:23-28 (1979).

24. R. Keil, Dr.-Ing. thesis (1981).

25. A. Neyer, private communication.

THE FINITE ELEMENT METHOD IN THE ANALYSIS OF OPTICAL WAVEGUIDES*

L. Manià, T. Corzani and E. Valentinuzzi

Instituto di Electtrotecnica e di Elettronica
Università di Trieste
34127 - Trieste, Italy

INTRODUCTION

During the last years a very great interest has been devoted
to the analysis of the propagation characteristics of dielectric
waveguides, on account of their possible use as the fundamental
building blocks of integrated optical circuits. These waveguides
serve not only as a transmission medium to confine and direct optical
signals, but also as the basis for circuits such as filters and directional couplers. Thus, it is important to have a thorough knowledge
of the properties of their modes.

Besides a few examples of dielectric waveguides which can be
investigated exactly in an analytical way (such as the circular and
the elliptical waveguide), there are many other cases that, for
their geometry, require an analysis based on some numerical procedure:
we mention, for example, the rectangular dielectric waveguide, the
channel guide, the optical strip-line and so on. Among these numerical procedures, the finite element method (FEM), to which these notes
are devoted, turns out to be one of the most general techniques
which can be successfully applied to the analysis of the surface
wave modes propagating along a dielectric waveguide. Today, this
numerical method is widely used in many fields of engineering[1].
In particular, its applications in electromagnetic problems concern
mostly the study of the wave propagation in uniform cylindrical waveguides, where the uniformity along the cylinder axis reduces the
original three-dimensional problem to a two-dimensional one, thus

---

*Presented by C.G. Someda, Università di Bologna (Italy).

obtaining drastic simplifications both from an analytical and from
a numerical point of view.

In these notes we present a formulation of the FEM suitable
for the analysis of dielectric waveguides. The most particular feature of these problems, with respect to those usually tackled by
the FEM, consists in the fact that the cross section of the waveguide,
that is, the domain of definition of the unknown functions, is in
theory unbounded. The application of the FEM to this class of problems requires some strategies in order to overcome this difficulty.
To this end, in these notes we make use of the so called "infinite
elements", first introduced by Bettess[2] for viscous flow problems.

Moreover, the formulation proposed here is based on the weighted
residual principle and Galerkin's criterion, and not on a variational
principle. This allows a very straightforward procedure for deriving
the finite element equations, especially when the waveguides include
lossy materials.

Finally, in order to show the performances which can be obtained
with this numerical method, we present a set of results relative
to some typical dielectric waveguides, such as the lossy dielectric
rod, the elliptical tube and the rectangular waveguides, the channel
guide and the optical strip-line.

STATEMENT OF THE PROBLEM

The problem we are dealing with is the evaluation of surface
wave modes which propagate along a uniform cylindrical dielectric
guide. The dielectric guide consists of a finite number of regions,
filled with linear, homogeneous, isotropic materials, with or without
losses, surrounded by an unbounded medium (typically air). The
wave is assumed to propagate along the positive z-direction, which
coincides with the axis of the guiding structure, with a z-dependence
of the type $\exp(-\gamma z)$ and with a time harmonic variation given by
$\exp(j\omega t)$, where ω is the angular frequency. Under these assumptions,
the fields may be completely specified in terms of two factors of
the longitudinal components of the electric and magnetic field
which depend on the transverse coordinates only. As it is well
known, over each homogeneous region of the waveguide cross section
E_z and H_z satisfy the following equations:

$$(\nabla^2 + k^2 + \gamma^2) \begin{Bmatrix} E_z \\ H_z \end{Bmatrix} = 0, \tag{1}$$

where $\gamma = \alpha + j\beta$ is the propagation constant along the z-axis and
$k = \omega\sqrt{\mu\varepsilon}$ represents the intrinsic wave number of the region. If it
is the case, complex scalar values must be considered for the con-
stants μ and ε of the different media, in order to take into account
various kinds of dissipation. One pair of equations like (1) need

to be solved within each homogeneous region, under the appropriate boundary conditions at the dielectric interfaces, and under the radiation condition at infinity.

The transverse field components $\overline{E}_t$ and $\overline{H}_t$ can be deduced from the knowledge of E_z and H_z: in fact, by means of Maxwell's equations, they can be expressed as:

$$\overline{E}_t = - \frac{j\omega\mu}{k^2+\gamma^2} \text{ grad } H_z \wedge \hat{z} - \frac{\gamma}{k^2+\gamma^2} \text{ grad } E_z \quad ,$$

$$\overline{H}_t = \frac{j\omega\varepsilon}{k^2+\gamma^2} \text{ grad } E_z \wedge \hat{z} - \frac{\gamma}{k^2+\gamma^2} \text{ grad } H_z \quad ,$$

(2)

where $\hat{z}$ is the unit vector in the z-direction.

As to the boundary conditions, the tangential components of $\overline{E}$ and $\overline{H}$ must be continuous at the interface between two different dielectric media. Thus along the contour line τ, obtained by inter-secting the interface with a plane normal to the z-axis, the conti-nuity of the tangential component of $\overline{E}$ and $\overline{H}$ is expressed by the following equations:

$$E_{z1} = E_{z2} , \quad H_{z1} = H_{z2} ,$$

(3)

$$E_{\tau 1} = E_{\tau 2} , \quad H_{\tau 1} = H_{\tau 2} ,$$

where the subscript τ denotes the components of the field tangential to the contour line τ, and the subscripts 1 and 2 refer to the dif-ferent media separated by τ. The components E_τ and H_τ may be con-veniently expressed in terms of the normal and the tangential deriva-tives of E_z and H_z, that is:

$$E_\tau = \overline{E}_t \cdot \hat{\tau} = \frac{j\omega\mu}{k^2+\gamma^2} \frac{\partial H_z}{\partial n} - \frac{\gamma}{k^2+\gamma^2} \frac{\partial E_z}{\partial \tau} \quad ,$$

(4')

$$H_\tau = \overline{H}_t \cdot \hat{\tau} = - \frac{j\omega\varepsilon}{k^2+\gamma^2} \frac{\partial E_z}{\partial n} - \frac{\gamma}{k^2+\gamma^2} \frac{\partial H_z}{\partial \tau} \quad .$$

(4")

In (4') and (4") reference is made to a clockwise set of three unit vectors $\hat{n}$, $\hat{\tau}$, $\hat{z}$, whose directions coincide with the normal, the tangent and the bi-normal to the considered contour curve, re-spectively. The two boundary conditions relative to surfaces of perfect electric and magnetic conductors are particularly interesting: in fact, the first can stimulate appropriately a good conductor, while both of them are very useful for taking advantage of the sym-metries of the structure. On such surfaces these conditions can be written respectively as:

$E_z = 0$, $E_\tau = 0$ (perfect electric conductor);

$H_z = 0$, $H_\tau = 0$ (perfect magnetic conductor).

(5)

For computational convenience, it is preferable to put (1) in a non-dimensional form. This can be done by choosing a suitable metric coefficient a, which in general is one of the typical dimensions of the corss section of the waveguide: equations (1) are then multiplied by a^2. Moreover, E_z and H_z are normalized with respect to the values $|E_{zo}|$ and $|H_{zo}|$ taken by thir moduli in an arbitrary reference point. Using a prime for denoting normalized quantities, one gets:

$$\nabla'^2 = a^2 \nabla^2, \quad k'^2 = a^2 k^2, \quad \gamma'^2 = a^2 \gamma^2;$$

(6)

$$E'_z = \frac{E_z}{|E_{zo}|}, \quad H'_z = \frac{H_z}{|H_{zo}|} \quad .$$

(6')

In addition, one can write:

$$k'^2 = \omega^2 \mu \varepsilon a^2 = \left(\frac{\omega a}{c}\right)^2 \mu_r \varepsilon_r \ ,$$

(7)

where c is the light speed in the vacuum and μ_r and ε_r are the relative permeability and permittivity of the medium, respectively. Therefore, it seems to be convenient to define a normalized frequency ω' as:

$$\omega' = \frac{\omega a}{c} \ ,$$

(8)

so obtaining:

$$k'^2 = \omega'^2 \mu_r \varepsilon_r \quad .$$

(9)

Substituting the normalized quantities in (1), one obtains:

$$(\nabla'^2 + k'^2 + \gamma'^2) \begin{Bmatrix} E'_z \\ H'_z \end{Bmatrix} = 0 \ .$$

(10)

In the sequel all quantities will be considered in normalized form, and, for the sake of simplicity, the prime will be neglected. Moreover, for mere mathematical reasons, it is convenient to rewrite expressions (10) in the following form, with reference to each homogeneous region:

$$\varepsilon_r \left(\frac{1}{k^2+\gamma^2} \nabla^2 E_z + E_z \right) = 0 \ ,$$

$$\mu_r \left(\frac{1}{k^2+\gamma^2} \nabla^2 H_z + H_z \right) = 0 \ .$$

(11)

FINITE AND INFINITE ELEMENTS

The application of the FEM to the solution of field problems
is based on a partition of the domain of definition W of the unknown
functions into a sufficient number of subdomains. The unknown functions are then approximated over each element by suitable functions,
expressed through a given number of parameters: the solution of
the numerical problem consists in determining the values of these
parameters in such a way that the corresponding functions yield a
good approximation of the true fields, according to a given citerion.

Before discussing the mathematical aspects relative to the
application of the FEM to the solution of surface wave propatagtion
we wish to stress that in our problem the domain W is unbounded:
this is an important fact, which makes our problem somewhat more
complicated than those usually tackled by means of the FEM. Actually,
the name "finite element method" comes from the fact that the subdomains used for discretizing the domain W have finite size. If W
is unbounded, an infinite number of such elements would be required
for its partition, and the resulting numerical problem would turn
out to be intractable. In the past various strategies have been
proposed to fit the FEM to unbounded geometry[3]. Generally, all
these strategies are based on a division of the domain of interest
into two parts; the first is a bounded domain, which is treated with
the classical FEM. Some tricks are then adopted simply for taking
into account the effect of the second part, which is unbounded, on
the unknown field associated with the bounded domain.

Also the method considered in these notes is based on such a
division of the domain W, but, in the unbounded region, we will adopt
the so called "infinite element method", which appears to be the
most natural extension of the classical FEM to this class of problems,
form an analytical as well as from a numerical point of view.

Therefore the starting point of the method described here consists in a subdivision of the cross section of the guide into two
adjacent domains, as schematically shown in Figure 1. The inner
domain is bounded by the contour C: it contains typically the dielectric guide, which can consist of different media, and a portion of
the surrounding space; the outer domain extends from C to the
infinity.

For the time being, let us consider the two domains separately.
According to the FEM, the inner domain is subdivided into a suitable
number T of subdomains, called precisely "finite elements", and the
unknown functions are then approximated over each element by complete
polynomials of order n. The problems regarding the choice of the
form of the finite elements and the order of the approximating polynomials are widely discussed in the literature (see for example
Zienkiewicz[1]). In particular, a widely adopted shape for the finite

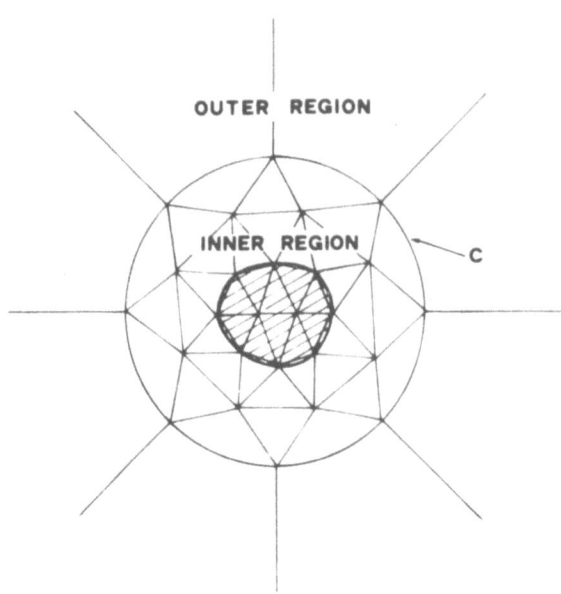

Fig. 1. Typical subdivision in the plane of the cross section of
         the guide.

elements is the triangular one, to which we will refer in the following. Over the general element t (t = 1,2,...,T) the n-order polynomials $\phi_E^t(x,y)$ and $\phi_H^t(x,y)$ which approximate E_z and H_z are written
in the form:

$$\phi_E^t(x,y) = \sum_1^{N_1} {}_k \phi_{Ek}^t \alpha_k^t(x,y) ,$$

$$\phi_H^t(x,y) = \sum_1^{N_1} {}_k \phi_{Hk}^t \alpha_k^t(x,y) ,$$

(12)

where N_1 = (n+1) (n+2)/2 and where ϕ_{Ek}^t and ϕ_{Hk}^t are the values taken
by ϕ_E^t and ϕ_H^t at the k-th point of a set of N_1 points, called "nodes",
which have been fixed on the above t-th element. The general func-
tion α_k^t, which is called "interpolating function at the k-th node",
is a complete polynomial of order n, whose coefficients can be expressed in terms of the cartesian coordinates of the N_1 nodes: α_k^t
is equal to one at the k-th node and zero at the other nodes of the
element; moreover, each function α_k^t is conventionally set equal to
zero outside the t-th element. If, for example, a value of 2 is
chosen for n, N_1 turns out to be equal to six, and in such a case
it is convenient to choose as nodes of each triangular element the
three vertices and the midpoints of its sides.

As indicated in Figure 1, the outer domain is subdivided into
a number (S, say) of adjacent elements which extend radially toward
the infinity (this is the reason why they are called "infinite elements"). A typical element of them is considered in Figure 2a,
with reference to an orthogonal coordinate system (x,y), called
"global system". Also upon this kind of elements, a set of N points
is fixed, which can be still called "nodes", or "nodal points", on
the analogy of what has been done in the case of a finite element.
As indicated in Figure 2a, the most exterior series of nodes are
allowed to tend to infinity.

The nodal points are partially used also as reference points
for establishing a mapping of the infinite element into a parent
element of more regular shape, over which the field representation
and all the subsequent operations of integration can be greatly simplified. The parent element consists of the rectangular open strip
shown in Figure 2b and is referred to a local coordinate system (u,v);
upon it n_u and n_v reference points are fixed in the u- and v-direc-
tion, respectively, being $n_u x n_v = N_2$. In the local coordinate system
the coordinates of the general reference point will be denoted by
u_i,v_j (i = 1,2,..,n_u; j = 1,2,..,n_v). The only difference between
the reference points (used in the element transformation) and the
nodal points is that the most external series of reference points
is assumed to have a large but not infinite u-coordinate: Figures 2a
and 2b clearify this assumption.

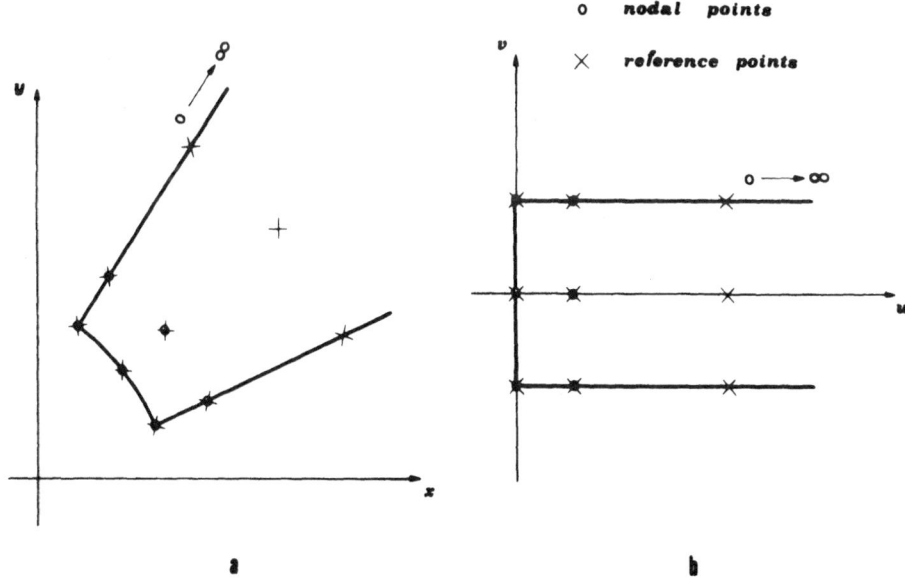

Fig. 2.   Infinite elements: a) original element; b) parent element.

The mapping between original and parent elements will have in general the form:

$$x = \sum_{k}^{N_2} F_k(u,v) x_k \ , \qquad\qquad y = \sum_{k}^{N_2} F_k(u,v) y_k \ , \qquad (13)$$

where x_k and y_k are the coordinates of the k-th reference point in the global system. The transformation functions $F_k(u,v)$ must meet the requirement that each reference point in the local system has to be mapped into the corresponding point in the global coordinate system. Therefore, each $F_k(u,v)$ must be equal to one in the k-th point and equal to zero in all the other reference points. Many kinds of functions can be chosen in order to satisfy these requirements: in these notes products of standard Lagrange polynomials are used for setting up the functions $F_k(u,v)$, that is:

$$F_k(u,v) = N_i(u) M_j(v) \ , \quad k = 1,2,..,N_2 \qquad (14)$$

where:

$$N_i(u) = \prod_{\substack{p=1 \\ p \neq i}}^{n_u} \frac{u_p - u}{u_p - u_i} \ ,$$

$$\qquad (15)$$

$$M_j(v) = \prod_{\substack{p=1 \\ p \neq j}}^{n_v} \frac{v_p - v}{v_p - v_j} \ .$$

In (15), i and j specify the k-th reference point in the u and v direction, respectively.

According to the procedure previously illustrated for the case of a finite element, also over the general infinite element s ($s = 1,2,..,S$) the fields E_z and H_z are approximated by a linear combination of interpolating fuctions, in the form:

$$\phi_E^S(x,y) = \sum_k^{N_2} \phi_{Ek}^S \ \psi_k^S(x,y) \ ,$$

$$\qquad (16)$$

$$\phi_H^S(x,y) = \sum_k^{N_2} \phi_{Hk}^S \ \psi_k^S(x,y) \ ,$$

where again ϕ_{Ek}^S (ϕ_{Hk}^S) are the nodal values of ϕ_E^S (ϕ_H^S), that is, the values assumed by ϕ_E^S (ϕ_H^S) at the nodal points.

The interpolating functions $\psi_k^S(x,y)$ can be described better in the local coordinate system (u,v); in fact, taking into account the transformation expressed by (13), and setting:

$$\psi_k'^S(u,v) = \psi_k^S\{x(u,v),y(u,v)\} \quad , \tag{17}$$

the approximations (16) can be transposed over the parent element, where one gets:

$$\phi_{E,H}'^S(u,v) = \sum_1^{N_2} \Phi_{Ek,Hk}^S \psi_k'^S(u,v) \quad . \tag{18}$$

The interpolating functions $\psi_k'^S(u,v)$ have to be chosen according to the following requirements: a) it must assume unit value at the k-th node and zero value at the other nodes; b) the continuity of the approximating functions must be assured along the contour lines between different infinite elements and between infinite and finite elements; c) the interpolating functions must be able to express the typical decay of the true fields at great distance along a radial direction (that is, along the u-direction in the parent elements).

The interpolating functions used here are similar to the transformation functions $F_k(u,v)$ indicated in (14). The functions $\psi_k'^S(u,v)$ relative to nodes placed at finite distance are built as product of two functions in the variable u and v, respectively:

$$\psi_k'^S(u,v) = U_i(u)V_j(v) \quad , \tag{19}$$

where the functions $V_j(v)$ coincide with $M_j(v)$ given by (15).

In order to satisfy item c), a special shape is used for the functions $U_i(u)$, of the general form:

$$U_i(u) = P_i(u) \exp(-u/L) \exp(-j\beta_T u) \quad , \tag{20}$$

where $P_i(u)$ is a polynomial, L is a parameter which can be called "decay length" and β_T is one more quantity which takes into account a possible progressive phase variation of the fields along the u-direction. The parameter β_T is useful when the analysis is performed on dissipative dielectric guides, while its use is not required when the guide is entirely loss free. The index i still specifies the k-th node along the u-direction.

For $P_i(u)$, standard Lagrange polynomials are used, that is:

$$P_i(u) = \prod_{\substack{q=1 \\ q \neq i}}^{n_u-1} \frac{u_q-u}{u_q-u_i} \quad , \tag{21}$$

in which the product is extended up to n_u-1, ie. the number of nodes at finite distance along the u-direction. Taking into account also item s), the final expression of $U_i(u)$ will be:

$$U_i(u) = \prod_{\substack{q=1 \\ q \neq i}}^{n_u-1} \frac{u_q-u}{u_q-u_i} \exp\left\{ (u_i-u)\left(\frac{1}{L} + j\beta_T\right)\right\} . \qquad (22)$$

As regards item b), it will be satisfied if the degrees of the polynomials $V_j(v)$ and of the polynomials used in the adjacent finite element are the same, and if the nodal points placed on the contour between the infinite and the finite elements coincide with those used for the finite element.

The interpolating functions relative to nodes placed at infinite distance are not necessary for the approximation of the filed. Nevertheless, they can be expressed conventionally as in (19), with $U_i(u)$ defined as:

$$U_i(u)_\infty = 1 - \sum_{1}^{n_u-1} U_j(u) . \qquad (23)$$

This definition makes the functions $\psi_k'^s(u,v)$ to satisfy the condition:

$$\sum_{1}^{N_2} \psi_k'^s(u,v) = 1 , \qquad (24)$$

required for the interpolating functions[1]. Moreover, each function $\psi_k'^s(x,y)$ is set equal to zero outside the s-th element.

Before continuing the algebraic formulation of our problem, we think it convenient to give some comments about the parameters L and β_T introduced in (21). The values of L and β_T have to be chosen in such a way that the behaviour of the true fields along a radial direction will result well approximated. This behaviour is not known in advance, but, fortunately, a great number of computations has shown that the solutions are in general not too sensitive to the variations of L and β_T around their optimum values. Anyway, a suitable choice of the values of L and β_T can be made on the basis of an estimation of the transversal separation constant (for example, through a preliminary solution of Eq.s (11), obtained by using finite elements only).

WEIGHTED RESIDUAL PRINCIPLE AND GALERKIN'S CRITERION

After introducing the concepts of finite and infinite elements and the associated approximating functions expressed by (12) and (16), respectively, we will try to give an expression for the approximation of E_Z and H_z valid on the whole domain W.

First of all it must be noted that when all elements (finite and infinite) are assembled together over the entire domain W, they will share certain sides and nodes: therefore, only N distinct nodes will be specified on the whole domain, with $N < N_1 \cdot T + N_2 \cdot S$. Then, in correspondence with the k-th of these distinct nodes, it is possible to introduce a function $f_k(x,y)$, so defined:

$$f_k(x,y) = \begin{cases} \alpha_k^t(x,y) & \text{if the point (x,y) belongs to the t-th} \\ & \text{finite element, assuming that this} \\ & \text{element includes the k-th node;} \\ \\ \psi_k^s(x,y) & \text{if the point (x,y) belongs to the s-th} \\ & \text{infinite element, assuming that this} \\ & \text{element includes the k-th node;} \\ \\ 0 & \text{if the point (x,y) belongs to an} \\ & \text{element which does not include the k-th} \\ & \text{node.} \end{cases} \qquad (25)$$

As it can be easily verified, the functions $f_k(x,y)$ are continuous on the whol domain W. With the introduction of the set of functions f_k, it is possible to give an approximation for E_Z and H_z similar to (12) and (16), which have been given for a single element: now the approximation holds over the whole domain W, ie.

$$\phi_E(x,y) = \sum_1^N {}_k \Phi_{Ek} f_k(x,y) \quad ,$$

$$\phi_H(x,y) = \sum_1^N {}_k \Phi_{Hk} f_k(x,y) \quad . \qquad (26)$$

The two sets of variable $\{\Phi_{Ek}\}$ and $\{\Phi_{Hk}\}$ must be determined in such a way that the resulting functions ϕ_E and ϕ_H give a "good approximation" of the solutions of (11). For an arbitrary choice of $\{\Phi_{Ek}\}$ and $\{\Phi_{Hk}\}$ the corresponding function ϕ_E and ϕ_H will not generally satisfy the two equations (11). Mathematically, this fact can be phrased as:

$$\varepsilon_r \left(\frac{1}{\chi^2} \nabla^2 \phi_E + \phi_E \right) = R_E(x,y) \quad ,$$

$$\mu_r \left(\frac{1}{\chi^2} \nabla^2 \phi_H + \phi_H \right) = R_H(x,y) \quad ,$$

(27)

in which R_E and R_H are called "residuals" and $\chi^2 = k^2 + \gamma^2$. One cannot hope to get $R_E(x,y)$ and $R_H(x,y)$ identically equal to zero over W, even by suitably choosing $\{\Phi_{Ek}\}$ and $\{\Phi_{Hk}\}$. Instead, by the weighted residual principle, a solution for $\{\Phi_{Ek}\}$ and $\{\Phi_{Hk}\}$ is sought in order to satisfy equations (11) in an average sense. More precisely, this principle consists in determining $\{\Phi_{Ek}\}$ and $\{\Phi_{Hk}\}$ in such a way that:

$$\int_W g_k(x,y) \; R_E(x,y) \; dw = 0 \quad ,$$

$$\int_W g_k(x,y) \; R_H(x,y) \; dw = 0 \quad ,$$

(28)

for every function $g_k(x,y)$ belonging to a set of N functions defined over W and called "weighing functions". Many different sets of weighting functions may be adopted; in particular, the Galerkin criterion suggests the choice $g_k \equiv f_k$ (k = 1,2..,N). This approach has been already widely adopted in engineering problems and offers various attractive features[4,5]. With the Galerkin choice of the weighting functions, Eq.s (28) become:

$$\int_W f_k \, \varepsilon_r \left(\frac{1}{\chi^2} \nabla^2 \phi_E + \phi_E \right) dw = 0 \quad ,$$

$$\int_W f_k \, \mu_r \left(\frac{1}{\chi^2} \nabla^2 \phi_H + \phi_H \right) dw = 0 \quad ,$$

(29)

with k = 1,2,..,N.

The following step consists in deriving, from the 2N set of equations like (29), an algebraic system of equations for the unknown vectors $\{\Phi_{Ek}\}$ and $\{\Phi_{Hk}\}$. If no otherwise stated, we will refer in the sequel by "element" both to a finite element or to an infinite element, indifferently. At first, we expand (29) by means of

Green's formula over each of the M homogeneous regions in W, so obtaining:

$$\sum_{1}^{M}{}_m \left(-\frac{\varepsilon_{rm}}{\chi_m^2} \int_{S_m} \nabla f_k \cdot \nabla \phi_E ds + \frac{\varepsilon_{rm}}{\chi_m^2} \oint_{\tau_m} f_k \frac{\partial \phi_E}{\partial n} d\tau \right.$$

$$\left. + \varepsilon_{rm} \int_{S_m} f_k \phi_E ds \right) = 0 ,$$

$$(30)$$

$$\sum_{1}^{M}{}_m \left(-\frac{\mu_{rm}}{\chi_m^2} \int_{S_m} \nabla f_k \cdot \nabla \phi_H ds + \frac{\mu_{rm}}{\chi_m^2} \oint_{\tau_m} f_k \frac{\partial \phi_H}{\partial n} d\tau \right.$$

$$\left. + \mu_{rm} \int_{S_m} f_k \phi_H ds \right) = 0 ,$$

where the subscript m denotes quantities relative to the m-th homogeneous region: in particular, $\chi_m^2 = k_m^2 + \gamma^2$, while S_m and τ_m denote the area and the contour of the m-th region.

In order to remove the normal derivatives, which are not continuous on crossing the boundary interface between different media, it is convenient to introduce two more functions, $\phi_{E\tau}$ and $\phi_{H\tau}$, which are an approximation of the tangential components of $\overline{E}_t$ amd $\overline{H}_t$ with respect to the contour τ. They can be obtained from (4), replacing E_z and H_z by the functions ϕ_E and ϕ_H. Therefore (30) can be rewritten as:

$$\sum_{1}^{M}{}_m \left(-\frac{\varepsilon_{rm}}{\chi_m^2} \int_{S_m} \nabla f_k \cdot \nabla \phi_E ds - \frac{\gamma}{j\omega} \frac{1}{\chi_m^2} \oint_{\tau_m} f_k \frac{\partial \phi_H}{\partial \tau} d\tau \right.$$

$$\left. - \frac{1}{j\omega} \oint_{\tau_m} f_k \phi_{H\tau} d\tau + \varepsilon_{rm} \int_{S_m} f_k \phi_E ds \right) = 0 , \qquad (31')$$

$$\sum_{1}^{M}{}_m \left(-\frac{\mu_{rm}}{\chi_m^2} \int_{S_m} \nabla f_k \cdot \nabla \phi_H ds + \frac{\gamma}{j\omega} \frac{1}{\chi_m^2} \oint_{\tau_m} f_k \frac{\partial \phi_E}{\partial \tau} d\tau \right.$$

$$\left. + \frac{1}{j\omega} \oint_{\tau_m} f_k \phi_{E\tau} d\tau + \mu_{rm} \int_{S_m} f_k \phi_H ds \right) = 0 . \qquad (31'')$$

By imposing that the functions ϕ_E and ϕ_H, as well as ϕ_{ET} and ϕ_{HT}, verify the same boundary conditions of the true electromagnetic fields along an interface line between different media, the integrals in (31') and (31") may be evaluated as the sum of the corresponding integrals over each element of W. This is possible because, when the line integrals are performed, the sides of the elements internal to each homogeneous region are traversed twice in opposite directions, so that their contributes cancel out.. In particular, denoting by τ_i the contour of the i-th element and by K the number of elements (K = T+S), one has:

$$\sum_{1}^{M} \frac{1}{j\omega} \oint_{\tau_m} f_k \phi_{HT} d\tau = \sum_{1}^{K} \frac{1}{ij\omega} \oint_{\tau_i} f_k \phi_{HT} d\tau \; ,$$

$$\sum_{1}^{M} \frac{1}{j\omega} \oint_{\tau_m} f_k \phi_{ET} d\tau = \sum_{1}^{K} \frac{1}{ij\omega} \oint_{\tau_i} f_k \phi_{ET} d\tau \; . \tag{32}$$

Moreover, expressions (32) are equivalent to the integrals of the same functions performed along the external contour which delimits the domain W. For an open dielectric waveguide, this contour is generally located at infinity, where the various field components must satisfy the radiation condition: therefore both the integrals (32) and zero. Sometimes, in order to take advantage of the symmetries of the cross section of the waveguide, part of this contour, although not located at infinity, is supposed to consit of a perfect electric (or magnetic) conductor: also in this case the above integrals are zero[5]. Therefore it is possible to conclude that the integrals (32) can always be dropped out from Eqn.s (31') and (31").

Taking into account the above considerations, Eqn.s (31') and (31") may be written in the following form, in which S_i and τ_i stand for the area and the contour of the i-th element, respectively:

$$\sum_{1}^{K} \left(-\frac{\varepsilon_{ri}}{\chi_i^2} \int_{S_i} \nabla f_k \cdot \nabla \phi_E ds - \frac{\gamma}{j\omega} \frac{1}{\chi_i^2} \oint_{\tau_i} f_k \frac{\partial \phi_H}{\partial \tau} d\tau + \right.$$

$$\left. + \omega^2 \varepsilon_{ri} \int_{S_i} f_k \phi_E ds \right) = 0 \; , \tag{33}$$

$$\sum_{1}^{K} {}_{i} \left(- \frac{\mu_{ri}}{\chi_i^2} \int_{S_i} \nabla f_k \cdot \nabla \phi_H ds + \frac{\gamma}{j\omega} \frac{1}{\chi_i^2} \oint_{\tau_i} f_k \frac{\partial \phi_E}{\partial \tau} d\tau + \right.$$

$$\left. + \omega^2 \mu_{ri} \int_{S_i} f_k \phi_H ds \right) = 0 \ .$$

From the two sets of equations (33) a set of algebraic equations in the unknown coefficients Φ_{Ek} and Φ_{Hk} is easily derived. First of all, according to the definition of f_k, one can state that all the integrals in (33) are zero if node k does not belong to the i-th element. Moreover, if node k belongs to the i-th element, let p be its index in the numbering of the N_1 (or N_2) nodes of the i-th element and q the current index of that numbering (q = 1,2,..,N_1 (or N_2)); then, if the i-th element is a finite element, these integrals can be expressed as:

$$\oint_{\tau_i} f_k \frac{\partial \phi}{\partial \tau} d\tau = \sum_{1}^{N_1} {}_q \Phi_q \oint_{\tau_i} \alpha_p^i \frac{\partial \alpha_q^i}{\partial \tau} d\tau = \sum_{1}^{N_1} {}_q \Phi_q \ r_{pq} \quad ,$$

$$\int_{S_i} \nabla f_k \cdot \nabla \phi ds = \sum_{1}^{N_1} {}_q \Phi_q \int_{S_i} \nabla \alpha_p^i \cdot \nabla \alpha_q^i ds = \sum_{1}^{N_1} {}_q \Phi_q \ s_{pq} \quad , \qquad (34)$$

$$\int_{S_i} f_k \phi \ ds = \sum_{1}^{N_1} {}_q \Phi_q \int_{S_i} \alpha_p^i \alpha_q^i \ ds = \sum_{1}^{N_1} {}_q \Phi_q \ t_{pq} \quad ,$$

where the function ϕ stands for ϕ_E or ϕ_H, indifferently. The quantities r_{pq}, s_{pq}, t_{pq} correspond to the (p,q) elements of appropriate matrices R, S and T of order $N_1 \times N_1$, which have been already evaluated in closed form[6] in the case of triangular elements and for n = 1, 2, 3 and 4.

When node k belongs to an infinite element, the various integrals are of the same form as in (34), with N_1 replaced by N_2 and the various functions α^i replaced by the functions ψ^i. However, in this case a closed expression for the required integrals does not exist and they have to be computed by means of a numerical procedure. To this end, they can be expressed in a more suitable form in the local coordinate system (u,v) by means of the transformation (13)

and its inverse. Over the parent elements the various integrals
will look as:

$$\int_{-1}^{+1} \int_{0}^{\infty} h(u,v) \exp(-u/L) \exp(-j\beta_T u) \ du \ dv \ , \tag{35}$$

where, obviously, the function $h(u,v)$ will have different expressions,
depending on what integral (34) we are dealing with.

Finally, inserting (34) and the parallel equations relative to
the infinite elements, in Eqn.s (33), a set of algebraic equations
in the unknowns $\{\Phi_{Ek}\}$ and $\{\Phi_{Hk}\}$ is obtained, which can be written
in compact form as:

$$[A] [x] - \omega^2 [B] [x] = 0 \ , \tag{36}$$

where $[A]$ and $[B]$ are two $2N \times 2N$-order matrices and $[x]$ is the
vector whose components are sequentially the coefficients Φ_{Ek} and
Φ_{Hk}.

Equation (36) represents a matrix eigenvalue equation, where
the parameter ω^2 has the meaning of an eigenvalue. It must be pointed
out that the parameter ω^2 affects also the entries of matrix $[A]$.
When lossy media are present in the waveguide, both matrices $[A]$
and $[B]$ will be complex. Therefore, starting from an initial tenta-
tive value for the propagation constant $\gamma = \alpha + j\beta$, the resulting
values of ω^2 will generally be complex, which is not a realistic
situation. It is then necessary to perform an iteration on α and β
until the imaginary part of ω^2 becomes negligible. The eigenvectors
associated with the eigenvalues, which have been made approximately
real in this way, represent the approximations of possible mode
fields in the waveguide.

NUMERICAL RESULTS

As it can be easily argued, the numerical method presented in
these notes for the analysis of the surface wave propagation is very
general and can be successfully applied to a large variety of problems with different geometrical and physical characteristics.
Accordingly, a computer program has been realized, which builds up
the matrices $[A]$ and $[B]$, computes the eigenvalues ω^2 and the associated eigenvectors and, for the various modes of propagation, plots
the pattern of E_z and H_z over the cross section of the waveguide.
The program makes use of the geometry of the mesh, consisting of
the finite and the infinite elements, drawn on the cross section of
the guide and of the electromagnetic constants relative to the media
composing the waveguide. If required, an automatic procedure performs

an iteration on α and β in order to make the eigenvalue approximately
real. This program has been used extensively for the analysis of
various dieletric eaveguides. In this section some examples are
presented, with the aim of illustrating the performances which can
be obtained by using this numerical technique.

The first example regards a simple structure, that is the lossy
circular dielectric guide. The dielectric loss is taken into account
by the dissipation factor tan δ, for which a value of 0.1 has been
assumed: therefore in this case a complex dielectric constant is
required, of the form $\varepsilon_r(1 - j0.1)$. The complex constant $\alpha + j\beta$
has been evaluated for the TM_{01} mode, for various values of frequency:
the corresponding results are shown in Figure 3, together with those
obtained by means of an approximated analytical method[7] (solid curve),
and with those obtained by using finite elements only, under the
same number of nodal points (dashed curve). As can be seen, in
this example the introduction of the infinite elements produces a
sharp improvement in the results.

We now describe the results for the elliptical dielectric-tube
guide. For this structure the analysis has been restricted to a
dominant mode, namely the even-HE_{11} mode. A lossy dielectric has
been assumed, characterized by ε_r = 2.26 and tan δ = 0.1. Figure 4
shows a quarter of the guide, together with the geometrical dimensions

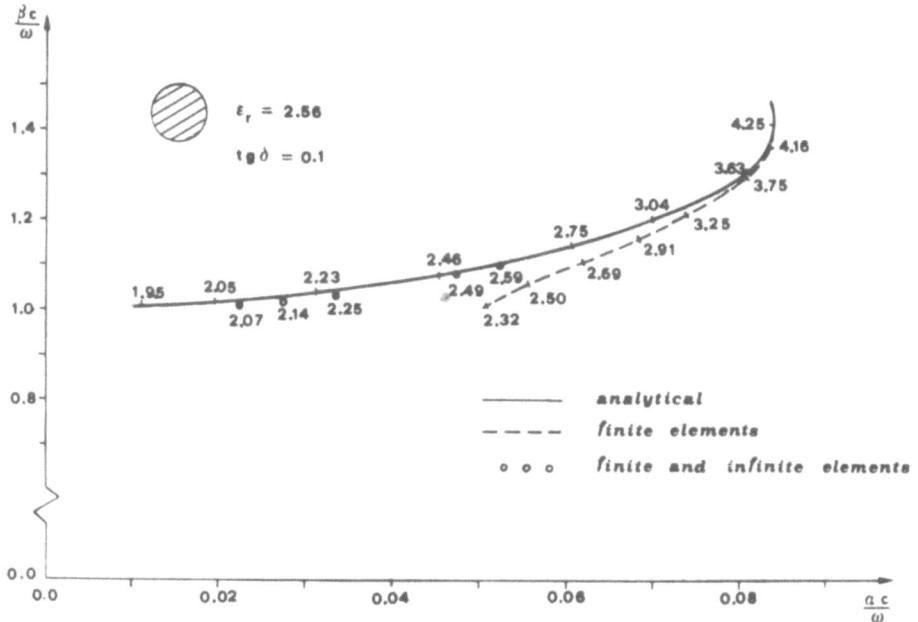

Fig. 3.   Phase constant versus attenuation constant for TM_{01} mode
          of a lossy circular dielectric guide.

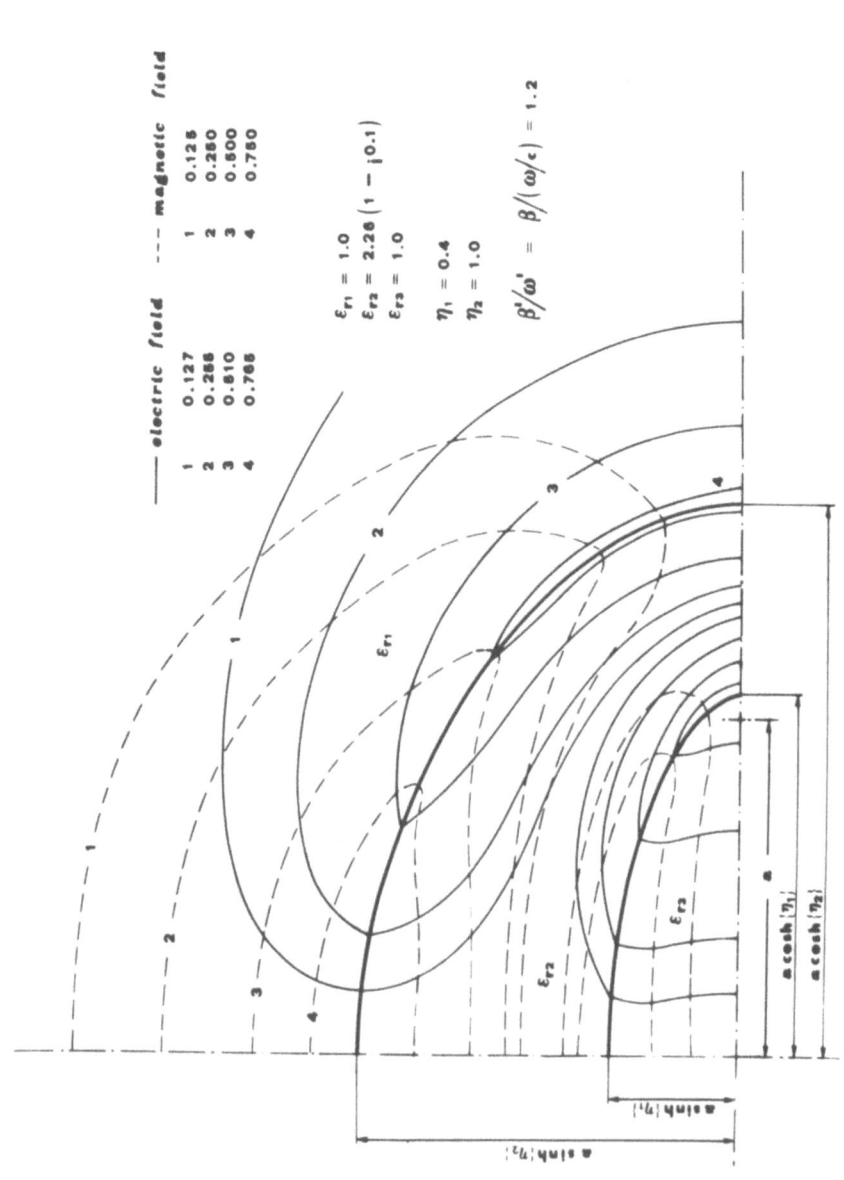

Fig. 4.   Patterns of the real parts of E_z and H_z in an elliptical dielectric-tube guide: even-HE_{11} mode.